The
Unweirding

The Unweirding

not just another quantum mechanics text

Richard A.W. Bradford

principia publishing unlimited

This edition first published 2020

A catalogue record for this book is available from the British Library

ISBN: 978-1-8380216-0-3

Cover design by rawb

principiapublishing.wordpress.com

Contents

Preface

This is not another quantum mechanics text on conventional lines. Schrodinger's equation in its usual form is used but rarely. You will not be treated to the configuration-space solutions of this equation in its usual differential form, whether as applied to a potential well, a harmonic oscillator or the hydrogen atom. There will be no Legendre, Laguerre or Hermite polynomials. There will be no wave packets, and perturbation theory and scattering amplitudes will put in only fleeting appearances. Even the wavefunction – as a function of spatial coordinates – is little used. And the canonical commutation relations between conjugate variables, which are usually regarded as the cornerstone of quantum mechanics, will be largely peripheral. Ideally, the reader of this book will have some familiarity with the aforementioned topics – or perhaps is destined shortly to become acquainted with them in an undergraduate physics course. This book is complementary to these standard treatments, not a replacement for them. It is, I flatter myself, the book I would have liked to have read as a student, though that would hardly have been possible as much of the material was then either unknown or not yet widely disseminated.

But this is not a book for the expert. Far from it. The emphasis of the book is upon understanding the logical structure of quantum mechanics. The mathematics deployed is not difficult – for the most part, at least. I cannot claim the same for the physical concepts. The physical concepts of quantum mechanics are famously counterintuitive. But that is not the same as being incomprehensible.

There has been a tendency in some quarters to pretend that quantum mechanics is not weird at all. Strictly this is correct. Physical reality can have no business being weird. The weirdness lies in us. Its origin is the perennial temptation to apply human intuition, derived from human-scale physics, in settings where it is inappropriate. What would have been truly weird would be the discovery that the fundamental basis of matter was indeed tiny little billiard balls obeying Newton's laws. That would have been like something dreamt up by Terry Pratchett.

The mission of this book is to unweird the seemingly weird.

One of the most surprising things about quantum mechanics is that its weirdness can be completely captured by a very simple mathematical formalism. Indeed, the simplicity of the maths is the key to the unweirding of the physics. A key message of the book is that the bulk of the apparent weirdness of quantum mechanics arises from the phenomenon of superposition, which also results in the phenomenon of entanglement.

The reader will require only to be conversant with Hilbert space formalism – essentially just a linear vector space with a complex inner product wherein superposition appears simply as vector addition. In fact, all the required mathematics will be explained in chapter two, so the reader may, in principle, tackle the book with no prior knowledge of quantum mechanics. However, most readers will gain more from the exercise if pursued in parallel with (or, preferably, following) a more conventional quantum mechanics course or text.

Finite dimensional Hilbert spaces and discrete variables are used rather than continuous variables, with only a few exceptions. Differential equations in spatial variables are therefore almost completely absent. Because low-dimensionality Hilbert spaces are to the fore, much of the development follows what would now be called quantum information theory, the technological motivation for which is quantum computation and quantum cryptography. However, this is incidental to the author's motivation, which is firmly rooted in elucidating the physical meaning of quantum mechanics. Consequently, there are many chapters discussing various configurations of interferometry, specifically to expose the apparent weirdness of such things as "delayed choice" and "quantum erasure" to the light of algebraic clarification.

I am tempted to say that this book will not help you pass exams, unless exams have become rather more interesting of late. However, that is probably untrue despite not being directed at that objective. In physics, one's ability to address problems hinges upon firm understanding, not upon rote learning. And this book is all about understanding. Nevertheless, I stress again that this book is not a substitute for a conventional quantum mechanics course, but should be regarded as an adjunct to a grounding in standard techniques and quantum calculations of practical importance. There are plenty of existing texts which do these jobs admirably. Notwithstanding Richard Feynman's famous dictum, this is a book for people who want to understand quantum mechanics, the key to which is to appreciate that quantum mechanics involves a shift of epistemology – a change in the nature of understanding itself. The physics of systems describable in terms of pure quantum states does not conform to our classical intuition. Get used to it. This book will help you do so.

The book is intended to be pedagogic rather than a work of deep scholarship. Consequently, whilst some references are included at the end of each chapter to enable the reader to pursue further studies, no attempt at completeness or rigorous literature review is intended. Exercises for the reader are randomly distributed throughout the text. Answers are provided at the end of each chapter. Finally, as the emphasis is on principles and understanding interpretational issues, the development is confined to the non-relativistic case. We shall not address relativistic equations nor quantum field theory.

I would be grateful if readers spotting errors would notify me at rawbradford@gmail.com.

Notation and Terminology

Symbol or Term	Meaning
adjoint (of an operator)	the Hermitian conjugate of the operator
Angstrom	10^{-10} m
basis	a set of vectors forms a basis for a given Hilbert space if any vector in the space can be expressed as a linear combination of the basis vectors, but that this would not remain true if any basis vector were discarded. By convention, a basis is usually assumed to be orthonormal.
Bell's inequality	an inequality which local, realistic hidden variables must obey, but Quantum Mechanics (QM) does not (see chapter 14)
Bell states	maximally entangled two-qubit states (see chapter 13)
BCS (theory)	Bardeen-Cooper-Schrieffer and/or these authors' theory of low temperature superconductivity
bipartite	consisting of two parts
BKS Theorem	Bell-Kochen-Specker theorem: one of the no-go theorems against hidden variables (see chapter 9)
Bose-Einstein Condensate or BEC	a state of many bosonic atoms in which all the atoms occupy the same quantum state (usually the ground state)
boson	a particle, possibly a compound particle such as a nucleus or an atom, with integral angular momentum (in units of $\hbar$)
commutable	two operators are commutable if the order in which they are applied does not matter, i.e., if $\hat{A}\hat{B} = \hat{B}\hat{A}$
commutator	the commutator of two operators is the operator $[\hat{A}, \hat{B}] = \hat{A}\hat{B} - \hat{B}\hat{A}$
conjugate variables	in both classical dynamics and QM, there are certain pairs of quantities which are

	closely allied, examples being position and momentum, time and energy, angular position and angular momentum, and in classical dynamics, generalised coordinates and their generalised momenta. Their product has the dimensions of action (same as $\hbar$). In QM they do not commute, and the independence of the Hamiltonian on one variable (symmetry) leads to the other being conserved. In operator form, one variable is the generator of changes in the other variable (see chapter 5)
decoherence	decoherence is the transition from a pure quantum state to a mixed state, usually due to interaction with the environment
density matrix	an Hermitian Hilbert space operator with unit trace whose eigenvalues lie in the range $[0,1]$ and sum to unity. The density matrix provides the most general description of a physical system within quantum mechanics.
δ_{ik}	Kronecker delta, $\delta_{ik} = 0$ if $i \neq k$, $\delta_{ik} = 1$ if $i = k$
einselection	"einselection" is a contraction of "environmental selection". It refers to the determination of a pointer basis through the interaction of a given system with a given environment
entangled	(of a bipartite or multipartite state) not separable. The physical inseparability of an entangled multipartite state is mirrored by the algebraic inseparability of its Hilbert space state, i.e., it cannot be expressed as a product of single-part vectors.
EPR	Einstein, Podolsky and Rosen: a famous 1935 publication (see chapter 13)
eV	electron-volt, 1.6 x 10^{-19} J

Exclusion Principle	no two or more identical fermions can be in the same quantum state. (The Exclusion Principle is a consequence of the spin-statistics theorem)
expectation value	the expectation value of an observable $\hat{Q}$ is the average outcome of measuring that observable many times starting with the system in the same state each time. In quantum mechanics the expectation value is $\langle \varphi | \hat{Q} | \varphi \rangle$
fermion	a particle, possibly a compound particle such as a nucleus or an atom, with half-integral angular momentum, i.e., $1/2$, $3/2$, etc. (in units of $\hbar$)
h	Planck's constant, 6.626 x 10^{-34} Js
$\hbar$	$h/2\pi = 1.055$ x 10^{-34} Js
Hamiltonian	the Hamiltonian is the operator whose eigenvectors are the system's states of well-defined energy
Hermitian conjugate of an operator, $\hat{Q}$	written as $\hat{Q}^+$, the Hermitian conjugate operator may be defined such that for every pair of vectors, $\langle \psi | \hat{Q}^+ | \varphi \rangle = \langle \varphi | \hat{Q} | \psi \rangle^*$. When represented as a matrix, the Hermitian conjugate is the combination of taking the complex conjugate (*) with the matrix transpose.
Hermitian operator	operators such that $\hat{Q}^+ = \hat{Q}$
Heisenberg's Uncertainty Principle	see Uncertainty Principle
Hilbert space	a Hilbert space is a linear vector space over the complex numbers with a conjugate-symmetric, linear inner product $\langle \psi | \varphi \rangle$ for which $\langle \varphi | \varphi \rangle$ is positive definite for non-null vectors. Chapter 2 treats Hilbert space in detail.
iff	if and only if

inner product	a complex number formed from two Hilbert space vectors and denoted $\langle\psi\vert\varphi\rangle$ with the property $\langle\psi\vert\varphi\rangle = \langle\varphi\vert\psi\rangle^*$ and $\langle\varphi\vert\varphi\rangle > 0$ for all non-null vectors.
inverse operator	the inverse $\hat{Q}^{-1}$ of an operator $\hat{Q}$ is such that, for any vector, $\hat{Q}\vert\varphi\rangle = \vert\omega\rangle$ implies $\vert\varphi\rangle = \hat{Q}^{-1}\vert\omega\rangle$. This means that $\hat{Q}^{-1}\hat{Q}$ is the unit operator. Not all operators have an inverse.
k_B	Boltzmann's constant, 1.38×10^{-23} J/K
log	logarithms are always natural (i.e., to base *e*) unless otherwise indicated. But quantum entropy is defined using log_2
measurement problem	the term "measurement problem" in QM alludes to the nature of the evolution of a system's state during measurement being profoundly different from the unitary evolution of pure states according to the Schrodinger equation, especially as regards the reduction of the state vector (see §2.6)
MeV	mega-electron volt, 10^6 eV $= 1.6 \times 10^{-13}$ J
mixed state	a quantum state which is not a pure state but which is describable by a density matrix with two or more non-zero eigenvalues.
multipartite	consisting of many parts
non-contextual	properties of a system itself, independent of the means of measurement (opposite of "contextual" which refers to results contingent upon measurement arrangements)
non-local	the property of quantum systems that the outcome of measurements on space-like separated parts of an extended system are correlated despite being indeterminate (usually arising from entanglement).

norm	the norm of a vector is the square-root of its inner product with itself, $\sqrt{\langle\psi\|\psi\rangle}$
normalised	a vector is normalised if its norm is unity
orthogonal	two vectors are orthogonal if their inner product is zero, $\langle\psi\|\varphi\rangle = 0$
orthonormal	a set of vectors is orthonormal if they are all normalised and they are all mutually orthogonal
outer product	the outer product of two vectors is the operator $\|\xi\rangle\langle\psi\|$, defined by its action on any vector $\|\mu\rangle$ being given by $\|\xi\rangle\langle\psi\|\mu\rangle \equiv \langle\psi\|\mu\rangle\|\xi\rangle$
partial trace, $Tr_{\mathcal{H}1}$	the partial trace of an operator is the sum of its expectation values over all the states in a basis of a given subspace, $\mathcal{H}1$, of the whole Hilbert space, $\mathcal{H}1 \subset \mathcal{H}$
pointer basis	the pointer basis consists of the preferred set of states into which a system decoheres in the sense that its density matrix is diagonal in the pointer basis. The pointer basis is determined by the combination of the system, the environment, and their interaction.
projection operator	a projection operator is an operator which is idempotent, i.e., $\hat{P}^2 = \hat{P}$. A common example is an operator of the form $\|\psi\rangle\langle\psi\|$
pure state	a pure state is a quantum state describable by a Hilbert space vector (strictly, a ray), or equivalently by a density matrix with one eigenvalue equal to 1 and all other eigenvalues zero.
purification	any single-part mixed state may be converted to a pure bipartite state by introducing an auxiliary part with which it is entangled (see §13.7)
QM	Quantum Mechanics

qubit	any system with a two-dimensional Hilbert space when the intended application relates to quantum information or quantum computation. The quantum equivalent of a "bit".
ray	a ray is a vector which is normalised
reduced density matrix, $\hat{\rho}_A$	for a bipartite Hilbert space, $A \otimes B$, the reduced density matrix $\hat{\rho}_A$ is the partial trace $Tr_B \hat{\rho}$ of the full density matrix
Schrodinger equation	the Schrodinger equation defines the temporal evolution of a pure quantum state; it identifies the time derivative of a pure quantum state (times $i\hbar$) with the action of the Hamiltonian.
self-adjoint operator	same as an Hermitian operator
separable	(of a bipartite or multipartite state) algebraically expressible as a product of single-part states (see chapter 13)
span	a set of vectors is said to span a subspace if any vector in the subspace can be expressed as a linear combination of the set of vectors. Thus, a basis spans the whole Hilbert space.
SPDC	spontaneous parametric down-conversion (see chapter 18)
spin-statistics theorem	this theorem states that bosons and fermions have different thermodynamic distributions of state occupancy, called Bose-Einstein and Fermi-Dirac statistics respectively. The latter is consistent with the Exclusion Principle, whereas the former implies that an arbitrarily large number of identical bosons can occupy the same quantum state (see BEC and chapter 24).

stationary state	a stationary state is an eigenstate of the Hamiltonian with time dependence given as in §2.4.1
superscript *	complex conjugate
superscript ^{T}	transpose (of a matrix)
superscript $^+$	Hermitian conjugation, i.e., the complex transpose, $\hat{H}^+ \equiv \hat{H}^{*T}$. When acting on a vector, Hermitian conjugation turns the ket-vector into the bra-vector, i.e., $\lvert\varphi\rangle^+ = \langle\varphi\rvert$, and vice-versa.
tensor product	given two Hilbert spaces, one of dimension N_1 spanned by basis $\{\lvert u_i\rangle\}$, and the other of dimension N_2 spanned by a $\{\lvert v_k\rangle\}$, the tensor product of these two Hilbert spaces is another Hilbert space of dimension $N_1 N_2$ and spanned by the tensor products $\{\lvert u_i\rangle \otimes \lvert v_k\rangle\}$.
trace, Tr	the trace of an operator is the sum of its expectation values over all the states in a basis. (The result is basis independent).
transpose	the transpose of a column matrix is the same array of elements written as a row matrix (and vice-versa). The transpose of a 2D matrix transposes every column, so that $(A^T)_{ik} = (A)_{ki}$
Uncertainty Principle	the Uncertainty Principle states that pairs of observables which do not commute cannot both be known with certainty at the same time. It is a consequence of the nature of quantum measurement.
Uncertainty Relation	the product of the uncertainty in two non-commuting observables must be at least as great as a certain minimum (see chapter 5)
unitary operator	operators whose inverse equals their adjoint such that $\hat{U}^+ = \hat{U}^{-1}$
wrt	with respect to

1

What Is To Be Unweirded

A brief preview is presented of each chapter of the book. I hope these applications entice you to digest the necessary mathematical formalism of chapter 2. The effort will be rewarded.

Chapter 2: Hilbert Space Formalism for Pure States

It will be necessary in this lengthy chapter to set up the mathematical formalism of quantum mechanics. Whilst rather dry, this is unavoidable. Ideally, for most readers, this material will be revision rather than being met for the first time. Lest the reader lose patience before getting to the meat of the book, in this short introductory chapter I present a foretaste of the weirdness that is to be tamed in the chapters which follow chapter 2.

Chapter 3: No Peeking

Readers will be familiar with double-slit experiments which demonstrate the "wave-like" nature of particles in producing interference fringes. Accounts of such experiments emphasise that if the procedure is modified to determine which slit the particle went through, the interference fringes will disappear. But exactly why is it that "which path" information destroys the interference effect, no matter the manner in which such information is obtained? In this chapter it is shown that the effect follows simply from the Hilbert space algebra and should not be regarded as due to "disturbance" by physical interaction *per se*. It is more to do with the nature of quantum measurement and how interference arises in terms of the state vectors involved. The same analysis shows that imperfect "which path" information only degrades the interference pattern rather than eliminating it entirely.

Chapter 4: The No-Cloning Theorem

In conventional information technology it is routine to produce copies of digital data, and this can be done with essentially perfect fidelity and any number of times. In stark contrast, it is not possible to create a device which will faithfully copy an arbitrary quantum state without destroying the original. This is a crucial feature of quantum cryptographic systems. The proof is remarkably simple and demonstrates how dramatically the quantum formalism developed in chapter 2 departs from conventional physical epistemology.

Chapter 5: Symmetry, Conservation Laws and Uncertainty

There is a deep connection between symmetry and conserved quantities. In classical physics this is shown by Noether's Theorem, appreciation of which requires knowledge of Lagrangian dynamics and familiarity with tensors. Remarkably, in quantum mechanics, the relationship between symmetry and conserved quantities is far easier to demonstrate: both relate to the commutation properties of the operators involved. As a by-product we obtain the commutation properties of conjugate variables which permit the Uncertainty Relation to be derived.

Chapter 6: The Elitzur-Vaidman Bomb Test

Can something that did not happen nevertheless influence a future outcome merely because it could have happened? In classical physics the answer is no. But in quantum mechanics things are otherwise. The Elitzur-Vaidman gedanken experiment is a dramatic illustration of this "observability of counterfactuals". Given a number of bombs whose sensitivity is such that any physical interaction with them whatsoever will detonate them, and given that some of the bombs are duds, how can one identify a definitely live bomb without exploding it? In classical physics it cannot be done. But in quantum mechanics it can: the fact that a bomb *might have* exploded is sufficient to identify it, despite the fact that it did not explode and therefore had experienced no physical interaction.

Chapter 7: State Vector Reduction Demonstrated

In this chapter I discuss Wheeler's delayed choice experiment in a modified form which has actually been carried out. The essence of delayed choice experiments is that a beam is split into two paths before the decision is made whether or not to later recombine the beams to examine interference. Yet in those cases where the beams are recombined, interference is indeed found. Such experiments appear to confirm that particles really do "go both ways at once" and, moreover, this must be the case also when the beams are not recombined and hence interference does not occur. This effectively demonstrates state vector reduction at the point of particle detection, as the superposed state in the latter case always collapses to detection in one detector or the other.

Chapter 8: A Watched Pot Never Boils

Notwithstanding the observability of counterfactuals demonstrated by the Elitzur-Vaidman bomb test, a completed quantum measurement on a system not in an eigenstate must perturb the system because state vector reduction will leave it in an eigenstate after measurement. An unexpected consequence of this is that the evolution of a system due to a perturbing interaction can be suppressed by continually measuring its energy. Hence, a watched pot never boils – but only if it's a pure quantum pot.

Chapter 9: There Are No Hidden Variables: Part 1

In the early days of QM physicists were disturbed by the claim that the outcomes of quantum measurements were irreducibly indeterminate. The unease remains to this day. The hope amongst some physicists was that a theory with some "hidden variables" might be found in terms of which measurement outcomes would be deterministic. This was essentially the motivation behind the 1935 paper by Einstein, Podolsky & Rosen ("EPR") which purported to demonstrate that QM is incomplete (for which read, "there must be hidden variables"). QM proponents like von Neumann early claimed to prove there are no hidden variables, but von Neumann's 1932 proof is fatally flawed. However, in 1966/67 a powerful no-go theorem was proved independently by Bell and by Kochen & Specker which ruled out a broad class of local, non-contextual hidden variable theories with deterministic measurement outcomes. A proof is given which is considerably simpler than Kochen & Specker's original.

Chapter 10: The General Quantum State is a Mixed State

This is the point in the book in which we accept that not all quantum states can be represented as Hilbert space vectors, i.e., not all states are pure. The more general description of a state must encode both the indeterminacy of quantum measurement (aleatory uncertainty) and also the ordinary uncertainty associated with incomplete knowledge of microstates (epistemic uncertainty). The vehicle for this description is the density matrix, which is an Hermitian operator on Hilbert space with eigenvalues in the range $[0,1]$ which sum to unity, and hence has unit trace. This chapter explains how the expectation value of an observable is found for a mixed state, and how the probabilities of a measurement returning the various eigenvalues are found from a generalised Born Rule.

Chapter 11: Schrodinger's Cat

It is rather distressing that so much has been written about Schrodinger's cat, causing quite unnecessary confusion. This diabolical gedanken experiment places a poor cat in a box with a vial of cyanide which a mechanism will release if a single radioactive nucleus decays. We are beguiled into believing that, since the unstable nucleus is in a superposed state of decayed and undecayed, so the cat is also in a superposed state – namely of being both dead and alive at the same time. But cats cannot be in pure quantum states, the sort of states for which superpositions are possible. This, in a nutshell, is the resolution of the Schrodinger's cat paradox. Moreover, following chapter 10, we have the formalism required to express the state of the cat within quantum mechanics without the false assumption that its state is pure. Actually, cats are always in mixed states. "Being alive" is a mixed state.

Chapter 12: Quantum Entropy

If there are two subjects which students perennially struggle to understand they are entropy and quantum mechanics. Putting the two together and investigating the entropy of quantum systems seems doomed to be difficult. But it really isn't. The entropy of a pure quantum state is zero. And this is the case whether you regard that pure quantum state as a superposition of other pure states, or not. How much simpler would you like that to be? On the other hand, mixed states – which the reader should by this point understand are quite different – do have non-zero entropy, and for precisely the same reason as in classical physics: their actual micro-state is epistemically uncertain. This chapter consists mostly of how entropy can be calculated when working with the quantum formalism for mixed states, i.e., the density matrix. As a by-product we see how measurement on a pure state provides information by first creating a mixed state, and hence increasing entropy (which can be viewed as information within the system).

Chapter 13: Entanglement of Bipartite Pure States

After a brief discussion of EPR (the 1935 paper by Einstein, Podolsky & Rosen) the concept of entanglement is defined and quantified for pure bipartite states, i.e., pure states drawn from the tensor product space of two different parts. Entanglement is seen to be an algebraic property of the Hilbert space vector, namely that a bipartite state is entangled if it is not separable. In contrast, a bipartite state is separable if it can be written as a product of states of the two parts separately. The concepts of a "reduced state" described by a "reduced density matrix" are introduced in this chapter. Entangled states can have different degrees of entanglement and it is shown that the entanglement may be quantified by the von Neumann entropy of the reduced state. This chapter does not address the far more problematical case of quantifying the entanglement of mixed states, which is delayed until chapter 17.

Chapter 14: There Are No Hidden Variables: Part 2

Chapter 9 has already discussed the powerful BKS no-go theorem against hidden variable theories. In this chapter the final nail is driven into the coffin of local, realistic, non-contextual hidden variable theories. Inspired by EPR type experimental arrangements, in 1964 John Bell derived an inequality which must be obeyed by the results of measurements on a pair of entangled particles if these results are to be made deterministic by some hidden variable (or set of hidden variables). In the case of measurements on two-state spins, the inequality is not respected by the QM prediction. This observation paved the way for experiment to discriminate which was right: QM or hidden variables. Many experiments have been conducted over the last 50 years or so, and with increasing precision and decreasing scope for loopholes. Except for discredited experiments, the outcomes clearly rule in favour of QM.

Chapter 15: Decoherence and Measurement

Decoherence theory posits that pure quantum states often degrade into mixed states due to unavoidable interactions with the environment. In this chapter the basic formulation is explained. Once again the concept of a reduced state is central, obtained by "tracing out the environment". In later chapters these ideas will be used to demonstrate decoherence in toy models. But decoherence can also be used to shed some (but limited) light on how measurement operates to produce a mixed system state from an initially pure state. The idea of the "pointer basis" is introduced as the preferred basis into which states decohere. The pointer basis depends upon the apparatus/environment and the strength of their interaction with the system in comparison with the system's eigenenergy spacings. The decoherence theory of measurement then requires that an apparatus to measure a certain observable must be contrived so that its pointer basis coincides with the eigenstates of the observable to be measured. In mathematical language this is what it means to devise a suitable apparatus for the desired observable.

Chapter 16: The Whole is Less than the Sum of Its Parts

The cryptic title refers, of course, to the well known aphorism that "the whole is greater than the sum of its parts". In this chapter we investigate inequalities obeyed by the entropy of bipartite states and ensembles. The two are very different. The entropy of an ensemble generally exceeds the "sum" of the entropies of its component elements (duly weighted by their relative frequency in the ensemble). But the manner in which the entropy of a bipartite state relates to the entropy of its parts is very different. In this case the entropy of the whole is generally less than the sum of the entropies of its parts, a result known as "subadditivity". This chapter also provides a plethora of other inequalities involving quantum entropy.

Chapter 17: The Entanglement of Mixed States

Chapter 13 defined entanglement in pure bipartite states and provided a means of both identifying and quantifying such entanglement. Mixed states are far more tricky. What is meant by an entangled mixed state is easy enough to define (once some pitfalls are dealt with) but actually identifying whether a given mixed state density matrix is entangled is not so simple in the general case. Moreover, there is no universally agreed quantification of entanglement for mixed states, though several different measures exist. The so-called "entanglement of formation" is a generalisation of that for pure states, but involves a highly non-trivial optimisation to compute and has no effective algorithm at present in the general case. Fortunately, in the case of mixed bipartite states of two qubits these problems have been solved and entanglement can be readily identified and quantified in such systems. This is a rather more demanding chapter than others and may be missed on first reading.

Chapter 18: Entangled Interference

What happens if we try an interference experiment with entangled particles? There is a lovely conundrum raised by this scenario. Recall that carrying out "which path" measurements will destroy interference. But for an entangled pair of particles, a measurement on one particle is effectively also a measurement on the other. So what if we set up equipment to observe interference with one particle, but carry out the "which path" measurement on the other particle? This would seem to destroy the interference. But the measurement on the second particle could be carried out at a remote distance.

The occurrence, or absence, of interference in the first particle would therefore seem to provide an opportunity for faster-than-light communication. What gives? All is explained in this chapter, and, as usual, this follows simply by tracking the state algebra. Rest assured that relativistic causality is not violated.

Chapter 19: Quantum Erasure

Chapter 20: Delayed Quantum Erasure

Quantum erasure is the phenomenon whereby a "measurement", which would destroy an interference pattern, can be erased and the interference regained. How can a measurement possibly be erased? Such erasures become even more intriguing when coupled with experiments on entangled pairs of particles because this provides the opportunity to carry out the erasure of the "measurement" **after** the interference data has already been obtained, so-called "delayed erasure". One particle, it is claimed in popular accounts, appears to "know" that a measurement will be carried out in the future on its entangled partner. Hence an interference pattern is re-established because of an erasure which will be carried out after the interference data has already been collected. This is often presented in popular accounts as posing a challenge to the correct temporal order of causality. It does not, of course, and these two chapters explain why. As usual it is the Hilbert state algebra which provides cogent clarification. The key issues are ambiguities in two things: how the word "measurement" is used, and how the claimed "interference" arises.

Chapter 21: How the Magic is Lost: A Simple Example of Decoherence

Chapter 22: How the Magic is Retained: Another Example of Decoherence

In chapters 21 and 22 the phenomenon of decoherence is illustrated by simple but explicit examples. The example in chapter 21 shows that if the interaction energy with the environment is strong compared with the spacing of the system's energy levels, an initially delocalised state decoheres into a spatially localised state. The pointer basis into which the system decoheres in this case is defined by the eigenstates of the interaction Hamiltonian. Conversely, if the interaction energy with the environment is weak compared with the spacing of the system's energy levels, the pointer basis into which the system decoheres is defined by the eigenstates of the system's free Hamiltonian. The example of chapter 22 illustrates this phenomenon. Since energy eigenstates are typically spatially delocalised, this delocalisation therefore remains indefinitely if the interaction with the environment is weak compared with the system's natural energy level spacings (atomic electron orbitals being an example). Only unitary evolution according to the Schrodinger equation is required for these demonstrations, but it is necessary to assume some suitable spectrum of environmental interaction strengths over which to average the system's density matrix. In these examples the time required for decoherence to take place is of the order $\hbar/V$, where V is characteristic of the strength of the interaction with the environment. This is typical but may not always be the case.

Chapter 23: Why Are Big Things Classical?

Why does quantum mechanics apply to small things, but classical physics applies to big things? Why does spatial delocalisation not apply to cats? And what do we mean by "small" in this context? The answer is that "small" means a system with a small number of degrees of freedom. Some simple examples illustrate how the minimum spacing between a system's energy levels tends to decrease as the number of degrees of freedom, N, increases, typically tending asymptotically towards zero as $N \to \infty$. But we know that if the strength of the interaction with the environment exceeds the system's minimum energy level spacing, the system will decohere into the pointer basis defined by the environmental interaction. This will inevitably be the case if the system's minimum energy level spacing becomes arbitrarily small for "large" systems, $N \to \infty$. This is the reason why "big" things tend to be classical, and "big" only means $N \gg 1$, which is not necessarily all that big. However, it is emphasised that this a general rule which sometimes does not apply.

Chapter 24: Big Things Behaving Quantumly

In chapter 23 we saw why it is that "big things" generally behave in a classical rather than a quantum manner. We also saw that "big" in this context is an absolute because it means a system with many degrees of freedom. However, I also warned that there are exceptions. Actually, you are surrounded by them. This chapter gives some examples of such exceptions and elucidates how quantum mechanical behaviour can be retained despite being "big". Unfamiliar behaviours resulting from uniquely quantum effects typically require specially contrived conditions to occur. That is why they are unfamiliar. However, one does not have to be a pedant to note that our everyday world is full of phenomena which are essentially quantum mechanical. The physical properties of matter are a case in point. Thus, whilst superconductivity is an unfamiliar phenomenon in the everyday world, and requires quantum mechanics to explain it, the fact that metals tend to be excellent conductors under ordinary room-temperature conditions, whilst most other materials are excellent insulators, is also a phenomenon explicable only within quantum mechanics. The apparently mundane nature of the familiar properties of matter under normal conditions may cause us to overlook their non-classical origins. I am tempted to say that "everything is quantum, dammit", but I doubt that engineers will start using Hilbert space to design bridges any time soon.

Chapter 25: Sins of Intention

In chapter 19, in the context of quantum erasure, we considered a case in which a measurement is not a measurement. Here we discuss a case in which a non-measurement *is* a measurement. Can such a null measurement reduce the state vector? Yes, but only those basis vectors that might have been detected, but aren't, are projected out, whilst the remaining states remain unreduced and coherent. Confused? You won't be.

Chapter 26: Superdense Coding and Quantum Security

There is a means of conveying two bits of classical information by the transmission of only one qubit from sender to receiver. This is known as "superdense coding" and relies on entanglement. Seemingly paradoxically, there is also a theorem that says the maximum amount of classical information that can be extracted from N qubits is N bits (the Holevo bound). Chapter 26 explains how superdense coding works and why it does not violate the Holevo bound. As a by-product, the scenario which implements superdense coding provides an example of quantum cryptographic security: an eavesdropper who intercepts the transmitted qubit can extract no information from it at all.

Chapter 27: Quantum Teleportation

This chapter explains what quantum teleportation is and what it is not. What it isn't is anything like Star Trek. The chapter explains, when carrying out quantum teleportation, what must be transported conventionally through space, and why the "teleportation" does not provide a means of faster-than-light travel or communication. As with superdense coding, the resource which makes quantum teleportation work is a bipartite entangled state shared by sender and recipient.

Chapter 28: Waves Making Tracks

The straight tracks of alpha or beta particles radiating out from a radioactive source in a cloud chamber give every impression that the agency causing them is particle-like. How is this consistent with the quantum description of particles in terms of Hilbert space vectors or wavefunctions? Why should a delocalised wave make a straight track? This is a puzzle to which every generation of physicists rediscovers the resolution anew.

Chapter 29: Algebraic Methods

In general, the state structure of a given system will not be easy to calculate, e.g., the eigenenergy states for an arbitrary Hamiltonian. However, there are some examples where the state structure can be derived in an elementary fashion which is purely algebraic rather than requiring the states to be expressed in the spatial basis, i.e., as wavefunctions requiring the solution of differential equations. All these cases can also be solved via wavefunctions, but the algebraic method is so elegant – and so important – that all students of quantum mechanics should be familiar with them. The examples given here are Fock space, angular momentum states, and the solution of the non-relativistic hydrogen atom.

Chapter 30: The Pitfalls of Popularisation

Popularisers of physics have a tough job. We must allow them considerable latitude as regards strict accuracy. Rigour is often the enemy of comprehensibility. Nevertheless, it is instructive to examine some sins that have been committed, because there is also an obligation not to mislead the public. Quantum mechanics is particularly easy to misrepresent in the effort to simplify – or, worse, in an endeavour to impress the audience with its weirdness. In this last chapter I examine some of the careless claims which have been made, and expose the unresolved nature of some of them.

2

Hilbert Space Formalism for Pure States

'The beginner should not be discouraged if he finds that he does not have the prerequisites for reading the prerequisites' (John Boccio).

2.1 Not All States Are Pure

Quantum mechanics recognises that atomic scale, or sub-atomic scale, phenomena behave differently from human-scale physics. On these very small scales a physical system often will not conform to our human-scale intuitions for describing the material world. When Galileo rolled balls down an inclined plane in order to study their trajectories, neither he nor his students had any problem understanding what was meant by 'a ball', nor 'an inclined plane', nor 'a trajectory'. We are not in the same luxurious position when it comes to systems on a scale far below our sensory perception. Particles and waves (or fields) are intuitive human-scale concepts, so we should not expect quantum scale systems to conform to either. Indeed, they do not, in general, and it is merely fortuitous when one of these conceptions happens to be a useful approximation and a prop to our intuition.

In certain circumstances, systems on a sufficiently small scale form "pure quantum states". A pure quantum state conforms to a particular mathematical description, which is the subject of this chapter. It is important to appreciate from the start that not all physical systems can be described in this way, i.e., not all physical systems are pure quantum states. Failure to appreciate this leads to confusion, such as that annoying cat of Schrodinger's (see chapter 11). The way in which a more general physical system is described mathematically, using so-called "mixed states", is addressed in chapter 10. In the current chapter, and in all chapters before chapter 10, we confine attention to pure quantum states. But I issue immediately the warning that not all physical states are pure quantum states, lest I encourage the same confusion I am attempting to alleviate.

The "certain circumstances" which result in a system being in a pure quantum state are not simple to state until the basic formulation of quantum mechanics has been set up and its implications examined. However, this will be addressed in chapters 21-24. For now note that being atomic, or sub-atomic, in scale certainly does not guarantee that the system will be in a pure quantum state. Moreover, scale is not strictly the arbiter of the possibility of being in a pure state, and macroscopic systems may behave "quantumly" (chapter 24). Nevertheless, as we shall see in chapters 21 and 23, the criteria for maintaining a pure state generally become increasingly hard to fulfil as the scale is increased.

2.2 Hilbert Space and Pure States

A physical system in a pure quantum state is represented by a ray in a Hilbert space, $\mathcal{H}$, where a ray is a vector of unit norm. The rest of this section unpacks what that opening sentence means.

A Hilbert space is a particular type of linear vector space. The concept of a linear vector space is inspired by ordinary space, but a linear vector space may have an arbitrary number of dimensions. However, our vectors have nothing to do with geometry. The vectors are abstract objects. In quantum mechanics their purpose is to represent the state of the physical system. Any two vectors may be added to form another vector, and any vector may be multiplied by a number to form another vector. A set of vectors exists (a "basis") such that every vector in the space may be expressed as a linear combination of these basis vectors. The smallest number of vectors in the basis which permits all vectors to be thus expressed is the dimension of the space. We will almost exclusively be concerned with spaces of finite dimension in this book. Standard quantum mechanics texts generally work with infinite dimensional spaces, though they tend not to make that explicit. (I refer to most texts' formulation in the spatial continuum). I will go over all that again more slowly in what follows and in the context of Hilbert spaces.

A Hilbert space is a linear vector space over the complex numbers equipped with a complex valued inner product (defined below). So any vector can be multiplied by a complex number to obtain another vector, and vectors can be added together. Writing vectors in the Dirac ket notation, $|\psi\rangle$, this means that for any vectors $|\psi\rangle$ and $|\varphi\rangle$ and any complex numbers a and b, it is possible to combine them to form another vector $a|\psi\rangle + b|\varphi\rangle$ which is within the Hilbert space (i.e., Hilbert space is closed with respect to these linear combinations). This innocent seeming property of vector spaces is actually the source of most quantum weirdness, namely superposition which is just another name for linear combination of Hilbert space vectors.

There is a unique zero vector, which can be obtained by multiplying any vector by the number zero, $|zero\rangle = 0|\psi\rangle$, and which leaves any vector unchanged when added to it: $|\psi\rangle + |zero\rangle = |\psi\rangle$.

NB: We shall often use the notation $|0\rangle$ to represent either the ground state of a system or perhaps a particular spin state of a particle. In neither case should this state be confused with the zero, or null, vector, $|zero\rangle$. A physical system cannot be represented by the zero vector, since a physical system is always normalisable – in fact it will be normalised to unit norm (see below).

The complex valued inner product between two vectors $|\psi\rangle$ and $|\varphi\rangle$ is written $\langle\psi|\varphi\rangle$ and is such that $\langle\psi|\varphi\rangle = \langle\varphi|\psi\rangle^*$ (where the asterisk denotes complex conjugation). It is linear in the sense that $\langle\chi|(a|\psi\rangle + b|\varphi\rangle) = a\langle\chi|\psi\rangle + b\langle\chi|\varphi\rangle$, from which linearity in the left-hand vector also follows. The inner product of a non-zero vector with itself is necessarily real and positive-definite, $\langle\psi|\psi\rangle > 0$, and this permits the norm (or magnitude) of a vector to be defined by $\|\psi\| = \sqrt{\langle\psi|\psi\rangle}$. The zero vector is the only vector with zero norm. The "norm" is the generalisation of the length of a vector in ordinary space. A ray is a vector with unit norm. Any vector can be "normalised" to be a ray by replacing it with $|\psi\rangle/\|\psi\|$.

The norm of the difference between two vectors can be considered as a "distance" measure from the "positions" in the Hilbert space represented by the two vectors, i.e., $d(x, y) \equiv \|x - y\|$. This is entitled to be regarded as a distance measure because it is positive, it is symmetric $d(x, y) = d(y, x)$, it is zero iff $x = y$, and it obeys the triangle inequality,

$$d(x, z) \le d(x, y) + d(y, z) \tag{2.2.1}$$

The proof of (2.2.1) is an exercise for the reader (see the last section of each chapter for solutions to reader exercises).

A Hilbert space is of dimension N if N is the smallest number of vectors, $\{|\varphi_i\rangle\}$, which is sufficient to span the space, in the sense that any vector can be expressed as $|\psi\rangle = \sum_i a_i|\varphi_i\rangle$ for some complex numbers a_i where the sum runs over all N basis vectors. The term "basis" will be reserved for a set of vectors $\{|\varphi_i\rangle\}$ which are "orthonormal". Orthonormality of the basis vectors means that $\langle\varphi_i|\varphi_k\rangle = \delta_{ik}$, where the Kronecker delta is defined such that $\delta_{ik} = 0$ if $i \ne k$ and $\delta_{ik} = 1$ if $i = k$. Orthonormality is the generalisation of the concept in ordinary space of vectors being orthogonal and of unit length. In general, any two vectors are said to be orthogonal if $\langle\xi|\mu\rangle = 0$.

A conjugate Hilbert space is defined, in which every vector $|\psi\rangle = \sum_i a_i|\varphi_i\rangle$ has a corresponding conjugate vector (a "bra", $\langle\psi|$, corresponding to the "ket", $|\psi\rangle$) which is written $\langle\psi| = \sum_i a_i^*\langle\varphi_i|$. If $|\xi\rangle = \sum_i b_i|\varphi_i\rangle$ this gives, $\langle\psi|\xi\rangle = \sum_i a_i^* b_i$ and $\langle\xi|\psi\rangle = \sum_i b_i^* a_i$. Hence the norm can be written in terms of "components" in the $\{|\varphi_i\rangle\}$ basis as $\|\psi\| = \sqrt{\sum_i a_i^* a_i}$.

In the particular basis $\{|\varphi_i\rangle\}$ we may think of $|\psi\rangle = \sum_i a_i|\varphi_i\rangle$ as being represented by a column matrix consisting of the numbers a_i, and the conjugate vector $\langle\psi| = \sum_i a_i^*|\varphi_i\rangle$ as represented by a row matrix containing the complex conjugate numbers, a_i^*. The inner product can then be regarded as the ordinary matrix multiplication of the vectors as represented by their row/column matrices, i.e.,

$$\langle\psi|\xi\rangle \equiv (a_1^* \quad a_2^* \quad a_3^* \ ..... \quad a_N^*) \begin{pmatrix} b_1 \\ b_2 \\ b_3 \\ etc. \end{pmatrix} = \sum_i a_i^* b_i \qquad (2.2.2)$$

We will sometimes use a notation in which a column matrix is written $\bar{b}$ and the combination of matrix transposition and complex conjugation is denoted by $^+$, so (2.2.2) can be written,

$$\langle\psi|\xi\rangle = \bar{a}^+ \bar{b} \qquad (2.2.3)$$

By this point anyone not already versed in quantum mechanics will be wondering what this Hilbert space formalism has got to do with the description of physical systems. Recall that we are dealing with systems for which we have no right to expect human-scale intuitions (such as particles moving on well-defined trajectories) to be applicable. The formalism is best regarded as a codification of a particular epistemology, free of classical intuitive concepts. That the formalism is appropriate, i.e., that it works, is far from obvious when presented in this manner. Historically it evolved over several decades, initially in particular forms in response to specific experimental findings before this general notation emerged. Contact between the Hilbert space formalism and the physical world will be made when we discuss measurement, very shortly. But first...

2.3 Operators

While Hilbert space vectors (more correctly, rays) will provide the specification of a system's state, operators in Hilbert space are essential to provide a means of extracting information from the system (measurements), and also to formulate how a system evolves in time (dynamics).

An operator on Hilbert space is a mapping which takes any vector to another vector, $\hat{Q}: \mathcal{H} \rightarrow \mathcal{H}$. Hence the result of acting the operator on any vector $|\psi\rangle$ is another vector in the Hilbert space, $\hat{Q}|\psi\rangle = |\varphi\rangle \in \mathcal{H}$. A linear operator preserves linear combinations after mapping. Thus, a linear operator is such that $\hat{Q}[a|\psi\rangle + b|\xi\rangle] \equiv a\hat{Q}|\psi\rangle + b\hat{Q}|\xi\rangle$. We shall be exclusively concerned with linear operators and so I shall be lazy and speak of "operators" but shall always mean "linear operators". Note that linearity of an operator implies linearity of all powers of the operator, i.e., linearity of $\hat{Q}$ implies linearity of $\hat{Q}^2, \hat{Q}^3$, etc. (A reader exercise is to prove this). Hence polynomials in $\hat{Q}$ are also linear operators, and, by extension so are infinite power series (subject to convergence).

Transcendental functions of any operator, such as $\exp(\hat{Q})$, are defined by their power series, e.g.,

$$\exp(\hat{Q}) \equiv 1 + \hat{Q} + \frac{1}{2}\hat{Q}^2 + \frac{1}{3!}\hat{Q}^3 + \frac{1}{4!}\hat{Q}^4 + \cdots \tag{2.3.1}$$

We will skirt over the mathematical niceties related to convergence, content in the knowledge that this will not be a problem in any of the applications addressed.

Any operator can be fully defined by its operation on a basis set of vectors, i.e., by the square matrix of complex numbers Q_{ik} which gives the result of the operation on an arbitrary vector in terms of the components of the vector in that basis, i.e.,

$$\hat{Q}|\varphi_k\rangle = \Sigma_i Q_{ik}|\varphi_i\rangle \tag{2.3.2}$$

Due to the assumed orthonormality of the basis $\{|\varphi_i\rangle\}$, by taking the inner product of both sides of (2.3.2) with $\langle\varphi_i|$ we can find the matrix representation of the operator to be,

$$Q_{ik} = \langle\varphi_i|\hat{Q}|\varphi_k\rangle \tag{2.3.3}$$

In general the matrix Q_{ik} will not be symmetric, i.e., $Q_{ik} \neq Q_{ki}$. For an arbitrary vector, $|\psi\rangle = \Sigma_i a_i|\varphi_i\rangle$, the operation of $\hat{Q}$ is,

$$\hat{Q}|\psi\rangle = \hat{Q}\Sigma_i a_i|\varphi_i\rangle = \Sigma_i a_i\hat{Q}|\varphi_i\rangle = \Sigma_i a_i \Sigma_k Q_{ki}|\varphi_k\rangle = \Sigma_k a'_k|\varphi_k\rangle$$

where $a'_k = \Sigma_i Q_{ki}a_i$. This means that if we consider Q_{ki} to be a square matrix, denoted (Q), then the action of the operator is to transform the components of a vector in accord with ordinary matrix multiplication, i.e., writing the components as a column matrix we have $\bar{a}' = (Q)\bar{a}$. Note that all the complex numbers a_i, a'_k and Q_{ki} are with respect to the specified basis $\{|\varphi_i\rangle\}$. For a given vector $|\psi\rangle$, these numbers (components) will change if the basis is changed.

Conversely, given any square matrix of complex numbers, Q_{ki}, we can define a linear operator by specifying the operator's action as being $\hat{Q}|\varphi_i\rangle = \sum_k Q_{ki}|\varphi_k\rangle$. One means of specifying a matrix is to form the product of a column matrix and a row matrix

$$(Q) \equiv \begin{pmatrix} b_1 \\ b_2 \\ b_3 \\ etc. \end{pmatrix} (a_1^* \quad a_2^* \quad a_3^* ..... \quad a_N^*) \tag{2.3.4}$$

so that $Q_{ki} = b_k a_i^*$. We can consider the column and row matrices as being the components of ket vector, $|\xi\rangle = \sum_j b_j |\varphi_j\rangle$ and the bra version of ket vector $|\psi\rangle = \sum_j a_j |\varphi_j\rangle$, so that we can write,

$$Q_{ki} \equiv \langle\varphi_k|\hat{Q}|\varphi_i\rangle = \langle\varphi_k|\xi\rangle\langle\psi|\varphi_i\rangle \tag{2.3.5}$$

From this there is an obvious notation for such operators as $\hat{Q} = |\xi\rangle\langle\psi|$ because forming the "matrix element", $\langle\varphi_k| ... |\varphi_i\rangle$, is consistent with (2.3.5). An expression like $|\xi\rangle\langle\psi|$ is said to be the "outer product" of the two vectors. Whilst the inner product of two vectors is a (complex) number, the outer product of two vectors is an operator. Whilst (2.3.5) is one way of defining the outer product, another is to note that any operator is defined by its mapping effect on the Hilbert space. For any vector $|\mu\rangle$ its image under the action of the operator $\hat{Q} = |\xi\rangle\langle\psi|$ is defined to be $\hat{Q}|\mu\rangle = |\xi\rangle\langle\psi|\mu\rangle \equiv \langle\psi|\mu\rangle|\xi\rangle$, which illustrates how natural is the notation.

However, not all operators can be written in this form, because an arbitrary matrix of components cannot always be written as (2.3.4). However, all operators can be written as a sum over "outer products" of the form $|\xi\rangle\langle\psi|$. In fact it follows from the above definitions that any operator can be written in terms of its components wrt the basis $\{|\varphi_i\rangle\}$ as,

$$\hat{Q} \equiv \sum_{i,j} Q_{ij}|\varphi_i\rangle\langle\varphi_j| \tag{2.3.6}$$

Operators of certain types play special roles in quantum mechanics. The definitions of these special types of operator follows.

2.3.1 The Adjoint Operator

For any operator, $\hat{Q}$, the "adjoint" operator, written $\hat{Q}^+$, can be defined via the adjoint matrix of elements, i.e., the complex conjugate transpose of the matrix. The complex conjugate transpose is also called the "Hermitian conjugate". Thus the adjoint of $\hat{Q} \equiv \sum_{i,k}|\varphi_i\rangle Q_{ik}\langle\varphi_k|$ is such that $\hat{Q}^+ \equiv \sum_{i,k}|\varphi_i\rangle Q_{ki}^*\langle\varphi_k|$. For example, $(|\psi\rangle\langle\xi|)^+ = |\xi\rangle\langle\psi|$. Note the useful identity,

$$\langle \xi | \hat{Q} | \psi \rangle^* \equiv \langle \psi | \hat{Q}^+ | \xi \rangle \tag{2.3.7}$$

Proof,

$$\langle \xi | \hat{Q} | \psi \rangle^* = \left[\Sigma_{i,k} \langle \xi | \varphi_i \rangle Q_{ik} \langle \varphi_k | \psi \rangle \right]^* = \Sigma_{i,k} \langle \psi | \varphi_k \rangle Q_{ik}^* \langle \varphi_i | \xi \rangle = \langle \psi | \hat{Q}^+ | \xi \rangle$$

2.3.2 Hermitian Operators

An operator is said to be Hermitian, or self-adjoint, if it equals its adjoint, $\hat{H}^+ = \hat{H}$, and this is equivalent to its matrix of components being Hermitian, i.e., $(H^+)_{ik} = H_{ki}^* = H_{ik}$, or simply $(H)^+ = (H)$ in matrix notation. An exercise for the reader is to demonstrate that being self-adjoint wrt one basis $\{|\varphi_i\rangle\}$ implies being self-adjoint in any other basis, so that being self-adjoint is a property of the operator, not of the basis.

2.3.3 Inverse Operator

The inverse of an operator reverses its action. Thus, if $\hat{Q} |\psi\rangle = |\xi\rangle$ then $\hat{Q}^{-1} |\xi\rangle = |\psi\rangle$ for any vector $|\psi\rangle$. In other words $\hat{Q}^{-1} \hat{Q} = \hat{Q} \hat{Q}^{-1} = 1$ (where we denote simply by 1 the unit operator which leaves all vectors unchanged). Not all operators have an inverse. For example, if a mapping is many-to-one then the original vector cannot be regained from its image. The components of an inverse operator are the inverse matrix of the components of the operator. The existence of an inverse operator is therefore equivalent to the determinant $\|(Q)\|$ being non-zero.

2.3.4 Projection Operators

An important class of operators which do not have an inverse (because they are many-to-one maps) are projection operators. These are operators which "project out" the part of a vector which is parallel to a given ray $|\psi\rangle$. Hence we can write,

$$\hat{P}_\psi = |\psi\rangle\langle\psi| \tag{2.3.8}$$

Thus, for an arbitrary vector $|\xi\rangle$ it is clear that $\hat{P}_\psi |\xi\rangle$ must be "parallel to" $|\psi\rangle$, which simply means some complex number times $|\psi\rangle$, because (2.3.8) shows that it equals $\langle\psi|\xi\rangle|\psi\rangle$. $\hat{P}_\psi$ is a many-to-one mapping because any vector with the same "component" in the direction of $|\psi\rangle$ will have the same projection, whatever its components in other "directions" of the Hilbert space. It follows immediately from (2.3.8) that,

$$\hat{P}_\psi{}^2 = \hat{P}_\psi \tag{2.3.9}$$

i.e., projecting again causes no further change. For a given set of orthonormal basis vectors, $\{|\varphi_i\rangle\}$, the particular projection operators $\hat{P}_i = |\varphi_i\rangle\langle\varphi_i|$ will project out the part of any vector in the direction of the ith basis vector. Since any vector equals the vector sum of its projections in all "directions", it follows that the sum of the basis projectors must be the unit operator,

$$\Sigma_i \hat{P}_i = \Sigma_i |\varphi_i\rangle\langle\varphi_i| = 1 \tag{2.3.10}$$

2.3.5 Unitary Operators

A unitary operator is an operator with an inverse and whose inverse equals its adjoint, $\hat{U}^+ = \hat{U}^{-1}$. The same expression holds for the matrix of components. Unitary operators have the property that they preserve the norm of a vector and so they are the generalisation of the concept of rotation in ordinary physical space. Thus, if $|\xi\rangle = \hat{U}|\psi\rangle$ then $\langle\xi|\xi\rangle = \langle\psi|\psi\rangle$. This follows simply from $\langle\xi|\xi\rangle = \langle\psi|\hat{U}^+\hat{U}|\psi\rangle = \langle\psi|\hat{U}^{-1}\hat{U}|\psi\rangle = \langle\psi|\psi\rangle$, because $\left(\hat{Q}|\psi\rangle\right)^+ = \langle\psi|\hat{Q}^+$.

An operator which transforms from one orthonormal basis to another must be unitary. Proof: put $|\psi_i\rangle = \hat{Q}|\varphi_i\rangle$ where both $|\psi_i\rangle$ and $|\varphi_i\rangle$ form orthonormal bases. It follows that, $\langle\psi_n|\psi_i\rangle = \delta_{ni} = \langle\varphi_n|\hat{Q}^+\hat{Q}|\varphi_i\rangle = \langle\varphi_n|\hat{Q}^+ \Sigma_k|\varphi_k\rangle\langle\varphi_k|\hat{Q}|\varphi_i\rangle = \Sigma_k(Q^+)_{nk}(Q)_{ki}$, so that the matrix of components of (Q^+) is seen to be the inverse of that of (Q), hence (Q) is unitary. Here we have made use of (2.3.10).

If $\hat{H}$ is an Hermitian operator, and α is a real number, then the operator formed by exponentiation, $\hat{U} = exp(i\alpha\hat{H})$, is unitary. Here $i = \sqrt{-1}$. This follows because equ.(2.3.1) shows that $\left[exp(i\alpha\hat{H})\right]^+ = exp(-i\alpha\hat{H})$ and also that $exp(-i\alpha\hat{H})exp(i\alpha\hat{H}) = exp(0) = 1$, and hence that $\hat{U}^+ = \hat{U}^{-1}$.

2.3.6 Expectation Values and Matrix Elements

So far we have discussed only abstract mathematical objects: vectors and operators in Hilbert space. But contact with the physical world will require numbers. The inner product is the route for obtaining numbers, though in generally it is a complex number. We will see that certain operators represent familiar physical quantities, such as energy or momentum. Such operators can be regarded as "machines" for extracting the corresponding physical quantity, the actual value of which depends upon the state of the system – which, recall, is being represented by a Hilbert space vector such as $|\psi\rangle$. One of the lessons of quantum mechanics is that not all physical quantities are uniquely defined even if the quantum state, $|\psi\rangle$, is specified. Instead we talk of the "expectation value" of an observable (represented by an operator, $\hat{Q}$) in the state $|\psi\rangle$.

Physically this means the average value that would be found for this observable quantity if it were measured repeatedly on a system prepared each time in the same state $|\psi\rangle$. Quantum mechanics posits that the expectation value is given by,

$$\langle \hat{Q} \rangle = \langle \psi | \hat{Q} | \psi \rangle \tag{2.3.11}$$

Physical quantities are required to be real. This requires physical quantities to be represented by Hermitian operators because,

$$\langle \hat{Q} \rangle^* = \langle \psi | \hat{Q} | \psi \rangle^* = \langle \psi | \hat{Q}^+ | \psi \rangle \tag{2.3.12}$$

and this is real for arbitrary $|\psi\rangle$ if and only if $\hat{Q} = \hat{Q}^+$, i.e., if $\hat{Q}$ is Hermitian.

The quantity $\langle \varphi | \hat{Q} | \psi \rangle$ is often called the "matrix element" of the operator $\hat{Q}$ between states $|\varphi\rangle$ and $|\psi\rangle$. In general, when the states $|\varphi\rangle$ and $|\psi\rangle$ are different, such matrix elements are complex valued, even for Hermitian operators. These "off diagonal" matrix elements do not represent observable quantities but their square-magnitude, $|\langle \varphi | \hat{Q} | \psi \rangle|^2$, is often related to experimentally measurable quantities. For example, when $\hat{Q}$ is some form of interaction energy, such as that between an atomic electron and electromagnetic radiation, this term can be shown to be proportional to the rate at which some transition takes place, for example an atomic electron moving to a different atomic energy level (look up Fermi's Golden Rules in a standard text).

2.4 Evolution (Dynamics): The Schrodinger Equation

Evolution in time can be formulated in different ways in quantum mechanics, depending upon whether the time dependence is attributed to the state vector or to the operators which represent observables (more of which later). These alternatives are referred to as the Schrodinger and Heisenberg pictures respectively. In this section we shall use the Schrodinger picture in which the state carries the time dependence. (In fact the Schrodinger picture will be used throughout this book unless otherwise stated).

For any system there will be an operator which represents energy, called the Hamiltonian, $\hat{H}$. As energy is an observable physical quantity the Hamiltonian must be an Hermitian operator. Those familiar with Noether's Theorem will know that time and energy are fundamentally related in that the conservation of energy results from translational invariance in time (the homogeneity of time), something that we will explore in the quantum mechanical context in chapter 5. Even before Emmy Noether proved the theorem in the second decade of the twentieth century, by the mid-nineteenth century William Rowan Hamilton had cast classical dynamics in a form which made explicit how energy as a function of generalised position and momentum variables was related to the time derivatives of those quantities. I mention this history so the fundamental quantum mechanical dynamical equation does not appear to have been pulled entirely out of the air. It is this,

$$\hat{H}|\psi\rangle = i\hbar\partial_t|\psi\rangle \qquad (2.4.1)$$

where ∂_t is the time derivative, $\frac{d}{dt}$. Equ.(2.4.1) is Schrodinger's equation.

In (2.4.1) the term $\hbar$ is the universally used notation for $h/2\pi$ and h is Planck's constant. It is merely historical accident that Planck's constant was originally defined via the quantum of energy associated with electromagnetic radiation of frequency f, namely $E = hf$. Later it was found that h almost always occurs in equations in the form $h/2\pi$ and so it became a natural shorthand to define $\hbar = h/2\pi$. It is a universal constant, $\hbar = 1.054$ x 10^{-34} Js. I shall be lazy and often refer to "Planck's constant" when I really mean $\hbar$. Note that the units (or dimensions) of Planck's constant can be regarded equivalently as energy ✕ time or as momentum ✕ length or simply as angular momentum. (The reader should check that these units are all the same).

The reader may be more familiar with Schrodinger's equation in its explicit form for a single particle in a potential field, in which the Hamiltonian operator appears as a differential operator acting on real physical space. We shall deploy this version of Schrodinger's equation, which is technically an infinite dimensional Hilbert space representation, only in chapter 28. Apart from that this book will always use finite dimensional Hilbert space representations of (2.4.1) which, in the author's opinion, leads to greater conceptual clarity and is sufficient for our purposes. In any case, equ.(2.4.1) is the most general form of Schrodinger's famous equation.

Schrodinger's equation tells us how a state vector develops in time. One can imagine finding the state at time t, which we will write $|\psi(t)\rangle$, by integrating (2.4.1) from some starting state at time zero, $|\psi(0)\rangle$. In principle such integration might be carried out numerically if algebraic integration is too hard, by incremental time stepping. However, in the case that the Hamiltonian operator has no explicit time dependence (which is usually the case) a formal solution of equ.(2.4.1) can be written down immediately,

$$|\psi(t)\rangle = exp\left\{-i\frac{\hat{H}t}{\hbar}\right\}|\psi(0)\rangle \qquad (2.4.2)$$

where the exponentiated operator is to be understood as in equ.(2.3.1). That (2.4.2) is a solution of (2.4.1) can be checked by direct substitution (confirmation of which is an exercise for the reader). In practice this formal solution may be some way from an explicit solution as the exponentiated operator will generally not be easy to evaluate, and its action on the state $|\psi(0)\rangle$ also needs to be calculated. Nevertheless, equ.(2.4.2) is important conceptually in specifying how the Hamiltonian operator drives the time evolution of the state.

Recall from §2.3.5 that the operator appearing in equ.(2.4.2) must be unitary by virtue of the Hamiltonian operator being Hermitian. Consequently, the evolution equation preserves normalisation, i.e.,

$$\langle\psi(t)|\psi(t)\rangle = \langle\psi(0)|\psi(0)\rangle \tag{2.4.3}$$

This is known as "unitarity" as it follows from the unitary nature of the operator $\hat{U}$ in $|\psi(t)\rangle = \hat{U}|\psi(0)\rangle$ where $\hat{U} = exp\left\{-i\frac{\hat{H}t}{\hbar}\right\}$. Physically the normalisation of the state vector refers to the fact that the total probability that the particle is at some location must always be unity (everything has to be somewhere), and unitarity is the requirement that the dynamical evolution of the state must maintain this unit total probability at all times.

2.4.1 Stationary States

An initial state which is an eigenvector of the Hamiltonian with energy E is called a "stationary state". However, despite the name, it retains a degree of time dependence as (2.4.2) becomes $|\psi(t)\rangle = exp\{-iEt/\hbar\}|\psi(0)\rangle$. Hence the state retains only the minimal time-dependence through the phase factor with a fixed frequency, $E/\hbar$. We shall have need of this in chapters 8 and 28.

2.5 Eigenvectors and Eigenvalues

For a given operator there will, in general, be some vectors for which the effect of the operator is simply to multiply the vector by a number, i.e.,

$$\hat{Q}|q_k\rangle = q_k|q_k\rangle \tag{2.5.1}$$

Here q_k is a number (complex in general) and the notation $|q_k\rangle$ refers to a vector such that equ.(2.5.1) holds. Any non-null vector such that (2.5.1) holds is said to be an eigenvector of the operator $\hat{Q}$ with q_k as its corresponding eigenvalue. The subscript k indicates that there will, in general, be many different eigenvectors with their associated eigenvalues. If $|q_k\rangle$ obeys (2.5.1) then so does any numerical multiple, $\alpha|q_k\rangle$. However we assume eigenvectors are normalised to unity and all such multiples are "the same" eigenvector. An eigenvalue which has more than one non-parallel eigenvector is said to be degenerate. The set of all eigenvalues of a given operator is called its spectrum. Trivially the zero vector obeys (2.5.1) but this is excluded from the definition of a valid eigenvector.

If an operator has been written in terms of components wrt an orthonormal basis, as in equ.(2.3.2), then the eigenvalues of the operator are the eigenvalues of the matrix of components, $(Q) = \{Q_{ik}\}$. The eigenvalues can be found by solving the "secular equation", which is the requirement that, for a non-zero eigenvector, the determinant of $(Q) - q_i(I)$ must be zero, where (I) is the unit matrix. In a Hilbert space of dimension N where (Q) will be an $N \times N$ matrix, there will therefore be N eigenvalues, defined as the roots of the Nth order polynomial $\|(Q) - q_i(I)\| = 0$. As a polynomial may have equal roots, some of these eigenvalues may be the same. However, they will correspond to distinct eigenvectors.

Some operators may have some eigenvalues which are zero, in which case the secular equation becomes $\|(Q)\| = 0$, which shows that only operators with no inverse can have a zero eigenvalue. (This generalises the fact that zero itself has no inverse).

Certain types of operator have eigenvalues and eigenvectors with particularly important properties which are of central importance in quantum mechanics, examples of which follow.

2.5.1 Hermitian Operators

Hermitian operators have real eigenvalues and eigenvectors which are mutually orthogonal. The proof is,

Put $\qquad \hat{H}|\psi_i\rangle = \lambda_i|\psi_i\rangle$

Then $\qquad \langle\psi_j|\hat{H}^+ = \langle\psi_j|\hat{H} = \lambda_j^*\langle\psi_j|$

Hence $\qquad \langle\psi_k|\hat{H}|\psi_i\rangle = \lambda_i\langle\psi_k|\psi_i\rangle = \lambda_k^*\langle\psi_k|\psi_i\rangle \qquad (2.5.2)$

In the case $i = k$ the norm $\langle\psi_i|\psi_i\rangle$ cannot be zero and hence (2.5.2) gives $\lambda_i = \lambda_i^*$, i.e., the eigenvalues are real. However, if the eigenvalues are different, and knowing now that they are real, (2.5.2) can hold only if $\langle\psi_k|\psi_i\rangle = 0$, i.e. the eigenvectors corresponding to different eigenvalues are orthogonal. QED.

A subtlety occurs for degenerate eigenvalues, i.e., when distinct eigenvectors have the same eigenvalue. In such a case, any linear combination of the degenerate eigenvectors is also an eigenvector with the same eigenvalue. But this "degenerate subspace" can be spanned by a mutually orthogonal set of eigenvectors, so the result is the essentially same. We merely assume that some (arbitrary) orthonormal basis of this subspace has been chosen as the eigenvectors in question.

We have already anticipated in §2.3.6 that Hermitian operators will play the part of physical observables because they have real expectation values, $\langle\psi|\hat{H}|\psi\rangle$. In the case that the state is an eigenstate, this becomes equivalent to the eigenvalues being real, i.e., $\langle\psi_i|\hat{H}|\psi_i\rangle = \lambda_i$ assuming normalisation.

2.5.2 Unitary Operators

Unitary operators have eigenvalues which are complex numbers of unit modulus, i.e., pure phase factors of the form $e^{i\theta}$ for real θ. Their eigenvectors are also mutually orthogonal. Proof,

Put $\qquad \hat{U}|\psi_i\rangle = \lambda_i|\psi_i\rangle$

Then $\qquad \langle\psi_j|\hat{U}^+ = \langle\psi_j|\hat{U}^{-1} = \lambda_j^*\langle\psi_j|$

Hence $\qquad \langle\psi_j| = \lambda_j^*\langle\psi_j|\hat{U}$

And thus $\qquad \langle\psi_i|\psi_k\rangle = \lambda_i^*\langle\psi_i|U|\psi_k\rangle = \lambda_i^*\lambda_k\langle\psi_i|\psi_k\rangle$ $\qquad$ (2.5.3)

Hence, in the case $i = k$ the norm $\langle\psi_i|\psi_i\rangle$ cannot be zero and hence we have $\lambda_i^*\lambda_i = 1$, i.e. the eigenvalues are of unit modulus as claimed. Where the eigenvalues are different we thus have $\lambda_k^*\lambda_i = e^{i(\theta_i-\theta_k)}$, and the exponent is non-zero (and not a multiple of 2π), so this cannot equal unity. Hence to satisfy (2.5.3) we must have $\langle\psi_i|\psi_k\rangle = 0$ when $i \neq k$, i.e., the eigenvectors are orthogonal. As for Hermitian matrices there is a subtlety for degenerate eigenvalues, but this is salvaged in the same manner by adopting a mutually orthogonal set of eigenvectors which span such degenerate subspaces.

It follows from the above observations that any Hermitian or unitary operator can be used to define a basis, i.e., its eigenvectors can be used to form an orthonormal basis.

2.5.3 Projection Operators

Projection operators were defined in §2.3.4. Projection operators have one eigenvalue equal to 1, and the remaining eigenvalues are zero. To see this write $\hat{P}|\psi_i\rangle = \lambda_i|\psi_i\rangle$ and recall from (2.3.9) that $\hat{P}_\psi^2 = \hat{P}_\psi$. Hence it follows that $\hat{P}_\psi^2|\psi_i\rangle = \lambda_i|\psi_i\rangle = \lambda_i^2|\psi_i\rangle$. Since eigenvectors cannot be the zero vector this means that $\lambda_i = \lambda_i^2$ from which it follows that $\lambda_i = 0$ or 1.

But we can write any projection operator as $\hat{P} = |\xi\rangle\langle\xi|$ for some vector $|\xi\rangle$. It is clear that this $|\xi\rangle$ has eigenvalue 1, i.e., we have $\hat{P}|\xi\rangle = |\xi\rangle$. It is also clear that any vector $|\mu\rangle$ orthogonal to $|\xi\rangle$ gives $\hat{P}|\mu\rangle = |\xi\rangle\langle\xi|\mu\rangle = 0$ and hence has zero eigenvalue. Consequently, a projection operator has just one unity eigenvalue, with eigenvector $|\xi\rangle$, and $N - 1$ zero eigenvalues whose eigenvectors are any basis spanning the $N - 1$ dimensional space orthogonal to $|\xi\rangle$.

2.5.4 The Spectral Representation

Any operator can be written $\hat{Q} \equiv \sum_{i,k} Q_{ik}|\varphi_i\rangle\langle\varphi_k|$ in terms of an arbitrary orthonormal basis $\{|\varphi_i\rangle\}$. If $\hat{Q}$ is an operator of a type whose eigenvectors form an orthonormal basis, e.g., Hermitian or unitary, then we can use these eigenvectors as the basis, i.e., use $\{|\varphi_i\rangle\}=\{|q_i\rangle\}$, where its eigenvalues are q_i and $|q_i\rangle$ are the corresponding eigenvectors. In which case $Q_{ik} = \langle q_i|\hat{Q}|q_k\rangle = q_i\delta_{ik}$ and hence the matrix of components is diagonal in this basis, and hence,

$$\hat{Q} \equiv \sum_i q_i|q_i\rangle\langle q_i| \tag{2.5.4}$$

This is called the "spectral representation" of the operator. It is valid if the operator's eigenvectors form an orthonormal basis. In this basis the matrix representation of the operator is diagonal.

2.5.5 Non-Commuting Operators Have Different Eigenvectors

In this section we show that non-commuting operators must have at least some eigenvectors which are different. If two operators, $\hat{P}$ and $\hat{Q}$, are applied in turn to a given vector, the result will depend, in general, upon the order in which they are applied, i.e., $\hat{P}\hat{Q}|\varphi\rangle$ will generally, though not always, produce a different vector from $\hat{Q}\hat{P}|\varphi\rangle$. The juxtaposition of two operators is just another operator (because a mapping from Hilbert space to itself, followed by another such mapping is just another mapping from Hilbert space to itself). Hence $\hat{P}\hat{Q}$ is another operator, as is $\hat{Q}\hat{P}$, but they are distinct operators if, for any vector, they produce different results. In this case $\hat{P}\hat{Q} \neq \hat{Q}\hat{P}$ and the difference between them is therefore non-zero. The "commutator" of the two operators is defined as this difference,

$$[\hat{P}, \hat{Q}] \equiv \hat{P}\hat{Q} - \hat{Q}\hat{P} \tag{2.5.5}$$

Theorem: Non-commuting Hermitian operators have at least some different eigenvectors (i.e., their set of eigenvectors is not the same). Proof: Suppose that $\hat{P}$ and $\hat{Q}$ have the same set of eigenvectors. We therefore have $\hat{P}|\varphi_i\rangle = p_i|\varphi_i\rangle$ and $\hat{Q}|\varphi_i\rangle = q_i|\varphi_i\rangle$ and $(\hat{P}\hat{Q} - \hat{Q}\hat{P})|\varphi_i\rangle = (p_i q_i - q_i p_i)|\varphi_i\rangle = 0$. As the operators are Hermitian their eigenvectors form an orthonormal basis and so any vector can be written $|\mu\rangle = \Sigma_i A_i|\varphi_i\rangle$. And so for any vector $(\hat{P}\hat{Q} - \hat{Q}\hat{P})|\mu\rangle = \Sigma_i A_i(p_i q_i - q_i p_i)|\varphi_i\rangle = 0$. But an operator which produces zero for every vector is the zero operator, i.e., $[\hat{P}, \hat{Q}] = (\hat{P}\hat{Q} - \hat{Q}\hat{P}) = 0$. So, if they have the same set of eigenvectors, $\hat{P}$ and $\hat{Q}$ must commute. So, if $\hat{P}$ and $\hat{Q}$ do not commute they cannot have the same set of eigenvectors.

There is a special case when the commutator $\hat{P}\hat{Q} - \hat{Q}\hat{P}$ is a non-zero number, Z (times the unit operator). This is an important special case as it applies between conjugate variable (see chapter 5). In this case for any common eigenvector we have $(\hat{P}\hat{Q} - \hat{Q}\hat{P})|\varphi_i\rangle = Z|\varphi_i\rangle = 0$. But if Z is non-zero this means $|\varphi_i\rangle = 0$, which is a contradiction, as an eigenvector cannot be the zero vector (by definition). It follows that when the commutator $\hat{P}\hat{Q} - \hat{Q}\hat{P}$ is a non-zero number, and hence for conjugate variables, **all** the eigenvectors of $\hat{P}$ and $\hat{Q}$ are different.

2.5.6 Commuting Operators Have the Same Eigenvectors

The complementary theorem is that operators which commute do have the same set of eigenvectors. Hence if $\hat{H}|\varphi_i\rangle = E_i|\varphi_i\rangle$ and $\hat{H}$ commutes with another operator, $\hat{P}$, then $\hat{P}\hat{H}|\varphi_i\rangle = \hat{H}\hat{P}|\varphi_i\rangle = E_i\hat{P}|\varphi_i\rangle$ and hence $\hat{P}|\varphi_i\rangle$ is also an eigenvector of $\hat{H}$ with the same eigenvalue, E_i. Assuming that this is a non-degenerate eigenvalue this implies that $\hat{P}|\varphi_i\rangle = p_i|\varphi_i\rangle$ for some number p_i, in other words that $|\varphi_i\rangle$ is also an eigenvector of $\hat{P}$. Note, however, that the two operators will generally have distinct eigenvalues.

In the case that E_i is a degenerate eigenvalue we can only conclude that $\hat{P}|\varphi_i\rangle$ lies in the degenerate subspace spanned by the different eigenvectors corresponding to E_i. An arbitrary vector in this subspace will not, in general, be an eigenvector of $\hat{P}$. For example, if the degenerate subspace is two dimensional and spanned by $|\varphi_1\rangle$ and $|\varphi_2\rangle$ then we can only conclude that $\hat{P}|\varphi_1\rangle = \alpha|\varphi_1\rangle + \beta|\varphi_2\rangle$ for some complex numbers α and β, and similarly $\hat{P}|\varphi_2\rangle = \gamma|\varphi_1\rangle + \mu|\varphi_2\rangle$. However, we can always find an alternative basis within the degenerate subspace, $|\tilde{\varphi}_1\rangle$ and $|\tilde{\varphi}_2\rangle$, which are eigenvectors of $\hat{P}$, such that $\hat{P}|\tilde{\varphi}_1\rangle = p_1|\tilde{\varphi}_1\rangle$ and $\hat{P}|\tilde{\varphi}_2\rangle = p_2|\tilde{\varphi}_2\rangle$. In general, the two operators can be taken to have the same eigenvectors, but only if the eigenvectors in degenerate subspaces are chosen appropriately. The proof is an exercise for the reader (see §2.11 for the solution).

The operators for energy (the Hamiltonian) and for momentum provide a common example. In many cases these operators commute. It is also common for energy eigenvalues to be degenerate. These degenerate states may correspond to a given magnitude of momentum, but with various different directions of the vectorial momentum being possible. Considering a 2D example, the energy E may uniquely define the magnitude of momentum, p, but this may correspond to motion in various directions in the (x, y) plane. As far as the operator for the x-component of momentum, $\hat{P}_x$, is concerned, there are therefore a number of energy-degenerate states with different p_x eigenvalues because only $\sqrt{p_x^2 + p_y^2}$ is constrained to equal p.

2.5.7 Solving for Eigenvalues via the Secular Equation

The use of the secular equation to find eigenvalues and eigenvectors is sufficiently important that it is worth providing an example. Consider the Hermitian matrix,

$$(H) = \begin{pmatrix} 1 & 0 & 0 \\ 0 & 1 & i \\ 0 & -i & 1 \end{pmatrix}$$

Its secular equation for the eigenvalues λ is,

$$\begin{Vmatrix} 1 - \lambda & 0 & 0 \\ 0 & 1 - \lambda & i \\ 0 & -i & 1 - \lambda \end{Vmatrix} = 0$$

The determinant is $(1 - \lambda)[(1 - \lambda)^2 - 1] = (1 - \lambda)[\lambda^2 - 2\lambda]$ whose roots are $\lambda = 1, \lambda = 0, \lambda = 2$. The eigenvectors which obey $(H)\bar{v} = \lambda\bar{v}$ are, for eigenvalues $\lambda = 1, \lambda = 0, \lambda = 2$ respectively,

$$\begin{pmatrix} 1 \\ 0 \\ 0 \end{pmatrix} \qquad \frac{1}{\sqrt{2}}\begin{pmatrix} 0 \\ 1 \\ i \end{pmatrix} \qquad \frac{1}{\sqrt{2}}\begin{pmatrix} 0 \\ 1 \\ -i \end{pmatrix}$$

These are normalised, but note that they are still arbitrary up to a factor $e^{i\theta}$, for example they could be multiplied by -1 or by i or by $(1 - i)/\sqrt{2}$.

2.6 Making Contact with the Physical World: Measurement

Now, at last, we can describe how quantum mechanics makes contact with physically measurable quantities. There are five key interpretational rules. The first three are,

[1] Pure quantum states are represented by Hilbert space vectors (strictly, rays);

[2] Physical observables are represented by Hermitian operators;

[3] The possible outcomes of a measurement of an observable $\hat{Q}$ are the eigenvalues of $\hat{Q}$.

Note that [2] ensures that the outcomes, [3], must be real numbers. Note also that where the Hilbert space is of finite dimension, N, the only possible measurement outcomes are the N discrete values $\{q_i\}$. This discreteness is the defining characteristic of quantum mechanics. Another key lesson that quantum mechanics teaches is that systems described by a pure quantum state are completely defined by a relatively small number of integers.

It remains only to say which of the eigenvalues $\{q_i\}$ is the outcome of a measurement. In the case that the system is in an eigenstate of $\hat{Q}$, i.e., $|q_i\rangle$, then obviously the outcome will be the corresponding eigenvalue, q_i (assuming perfectly efficient, error-free measurements). More generally, because the eigenvectors of an observable form a complete set, an arbitrary state vector can be written $|\psi\rangle = \sum_i a_i |q_i\rangle$ for some complex numbers a_i. In this, most general, case the outcome of the measurement of $\hat{Q}$ is indeterminate. Instead of being able to specify the outcome uniquely, quantum mechanics holds that there is an irreducible randomness in the outcome obtained. However, there is a rule, known as the Born Rule, which determines the probability that the outcome will be q_i for each eigenvalue. So our fourth interpretational principle is,

[4] (Born Rule): The probability of the measurement outcome q_i is $|a_i|^2$

Note that the normalisation of the state vector to unity ensures that the total probability of all possible outcomes is one, i.e., $\sum_i |a_i|^2 = 1$, as it should be. Thus, if the same observable is measured repeatedly on a system prepared in the same state, $|\psi\rangle$, each time, the fraction of the measurements resulting in outcome q_i will be $|a_i|^2$ in the limit of a very large number of measurements.

As an example, if the initial state were $0.9|q_1\rangle + 0.4359|q_2\rangle$, and assuming $|q_1\rangle$ and $|q_2\rangle$ are normalised and orthogonal, then the measured value would be found to be q_1 in 81% of occasions, and q_2 in 19% of occasions, but never found to be anything else (assuming perfect measurement).

Because the probability of $|\psi\rangle = \sum_i a_i |q_i\rangle$ being found to be in state $|q_i\rangle$ is $|a_i|^2$, the coefficients a_i may be regarded as the **amplitude** of the probability. This is suggested by analogy with wave phenomena in which intensity is related to the square of the wave amplitude. Since the original state, $|\psi\rangle$, is linear in the coefficients, it too is a sort of probability amplitude. Thus, if the observable $\hat{Q}$ represents a position measurement, and we make the replacement $|q_i\rangle \rightarrow |x\rangle$ for a continuous position variable, x, then the conventional "Schrodinger wavefunction" is simply $\psi(x) \equiv \langle x|\psi\rangle = a_x$. So the wavefunction is a probability amplitude, not a probability. It is the absolute square of the wavefunction, $|\psi(x)|^2$, which represents the probability density of finding the particle at the given x-position.

Many people have attempted to derive the Born Rule, rather than merely assert it as an axiom. We will not be so ambitious here, being content to regard it as axiomatic.

The nature of quantum measurement as described here disturbed many people when quantum mechanics was in its infancy and continues to do so today. The irreducible indeterminacy inherent in quantum measurement is unique in physics, and perturbs many people raised on classical, strictly deterministic physics. This has led many physicists in the past to posit "hidden variables" which, if known, would render the outcome of every individual measurement deterministic. But there are very strong reasons to believe there can be no such hidden variables, and we shall examine these arguments in detail in chapters 9 and 14.

The final interpretational postulate of the quantum formalism is,

[5]   If the measurement outcome is q_i then the system is left in state $|q_i\rangle$.

Recall that the system is assumed to have been in state $|\psi\rangle = \sum_i a_i |q_i\rangle$ prior to the measurement. Hence the measurement process changes the state of the system, in general. This is referred to as "reduction of the state vector" (or "collapse of the wavefunction"). This is another feature which disturbs many people, and with good reason. The alert reader will notice that the time-evolution specified by the solution of the Schrodinger equation, (2.4.2), seems to have been suspended during a measurement, because the reduction of the state vector appears to involve a completely different evolution. This is clear because repeat measurements on the same initial state will generally produce differing results, according to postulate [4]. In contrast, the solution to Schrodinger's equation, (2.4.2), defines a unique state to result from the evolution of a given state, i.e., the same outcome each time. Thus, the coherent evolution of a pure quantum state obeys equ.(2.4.2) whereas the interaction of the same system with a "large" (classical) measuring device is quite different. That there are two incompatible types of evolution in quantum mechanics is rather a running sore which has never been resolved. It is known as "the measurement problem".

Our purpose in this book, however, is not to attempt to solve these foundational problems, but to take quantum mechanics as it is and examine its consequences in the situations summarised in chapter 1. However, it is worth noting that there is a perspective on quantum measurement which helps explain what lies behind the measurement problem, namely decoherence theory. We take a look at this in chapter 15, but the reader should expect only a partial, not a complete, resolution of the problem.

Another feature of quantum measurement worth noting is that it is an irreversible process. After measurement the original state vector cannot be regained. That is, for any single measurement, all knowledge of the expansion coefficients a_i is lost after the measurement. (More precisely, that should perhaps read "all possibility of obtaining knowledge of the expansion coefficients a_i is lost"). Only if the identical state can be reproduced many times will it become possible to deduce, at least within some statistical error, the relative magnitudes of these expansion coefficients, $|a_i|$. This is another key distinction between the type of evolution which occurs in measurement and the evolution controlled by the Schrodinger equation, namely (2.4.2), as the latter is reversible (because the operator which occurs on the RHS of (2.4.2) is unitary, and hence has an inverse).

That the expansion coefficients a_i cannot, in general, be measured means that the state vector (or wavefunction) is not itself an observable. That the central descriptor of a system's state is formulated in quantum mechanics so as to be empirically unknowable marks a profound departure from classical physics' epistemology.

2.7 Multipartite Systems

It is commonly the case that a system can be considered as comprising two or more distinct parts, e.g., two "particles". If one part, in isolation, is described by a Hilbert space of dimension N_1 and spanned by a basis set of vectors $\{|u_i\rangle\}$, and a second part, in isolation, is described by a Hilbert space of dimension N_2 and spanned by a basis set of vectors $\{|v_k\rangle\}$, then the combined system is described by the tensor product of the two Hilbert spaces, which is of dimension $N_1 N_2$ and is spanned by $\{|u_i\rangle \otimes |v_k\rangle\}$. The tensor product symbol $\otimes$ does not imply any computation and is little more than an ordered pair except that, unlike a mere ordered pair, the tensor product obeys $\alpha|a\rangle\otimes\beta|b\rangle = \alpha\beta|a\rangle\otimes|b\rangle$ for any complex numbers α, β. However, we shall generally omit the tensor product symbol $\otimes$ and write the basis states simply as the juxtaposition $|u_i\rangle|v_k\rangle$.

With this notation the order of the kets is important as this specifies which particle is in which state. The bra formed by taking the conjugate of the ket can either be written as $[|u_i\rangle|v_k\rangle]^+ = \langle v_k|\langle u_i|$ or as $[|u_i\rangle|v_k\rangle]^+ = \langle u_i|\langle v_k|$. If it is clear which state refers to which particle, this is a matter of convention. However, one convention should be chosen and adopted consistently or errors will surely follow. Here we shall use $[|u_i\rangle|v_k\rangle]^+ = \langle u_i|\langle v_k|$. This has the advantage of consistency with an alternative, more compact, notation in which we write $|u_i\rangle|v_k\rangle \equiv |u_i v_k\rangle$ and also $[|u_i\rangle|v_k\rangle]^+ \equiv \langle u_i v_k|$. The extension to three or more parts is obvious, e.g., $|u_i\rangle|v_k\rangle|w_m\rangle \equiv |u_i v_k w_m\rangle$, etc.

Note that the fact that the basis vectors of a multipartite system can be written as Cartesian products (i.e., ordered pairs) of the basis states of the parts should not lead the reader to erroneously assume that all multipartite states can be regarded as Cartesian products of states of its parts. The general state of a bipartite system is an element of the tensor product of the two Hilbert spaces, $\mathcal{H}_1 \otimes \mathcal{H}_2$. The elements of the tensor product space can be written as linear combinations of the product basis vectors, $\sum_{i,k} C_{ik} |u_i v_k\rangle$. This contrasts with the most general Cartesian product of vectors drawn from the two Hilbert spaces which is $\sum_i A_i |u_i\rangle \times \sum_k B_k |v_k\rangle$. The difference is clear when you note that the number of coefficients in the latter is only $N_1 + N_2$ whereas the general element of the tensor product space has $N_1 N_2$ coefficients (ignoring normalisation in both cases).

A simple example of a bipartite state which cannot be expressed as a Cartesian product of part-states is $a|u_1\rangle|v_2\rangle + b|u_2\rangle|v_1\rangle$ if a and b are both non-zero.

The reader should pause to ensure he has grasped firmly the distinction between states within the tensor product space and Cartesian products. We will see that states like $a|u_1\rangle|v_2\rangle + b|u_2\rangle|v_1\rangle$ are the key to entanglement as will be discussed further in chapter 13 and many chapters thereafter. Entanglement is the source of the most bewildering aspects of quantum mechanics, and the essential resource for many quantum technologies (as we will see in chapters 26 and 27). It is remarkable that such profound, and confounding, physical effects follow from such a simple-seeming algebraic property. If a bipartite state is not mathematically separable (i.e., into a product of single particle states) then the physical state is not separable either.

2.8 Including the Measuring Device

One common situation in which we inevitably must deal with a two-part system is in making a measurement, the two parts being the system to be measured and the measuring apparatus itself. Suppose we have a system in which some physical observable can take values a or b and that the corresponding orthogonal states of the system alone are $|a\rangle$ and $|b\rangle$. Suppose the system is in state $\alpha|a\rangle + \beta|b\rangle$. Suppose the state of the measuring device prior to interacting with the system can be written $|M, 0\rangle$ where the 0 represents the device in its "ready to measure" state. Imagine the device to have a dial or display of some sort which can register the result of a measurement being $|a\rangle$ or $|b\rangle$. The state of the device when registering a will be written $|M, a\rangle$ and its state when registering b will be written $|M, b\rangle$.

If the measuring device is perfectly reliable, we must be able to distinguish between states $|M, a\rangle$ and $|M, b\rangle$ with certainty. This means they must be orthogonal $\langle M, a|M, b\rangle = 0$ because otherwise there would be a probability of $|\langle M, a|M, b\rangle|^2$ that an outcome a would be registered when the system state was actually $|b\rangle$ (or vice-versa). This attribute of a perfect measuring device will be important in the chapters which follow.

The combined system-plus-apparatus state, prior to any interaction between them, is simply the product, $(\alpha|a\rangle + \beta|b\rangle)|M, 0\rangle$. After interaction, i.e., after the apparatus has done its job as a measuring device, the state evolves to a state in the tensor product space, $(\alpha|a\rangle|M, a\rangle + \beta|b\rangle|M, b\rangle)$.

I must immediately correct two sleights of hand in that description of measurement. The first is that it assumes the measurement apparatus to be describable as a pure quantum state, something which will almost always be false. However we shall not meet the more general description of states until chapter 10, and shall delay until chapter 15 how this impacts on the formulation of the measurement process, via decoherence theory. So, for pedagogic convenience, I shall maintain this fiction for now. However, there is another crucial issue I have also glossed over.

The apparatus has not really done its job of measuring until the system is left in either state $|a\rangle$ or state $|b\rangle$, rather than remaining in a superposition of the two. In contrast, the state $(\alpha|a\rangle|M, a\rangle + \beta|b\rangle|M, b\rangle)$ has not yet undergone reduction of the state vector. However, it is sometimes useful to consider a measurement to proceed in two steps: firstly by the formation of a non-product bipartite state like $(\alpha|a\rangle|M,a\rangle + \beta|b\rangle|M,b\rangle)$, in which the system and the apparatus have become entangled (see chapter 13), and then followed by a reduction of the combined-state vector. The first of these steps (but not the second) can be considered to occur due to evolution controlled by the Schrodinger equation, (2.4.1). Thus, if the Hamiltonian involves the total energy of the combined system, crucially including the energy due to interaction between the system and the apparatus, then the evolution from the initial state $(\alpha|a\rangle + \beta|b\rangle)|M,0\rangle$ to the state $(\alpha|a\rangle|M,a\rangle + \beta|b\rangle|M,b\rangle)$ arises from,

$$exp\left\{-i\frac{\hat{H}t}{\hbar}\right\}[(\alpha|a\rangle + \beta|b\rangle)|M,0\rangle] = (\alpha|a\rangle|M,a\rangle + \beta|b\rangle|M,b\rangle) \quad (2.8.1)$$

Evolution according to the Schrodinger equation, (2.4.1), is said to be "unitary evolution" because the operator $exp\{-i\hat{H}t/\hbar\}$ is unitary. Unitary evolution preserves normalisation, equ.(2.4.3), and this can be regarded as maintaining the total probability of all outcomes at unity. State vector reduction, however, is irreversible. Hence, in this two-stage view of measurement, the first stage, entanglement with the measuring device, (2.8.1), is unitary – and therefore reversible – whilst it is the second stage, the reduction of the state vector, which is irreversible. A completed measurement involves both parts. The distinction between the first part of a measurement and a completed measurement will be key to understanding some famous brain-twisters such as quantum erasure (chapters 19, 20).

2.9 The Two-Dimensional Hilbert Space

We can call a pair of orthogonal vectors which span a 2-dimensional Hilbert space, say, $|\uparrow\rangle$ and $|\downarrow\rangle$. The symbols which distinguish them are, in principle, of no significance, but have been chosen to be of significance in a certain interpretation, namely spin, as we shall see. In another interpretation we might have written them as $|0\rangle$ and $|1\rangle$. In a matrix representation, these chosen basis vectors are,

$$\begin{pmatrix} 1 \\ 0 \end{pmatrix} \quad \text{and} \quad \begin{pmatrix} 0 \\ 1 \end{pmatrix} \quad (2.9.1)$$

We are particularly interested in Hermitian operators as these are the observables. The most general Hermitian operator in a 2-dimensional Hilbert space, represented as a 2 x 2 Hermitian matrix, is,

$$(H) = \begin{pmatrix} t+z & x-iy \\ x+iy & t-z \end{pmatrix} \quad (2.9.2)$$

where t, x, y, z are arbitrary real numbers. This can be written,

$$(H) = t \begin{pmatrix} 1 & 0 \\ 0 & 1 \end{pmatrix} + +x \begin{pmatrix} 0 & 1 \\ 1 & 0 \end{pmatrix} + y \begin{pmatrix} 0 & -i \\ i & 0 \end{pmatrix} + z \begin{pmatrix} 1 & 0 \\ 0 & -1 \end{pmatrix} \qquad (2.9.3)$$

The unit matrix is not an interesting observable as every vector is an eigenvector of the unit matrix. The three independent Hermitian operators which represent non-trivial observables, up to an arbitrary real multiplicative constant, are,

$$\sigma_x = \begin{pmatrix} 0 & 1 \\ 1 & 0 \end{pmatrix}, \ \sigma_y = \begin{pmatrix} 0 & -i \\ i & 0 \end{pmatrix}, \ \sigma_z = \begin{pmatrix} 1 & 0 \\ 0 & -1 \end{pmatrix} \qquad (2.9.4)$$

These are known as the Pauli matrices and invariably occur extensively throughout quantum mechanics texts, of which this one will be no exception. They represent the independent observables of the simplest non-trivial Hilbert space. This would sufficiently account for their ubiquity. But that they also represent the observables of the quantum mechanical phenomenon of spin, in its minimum non-zero manifestation, is further reason for their universal occurrence.

Note that squaring each of the Pauli matrices produces the unit matrix.

I shall avoid a long digression into how the Pauli matrices represent spin (i.e., the intrinsic angular momentum of a particle). This topic is addressed very well in every standard quantum text, to which I refer the reader who has not already had this drummed into him. I shall just make a few allusions to tantalise those who have yet to be enlightened. Firstly, define quantities I, J, K by $I = -i\sigma_x, J = -i\sigma_y, K = -i\sigma_z$. The reader can readily confirm that these quantities have the algebraic properties,

$$I^2 = J^2 = K^2 = IJK = -1 \qquad (2.9.5)$$

These are the defining properties of the quaternions, as discovered by William Rowan Hamilton and carved by him into the stone of Brougham Bridge in Dublin on 16th October 1843, immediately upon realising their significance. Hamilton had sought the quaternions for some time. His motivation was to find an algebraic means of implementing rotations in three-dimensional space. In the two-dimensional plane, a position can be represented by a complex number, and multiplying by the complex number $e^{i\theta}$ causes a rotation about the origin by θ. Hamilton had been looking for a corresponding means of representing positions and rotations in three-dimensional space. The quaternions are the answer. Thus, a position in 3D space can be represented by the (pure) quaternion $p = xI + yJ + zK$ and a 3D rotation by angle θ about an axis defined by unit vector $\hat{n}$ is implemented by the quaternion $q = exp\{i\theta(n_x I + n_y J + n_z K)/2\}$, the quaternion representing the rotated position being $\tilde{p} = qpq^{-1}$.

Mathematically, the novel thing about the quaternions is that they are non-commutative; that is, for arbitrary quaternions, pq is not generally the same as qp. In fact, the quaternions were the first non-commutative division algebra (i.e., an algebra in which division is always possible, other than for the zero element). This non-commutative property is not incidental. It is essential for quantities to represent 3D rotations, as Hamilton required, because 3D rotations are themselves non-commutative. If the reader has never checked this for himself, I suggest you do so now. Take an object and perform two rotations about different axes in 3D space. Now carry out those rotations in the reverse order. In general, the final orientation of the object will be different.

The connection between the Pauli matrices and quaternions leads naturally to their connection also with angular momentum – provided the reader is already familiar with the connection between rotations and angular momentum, though at this point we anticipate chapter 5. The way in which the connection between the Pauli matrices and spin (= angular momentum) is usually accomplished, however, is via their commutation properties. To be explicit, the angular momentum operators for a (so-called) spin-half particle (generally simply referred to as the "spin operators") are,

$$\hat{S}_x = \frac{\hbar}{2}\sigma_x, \quad \hat{S}_y = \frac{\hbar}{2}\sigma_y, \quad \hat{S}_z = \frac{\hbar}{2}\sigma_z \tag{2.9.6}$$

These have the commutators, as defined by (2.5.5), as follows,

$$[\hat{S}_x, \hat{S}_y] = i\hbar\hat{S}_z \quad [\hat{S}_y, \hat{S}_z] = i\hbar\hat{S}_x \quad [\hat{S}_z, \hat{S}_x] = i\hbar\hat{S}_y \tag{2.9.7}$$

The terminology "spin half" is merely historical and arises from the factor of ½ in (2.9.6), which in turn results from Planck's constant having been defined (via $E = hf$) prior to spin being discovered. The commutators of the spin operators for any magnitude of spin, not just spin-half, are identical to (2.9.7). In fact, angular momentum generally, including "ordinary" orbital angular momentum, has this same commutator algebra. This is not coincidence. In mathematical terms, (2.9.7) defines the Lie algebra of the Lie group of rotations in 3D Euclidean space. So we come back to Hamilton and recognise the deep connection between the rotational transformation properties of 3D space, their non-commutativity and the specific commutators given by (2.9.7), and how these relate to the quantum mechanical description of angular momentum and the discovery of discrete spin. Yet it is curious that the Pauli matrices arise simply as the most general expression of an Hermitian operator in the lowest non-trivial (i.e. two-dimensional) Hilbert space. Those of a metaphysical bent may pontificate on how or why the properties of 3D Euclidean space arise from this purely algebraic starting point.

In chapter 29 we shall derive the eigenvalues and the eigenvector structure of general angular momentum / spin states using a purely algebraic approach (i.e., without appeal to specific representations). The method is so mathematically elegant that every student of quantum mechanics should be familiar with it. However, for now, let's look at the eigenstates of the Pauli matrices which characterise the general observable in any 2D Hilbert space (perhaps a spin-half particle). The eigenvectors of σ_z are just (2.9.1), with eigenvalues +1 and -1 respectively. (Note that this means the measured spin components of a spin half particle, i.e., the eigenvalues of $\hat{S}_z$, are $+\frac{\hbar}{2}$ and $-\frac{\hbar}{2}$). The possible values of the spin component cannot depend upon which direction in space we choose to measure it, so we expect the eigenvalues of the operators σ_x and σ_y must also be +1 and -1. Indeed they are, and the reader should ensure that he can check the fact. But what about the eigenvectors of these operators? Suppose we denote the state with σ_x eigenvalue +1 as $|\uparrow\rangle_x$ and that with x-eigenvalue -1 as $|\downarrow\rangle_x$, the reader should also confirm that he can derive,

$$|\uparrow\rangle_x = \tfrac{1}{\sqrt{2}}\begin{pmatrix}1\\1\end{pmatrix} = \tfrac{1}{\sqrt{2}}(|\uparrow\rangle_z + |\downarrow\rangle_z) \qquad (2.9.8a)$$

$$|\downarrow\rangle_x = \tfrac{1}{\sqrt{2}}\begin{pmatrix}1\\-1\end{pmatrix} = \tfrac{1}{\sqrt{2}}(|\uparrow\rangle_z - |\downarrow\rangle_z) \qquad (2.9.8b)$$

$$|\uparrow\rangle_y = \tfrac{1}{\sqrt{2}}\begin{pmatrix}1\\i\end{pmatrix} = \tfrac{1}{\sqrt{2}}(|\uparrow\rangle_z + i|\downarrow\rangle_z) \qquad (2.9.8c)$$

$$|\downarrow\rangle_y = \tfrac{1}{\sqrt{2}}\begin{pmatrix}1\\-i\end{pmatrix} = \tfrac{1}{\sqrt{2}}(|\uparrow\rangle_z - i|\downarrow\rangle_z) \qquad (2.9.8d)$$

Note that the two x-eigenstates are orthogonal, as are the two y-eigenstates and the two z-eigenstates, but they are not orthogonal to each other. For example, $\langle\downarrow_z|\uparrow\rangle_x = 1/\sqrt{2}$ and $\langle\downarrow_z|\downarrow\rangle_x = -1/\sqrt{2}$. Thus, for example, a state observed to be spin-down wrt the z-axis, if then measured wrt the x-axis will produce positive x-spin half the time and negative x-spin half the time – recalling that probabilities are given by the squares of these inner products. The expectation value of the x-spin, however, is zero.

2.10 The States of Two Spin-Half Particles

As before, we shall work with the Pauli matrices directly. It should be recalled that, if interpreted for spin-half, a factor of $\hbar/2$ should be applied, as per equs.(2.9.6). It will be understood that we are working in the z-basis throughout, and the subscripts to this effect will be dropped. The key action of the Pauli operators on the z-basis states are summarised thus,

$$\sigma_z|\uparrow\rangle = |\uparrow\rangle \qquad \sigma_z|\downarrow\rangle = -|\downarrow\rangle$$
$$\sigma_x|\uparrow\rangle = |\downarrow\rangle \qquad \sigma_x|\downarrow\rangle = |\uparrow\rangle$$
$$\sigma_y|\uparrow\rangle = i|\downarrow\rangle \qquad \sigma_y|\downarrow\rangle = -i|\uparrow\rangle \qquad (2.10.1)$$

It is immediately clear that four orthogonal states which span the tensor product space which describes the bipartite Hilbert space of two spin-half particles are $|\uparrow\uparrow\rangle, |\uparrow\downarrow\rangle, |\downarrow\uparrow\rangle, |\downarrow\downarrow\rangle$, recalling that $|\uparrow\downarrow\rangle$ is shorthand for $|\uparrow\rangle|\downarrow\rangle$. The total angular momentum about the z-axis is defined by the operator $\hat{\sigma}_{Az} + \hat{\sigma}_{Bz}$, where the A and B subscripts indicate that the operator acts only on the first, or only on the second, particle. Thus we find, for example,

$$(\hat{\sigma}_{Az} + \hat{\sigma}_{Bz})|\uparrow\uparrow\rangle = (\hat{\sigma}_{Az}|\uparrow\rangle)|\uparrow\rangle + |\uparrow\rangle(\hat{\sigma}_{Bz}|\uparrow\rangle) = 2|\uparrow\rangle|\uparrow\rangle \qquad (2.10.2)$$

The state $|\uparrow\uparrow\rangle$ is found to have 2 units of angular momentum about the z-axis (where the unit is $\hbar/2$). This is only reasonable as each particle contributes one unit, and the two particles spins are aligned in the z-direction. In the same way we find that $|\downarrow\downarrow\rangle$ has z-angular momentum of -2 units. But for state $|\uparrow\downarrow\rangle$ it works out like this, $\qquad (2.10.3)$

$$(\hat{\sigma}_{Az} + \hat{\sigma}_{Bz})|\uparrow\downarrow\rangle = (\hat{\sigma}_{Az}|\uparrow\rangle)|\downarrow\rangle + |\uparrow\rangle(\hat{\sigma}_{Bz}|\downarrow\rangle) = |\uparrow\rangle|\downarrow\rangle - |\uparrow\rangle|\downarrow\rangle = 0$$

Again, this is only reasonable since the two spins are opposed and so cancel, giving zero angular momentum about the z-axis. The same applies, of course, to state $|\downarrow\uparrow\rangle$.

So far, so good. We have found that one of the four basis states has total z-spin +2, one has -2, and two have zero. But there are two other components of angular momentum, about the x and y axes. However, since the operators for the x and y components of angular momentum do not commute with that for z, i.e., (2.9.7), their values cannot be deterministically specified in the z-basis, i.e., $|\uparrow\uparrow\rangle, |\uparrow\downarrow\rangle, |\downarrow\uparrow\rangle, |\downarrow\downarrow\rangle$ are not eigenstates of the x or y spins. However, there is another spin observable which does commute with the z-component of spin and this is the scalar magnitude of the spin. This is defined by its square,

$$\hat{S}^2 = \hat{S}_x^2 + \hat{S}_y^2 + \hat{S}_z^2 \qquad (2.10.4)$$

An exercise for the reader is to confirm that the commutation relations (2.9.7) imply that $\hat{S}^2$ commutes with each component, $\hat{S}_i$, in particular $\hat{S}_z$. This is trivial in terms of the Pauli matrix representation because each of the squared operators σ_i^2 is the unit matrix. However, $[\hat{S}^2, \hat{S}_z] = 0$ follows purely from the commutation relations, (2.9.7), irrespective of representation or magnitude of spin. It would hold just as well if we were dealing with an object with 20 spin units.

For our two-particle states, the total scalar spin-squared operator is,

$$\Sigma_{tot}^2 = (\sigma_{Ax} + \sigma_{Bx})^2 + (\sigma_{Ay} + \sigma_{By})^2 + (\sigma_{Az} + \sigma_{Bz})^2 \qquad (2.10.5)$$

Calculation is simplified by noting that, for each component, the Pauli representation gives $(\sigma_{Ai} + \sigma_{Bi})^2 = 2(1 + \sigma_{Ai}\sigma_{Bi})$. Thus we find, using (2.10.1),

$$\Sigma_{tot}^2 |\uparrow\uparrow\rangle = \left[6 + 2(\sigma_{Ax}\sigma_{Bx} + \sigma_{Ay}\sigma_{By} + \sigma_{Az}\sigma_{Bz})\right]|\uparrow\rangle|\uparrow\rangle$$

$$= 6|\uparrow\rangle|\uparrow\rangle + 2(|\downarrow\rangle|\downarrow\rangle + i^2|\downarrow\rangle|\downarrow\rangle + |\uparrow\rangle|\uparrow\rangle) = 8|\uparrow\uparrow\rangle \qquad (2.10.6)$$

In the same way we find $\Sigma_{tot}^2 |\downarrow\downarrow\rangle = 8|\downarrow\downarrow\rangle$. So we have confirmed that $|\uparrow\uparrow\rangle$ and $|\downarrow\downarrow\rangle$ are also eigenstates of the total scalar spin-squared operator, with eigenvalue 8. For readers who are surprised to see 8 appearing here, note that if we reintroduce the omitted factor of $\hbar/2$, which is squared in the spin-squared operator, then the eigenvalue becomes $8(\hbar/2)^2 = 2\hbar^2$. More generally, as we will show in chapter 29, the spin-squared operator takes eigenvalues $S(S+1)\hbar^2$ where $S = 0, \frac{1}{2}, 1, \frac{3}{2}, 2, \frac{5}{2}, ...$ For the states $|\uparrow\uparrow\rangle$ and $|\downarrow\downarrow\rangle$ the total spin scalar, formed from two half-spins, is just $S = 1$, and hence the spin-squared operator has eigenvalue $S(S+1)\hbar^2 = 2\hbar^2$.

However, we have yet to address $|\uparrow\downarrow\rangle$ and $|\downarrow\uparrow\rangle$. We find,

$$\Sigma_{tot}^2 |\uparrow\downarrow\rangle = \left[6 + 2(\sigma_{Ax}\sigma_{Bx} + \sigma_{Ay}\sigma_{By} + \sigma_{Az}\sigma_{Bz})\right]|\uparrow\rangle|\downarrow\rangle$$

$$= 6|\uparrow\rangle|\downarrow\rangle + 2(|\downarrow\rangle|\uparrow\rangle + i(-i)|\downarrow\rangle|\uparrow\rangle - |\uparrow\rangle|\downarrow\rangle)$$

$$= 4(|\uparrow\rangle|\downarrow\rangle + |\downarrow\rangle|\uparrow\rangle) \qquad (2.10.7)$$

And,

$$\Sigma_{tot}^2 |\downarrow\uparrow\rangle = \left[6 + 2(\sigma_{Ax}\sigma_{Bx} + \sigma_{Ay}\sigma_{By} + \sigma_{Az}\sigma_{Bz})\right]|\downarrow\rangle|\uparrow\rangle$$

$$= 6|\downarrow\rangle|\uparrow\rangle + 2(|\uparrow\rangle|\downarrow\rangle + i(-i)|\uparrow\rangle|\downarrow\rangle - |\downarrow\rangle|\uparrow\rangle)$$

$$= 4(|\uparrow\rangle|\downarrow\rangle + |\downarrow\rangle|\uparrow\rangle) \qquad (2.10.8)$$

So, neither $|\uparrow\downarrow\rangle$ nor $|\downarrow\uparrow\rangle$ are eigenvectors of Σ_{tot}^2, and they give the same result when acted upon by this operator. We therefore form the following alternative pair of orthogonal vectors,

$$|0,0\rangle = \tfrac{1}{\sqrt{2}}\left(|\uparrow\downarrow\rangle - |\downarrow\uparrow\rangle\right) \quad \text{and} \quad |1,0\rangle = \tfrac{1}{\sqrt{2}}\left(|\uparrow\downarrow\rangle + |\uparrow\downarrow\rangle\right) \qquad (2.10.9a)$$

From (2.10.7,8) we see that $|0,0\rangle$ is an eigenstate of Σ_{tot}^2 with eigenvalue zero, and $|1,0\rangle$ is an eigenstate of Σ_{tot}^2 with eigenvalue 8. We now also introduce the alternative notation for the other pair of basis vectors,

$$|1,1\rangle = |\uparrow\uparrow\rangle \quad \text{and} \quad |1,-1\rangle = |\downarrow\downarrow\rangle \qquad (2.10.9b)$$

The logic of this notation becomes clearer if the general case is written $|S,m\rangle$, so we see that, in terms of the spin operators with the dimensioned factors reintroduced,

$$\hat{S}^2|S,m\rangle = S(S+1)\hbar^2|S,m\rangle \quad \text{and} \quad \hat{S}_z|S,m\rangle = m\hbar|S,m\rangle \qquad (2.10.10)$$

So m is the z-component of the total spin, in units of $\hbar$. What we have shown by this rather pedestrian derivation is that two spin-half particles form a 4-dimensional Hilbert space of bipartite states which may be spanned by three states with total spin $S = 1$, corresponding to the three possible z-spin components, $m = -1,0,+1$, plus a single spin zero state, $S = m = 0$. Thus, the two states in (2.10.9a) both have zero z-component of spin, but the first has zero total spin whereas the second has $S = 1$ but happens to be aligned with zero component in the z direction.

2.11 Reader Exercises

(1) Proof of Equ.(2.2.1), the triangle inequality for the distance measure defined by the Hilbert inner product, $d(x,z) \le d(x,y) + d(y,z)$.

Recall that $d(x,z) \equiv \||x\rangle - |z\rangle\| = [(\langle x| - \langle z|)(|x\rangle - |z\rangle)]^{1/2}$

Also recall that for two real vectors, of arbitrary dimension, we have a scalar product which can be expressed in terms of the angle between the vectors, i.e., $\bar{a} \cdot \bar{b} = \sum_i a_i b_i = ab\cos\theta_{ab}$, where $a = |\bar{a}|$.

Consider $|A\rangle = |x\rangle - |y\rangle$ and $|B\rangle = |y\rangle - |z\rangle$

so $|C\rangle = |x\rangle - |z\rangle = |A\rangle + |B\rangle$

Hence we require to show $\|C\| \le \|A\| + \|B\|$.

But $\|C\|^2 = (\langle A| + \langle B|)(|A\rangle + |B\rangle) = \|A\|^2 + \|B\|^2 + \langle B|A\rangle + \langle A|B\rangle$

And $\langle B|A\rangle + \langle A|B\rangle = \sum_i(A_i^* B_i + B_i^* A_i) = 2\sum_i|A_i||B_i|\cos(\theta_A - \theta_B)$

where we have written the complex components as $A_k = |A_k|e^{i\theta_A}$. It follows that $\langle B|A\rangle + \langle A|B\rangle \le 2\sum_i |A_i||B_i|$ and hence that

$$\|C\|^2 \le \|A\|^2 + \|B\|^2 + 2\sum_i |A_i||B_i| \quad \text{where, } \|A\|^2 = \sum_i |A_i|^2, \text{ etc.}$$

But $\sum_i |A_i||B_i| = \|A\|\|B\|\cos\theta_{AB} \le \|A\|\|B\|$

Hence, $\|C\|^2 \le \|A\|^2 + \|B\|^2 + 2\|A\|\|B\| = (\|A\| + \|B\|)^2$

And so $\|C\| \le \|A\| + \|B\|$ as both are positive. **QED**

(2) Prove that linearity of $\hat{Q}$ implies linearity of $\hat{Q}^n$.

Proof:

$$\hat{Q}^n[a|\psi\rangle + b|\xi\rangle] = \hat{Q}^{n-1}\hat{Q}[a|\psi\rangle + b|\xi\rangle] = \hat{Q}^{n-1}[a\hat{Q}|\psi\rangle + b\hat{Q}|\xi\rangle]$$

Repeating this process leads to,

$$\hat{Q}^n[a|\psi\rangle + b|\xi\rangle] = a\hat{Q}^n|\psi\rangle + b\hat{Q}^n|\xi\rangle \qquad \textbf{QED}$$

(3) Show that being Hermitian is basis independent.

The operator $\hat{H} = \sum_{ij} H_{ij}|i\rangle\langle j|$ is Hermitian iff the matrix of components is Hermitian, i.e., $(H)^+ = (H)$.

Hence $\hat{H}^+ = \sum_{ij} H_{ij}{}^*|j\rangle\langle i| = \sum_{ij} H_{ji}{}^*|i\rangle\langle j| = \sum_{ij} H_{ij}|i\rangle\langle j| = \hat{H}$.

A change of basis is defined by $|i\rangle = \sum_k U_{ik}|\tilde{k}\rangle$ hence,

$$\hat{H} = \sum_{ij} H_{ij} \sum_{kn} U_{ik}|\tilde{k}\rangle\langle\tilde{n}| U_{jn}{}^*$$

$$= \sum_{ij}\sum_{kn} U^T{}_{ki}H_{ij}U_{jn}{}^*|\tilde{k}\rangle\langle\tilde{n}| = \sum_{kn}\tilde{H}_{kn}|\tilde{k}\rangle\langle\tilde{n}|$$

where the matrix of components is $(\tilde{H}) = (U)^T(H)(U)^*$ from which we immediately have $(\tilde{H})^+ = (\tilde{H})$, using the identity $(ABC)^+ = C^+B^+A^+$.

(4) Show that (2.4.2) is the formal solution of Schrodinger's equ.(2.4.1).

That the solution (2.4.2) obeys the Schrodinger equation, (2.4.1), is established immediately by acting $i\hbar\partial_t$ on it. This brings down a factor of $\hat{H}$ and so (2.4.1) results. To establish the reverse, that (2.4.2) is necessarily the solution of (2.4.1) we have to work a little harder, but not much. Consider a small time increment, δt, from datum time zero. (2.4.1) tells us that, $\hat{H}|\psi(0)\rangle\delta t \approx i\hbar(|\psi(\delta t)\rangle - |\psi(0)\rangle)$ and this can be re-arranged as,

$$|\psi(\delta t)\rangle = (1 - i\hat{H}\delta t/\hbar)|\psi(0)\rangle$$

A second small time increment gives us,

$$|\psi(2\delta t)\rangle = (1 - i\hat{H}\delta t/\hbar)|\psi(\delta t)\rangle$$

and so on. A finite time interval, given by $t = N.\delta t$ as $N \to \infty$, therefore has,

$$|\psi(t)\rangle = \left(1 - i\frac{\hat{H}t}{N\hbar}\right)^N |\psi(0)\rangle$$

But recall the definition of the Naperian exponential is,

$$e^x \equiv LIM_{N\to\infty}\left(1 - \frac{x}{N}\right)^N$$

and hence $\qquad |\psi(t)\rangle = exp\left\{-i\frac{\hat{H}t}{\hbar}\right\}|\psi(0)\rangle$

If you are concerned about this in the context of operator valued quantities, note that it follows simply from the binomial expansion, and that what we mean by e^x when x is an operator is that is simply a short-hand for the familiar power series, $e^x = 1 + x + \frac{x^2}{2!} + \frac{x^3}{3!} + \cdots$. **QED** except that we have skirted around issues of convergence. Whilst e^x is absolutely convergent for numerical x we have not demonstrated this for operator valued x.

(5) If Hermitian operators $\hat{H}$ and $\hat{P}$ commute, and if E_i is a degenerate eigenvalue of $\hat{H}$, then there always exists a basis $\{\varphi_i\}$ of the corresponding degenerate subspace such that the basis vectors are eigenvectors of $\hat{P}$.

Proof: Because $\hat{P}\hat{H}|\varphi_i\rangle = \hat{H}\hat{P}|\varphi_i\rangle = E_i\hat{P}|\varphi_i\rangle$ we can conclude that, $\hat{P}|\varphi_i\rangle = \sum_k A_{ik}|\varphi_k\rangle$ for some basis of the degenerate subspace and for some complex matrix of coefficients A_{ik} where the summation over k extends over the degenerate subspace, all corresponding to the same eigenvalue E_i . We can write $A_{ik} = \langle\varphi_k|\hat{P}|\varphi_i\rangle$ from which it follows that the matrix (A) is Hermitian and hence has orthonormal eigenvectors. Arranging these eigenvectors as columns in a square matrix (B), the matrix $(p_{diag}) = (B)^+(A)(B)$ is diagonal. Call its diagonal elements $p_1, p_2, p_3, \ldots$ Writing $|\varphi_1\rangle, |\varphi_2\rangle, |\varphi_3\rangle \ldots$ as a column matrix $\bar{\varphi}$, we have $\hat{P}\bar{\varphi} = (A)\bar{\varphi}$ hence,

$$(B)^+\hat{P}\bar{\varphi} = \hat{P}(B)^+\bar{\varphi} = (B)^+(A)\bar{\varphi} = (p_{diag})(B)^+\bar{\varphi}$$

Hence, defining the alternative basis of the degenerate subspace by $\bar{\omega} = (B)^+\bar{\varphi}$ this means $\hat{P}\bar{\omega} = (p_{diag})\bar{\omega}$, i.e., $\hat{P}\omega_k = p_k\omega_k$ and so the new basis vectors are eigenvectors of $\hat{P}$. QED.

(6) $[\hat{S}^2, \hat{S}_z] = 0$ follows purely from the commutation relations, (2.9.7).

Proof: It suffices to prove $[\hat{S}_x^2 + \hat{S}_y^2, \hat{S}_z] = 0$. From (2.9.7) we note that,

$\hat{S}_x\hat{S}_z = \hat{S}_z\hat{S}_x - 2i\hat{S}_y$ and hence,

$$\hat{S}_x^2\hat{S}_z = \hat{S}_x\hat{S}_z\hat{S}_x - 2i\hat{S}_x\hat{S}_y = (\hat{S}_z\hat{S}_x - 2i\hat{S}_y)\hat{S}_x - 2i\hat{S}_x\hat{S}_y$$
$$= \hat{S}_z\hat{S}_x^2 - 2i(\hat{S}_x\hat{S}_y + \hat{S}_y\hat{S}_x)$$

Similarly, $\hat{S}_y\hat{S}_z = \hat{S}_z\hat{S}_y + 2i\hat{S}_x$ and hence,

$$\hat{S}_y^2\hat{S}_z = \hat{S}_y\hat{S}_z\hat{S}_y + 2i\hat{S}_y\hat{S}_x = (\hat{S}_z\hat{S}_y + 2i\hat{S}_x)\hat{S}_y + 2i\hat{S}_y\hat{S}_x$$
$$= \hat{S}_z\hat{S}_y^2 + 2i(\hat{S}_x\hat{S}_y + \hat{S}_y\hat{S}_x)$$

Adding these two results gives,

$$(\hat{S}_x^2 + \hat{S}_y^2)\hat{S}_z = \hat{S}_z(\hat{S}_x^2 + \hat{S}_y^2) \qquad \textbf{\underline{QED}}$$

3

No Peeking

Why is interference destroyed by "which path" information? And if we only obtain partial information, is the interference only degraded not completely destroyed?

3.1 The Double Slit Experiment

The Double Slit experiment is archetypal in quantum mechanics. It provides an excellent illustration of quantum weirdness. In its original form, as conducted by Thomas Young in 1801 using light, its significance was to demonstrate the wave nature of light. Light incident on a barrier, opaque apart from containing two, very narrow and very closely spaced parallel slits, was shown to produce an interference pattern on a distant screen. The interference pattern is a series of light and dark parallel bands. Familiar from school physics, the formula for the angular position, θ, of the fringes is $n\lambda = d\sin\theta$, where λ is the wavelength of the light, d is the distance between the slits, and n is the fringe order (1, 2, 3…).

But the quantum mechanical behaviour of entities which are generally regarded as particles includes cases where they behave like waves. Thus, a particle with momentum p can be associated with a de Broglie wavelength of $\lambda = h/p$, where h is Planck's constant (without the denominator of 2π in this case). For example, an electron with an energy of 54eV has a wavelength of 1.67 Angstroms, and hence comparable with interatomic spacings in crystal lattices. This makes possible the use of a crystal as a diffraction grating to diffract "particles" thus demonstrating their dual wave-like nature. In the 1920s this was confirmed in the case of electrons scattering off a crystal of nickel by Davisson and Germer, Ref.[3.1]. This has now become a common technique for investigating materials. For example, the diffraction of neutrons is commonly used by engineers to measure residual stresses within steel structures. The first electron diffraction pattern from an artificially produced screen contain slits was much later, in 1961, Ref.[3.2].

The interpretation of "particle" diffraction experiments took a major leap forwards in 1976 when Merli, Missiroli, and Pozzi reported a diffraction pattern being created when only one electron at a time passed through their apparatus, Ref.[3.3]. More precisely, the pattern builds up as many electrons pass through the apparatus, though only one electron is ever present at one time. This establishes the fundamental feature that it is single particles which can behave in a "wave-like" manner, rather than the interference pattern being the result of interactions between different electrons. In 2002, this single-electron interference experiment was voted the most beautiful physics experiment of all time by the readers of *Physics World*, beating a stupendous cast of competitors for the title, including Newton and Galileo (twice), Ref.[3.4]. With the benefit of hindsight, single particle interference was first demonstrated in 1909 by Geoffrey Taylor, at least if one is willing to regard photons as particles. Taylor demonstrated interference with light so weak that only one photon was within the apparatus for most of the time, Ref.[3.5].

The two slits in a double-slit experiment are just a convenient way of splitting a beam into two paths. There are many ways of accomplishing such beam splitting and subsequently bringing the two beams back together to create interference. We will make extensive use in other chapters of the Mach-Zehnder interferometer, which is another way of achieving beam splitting and subsequent interference. The discussion here will be expressed in terms appropriate for a double-slit experiment, but the reader will readily see that it could apply to any arrangement based on beam splitting and subsequent reuniting to (potentially) create interference.

In a two-slit arrangement an interference pattern is developed on a screen. In terms of the Hilbert space states this can be understood as follows. The state of the "particle" prior to hitting the screen is a coherent superposition of the two states representing passage through each slit, thus,

$$|\psi\rangle = \frac{1}{\sqrt{2}}\left(|U\rangle + |L\rangle\right) \tag{3.1.1}$$

where U and L stand for the upper and lower slits. I have slipped the phrase "coherent superposition" into that last sentence, because that is the language used for waves in classical physics. But note that what it means – and what (3.1.1) expresses – is that there is a single particle which might have passed through either slit and which is in a pure quantum state. The "coherent" bit translates into "pure state", while the "superposition" bit translates into the summation of vectors in (3.1.1).

Each of the states $|U\rangle$ and $|L\rangle$ is normalised to unity, representing a single particle. The factor of $1/\sqrt{2}$ in (3.1.1) is so that $|\psi\rangle$ is also normalised to unity. Consider a given position, x, on the screen. The coherence of two waves means that they differ only by a constant phase difference. I won't drag you through the detailed solution of the Schrodinger equation which reproduces this for the two quantum states. But the result is that we can consider the upper vector as differing from the lower vector only by some phase difference which depends upon x, say $\theta(x)$, so that $|U\rangle = e^{i\theta(x)}|L\rangle$. Consequently (3.1.1) becomes,

$$|\psi\rangle = \frac{1}{\sqrt{2}}\left(e^{i\theta(x)} + 1\right)|L\rangle \qquad (3.1.2)$$

The intensity of the pattern at position x on the screen is proportional to the probability density of the particle reaching that point, which is given by the square magnitude of the wavefunction (see §2.6), $|\psi|^2 = \langle\psi|\psi\rangle$, which (3.1.2) gives to be,

$$\langle\psi|\psi\rangle = 1 + \cos\big(\theta(x)\big) \qquad (3.1.3)$$

If this is not obvious, the calculation is carried out explicitly as follows,

$$\langle\psi|\psi\rangle = \langle L|\frac{1}{\sqrt{2}}\left(e^{-i\theta(x)} + 1\right)\frac{1}{\sqrt{2}}\left(e^{i\theta(x)} + 1\right)|L\rangle$$

$$= \langle L|\frac{1}{2}\left(1 + 1 + e^{-i\theta(x)} + e^{i\theta(x)}\right)|L\rangle$$

$$= 1 + \cos\big(\theta(x)\big)$$

Equ.(3.1.3) produces the familiar interference pattern, with alternating dark bands when θ is an odd multiple of π (zero intensity), and bright bands where θ is an even multiple of π (intensity 2 units). The average intensity is unity, as it should be to consistently represent a single particle.

The reader may worry that, for a single particle state, $\langle\psi|\psi\rangle$ should simply be unity, contrary to (3.1.3). We have taken some liberties with the notation. Strictly, by introducing an explicit spatial dependence in equ.(3.1.2) we are really dealing with the wavefunction, defined as $\psi(x) = \langle x|\psi\rangle$ where $|x\rangle$ is the state defined by a specific spatial location, x. The LHS of (3.1.3) is therefore strictly $|\psi(x)|^2 = \langle\psi|x\rangle\langle x|\psi\rangle$ which is the probability per unit distance of finding the particle at the position x. Normalisation applies only when integrating over all possible spatial locations, $\int|\psi(x)|^2 dx = \int\langle\psi|x\rangle\langle x|\psi\rangle dx = 1$. But $\int|x\rangle\langle x|dx$ is the unit operator as $|x\rangle$ forms a complete set, so we regain $\langle\psi|\psi\rangle = 1$, as we should.

3.2 Why "Which Path" Information Destroys Interference

Now let us introduce a two-state device, M, in the beam path of the lower slit. This device is set up in state $|M{:}U\rangle$ but it will change to state $|M{:}L\rangle$ if a particle passes through it. In other words, M is a device which records which path the particle takes.

Readers will recall that such a measurement is expected to destroy the interference pattern. But why should this always be the case – for any sort of measurement, M, however we choose to contrive it? The impression that is sometimes given is that the wave/particle is physically disturbed by the measurement, and that it is this physical disturbance which destroys the interference. Indeed, this was Heisenberg's original argument. But this is seriously misleading and misrepresents the nature of quantum mechanics. The vanishing of the interference arises from the very possibility of distinguishing paths and is a result of the state algebra alone (or, if you will, the quantum logic). The means of obtaining the "which path" information, however subtle, is irrelevant. (Arrangements which seem to refute this claim will be discussed in chapter 18). The derivation is as follows.

Suppose that M is a perfect measuring device, and suppose that we are certain that exactly one particle has passed through our apparatus. This means that M will infallibly switch to state $|M:L\rangle$ if the particle takes the lower path and infallibly be left in state $|M:U\rangle$ if the particle takes the upper path. However, to be a perfect measurement we must be able to distinguish with certainty between the two states $|M:L\rangle$ and $|M:U\rangle$. In other words, these two states must be orthogonal,

Perfect Measurement: $$\langle M:L|M:U\rangle = 0 \qquad (3.2.1)$$

Prior to any interaction between the particle and the measuring device, their combined state is just the tensor product state $|\Psi\rangle = |\psi\rangle|M:U\rangle = (1/\sqrt{2})(|U\rangle + |L\rangle)|M:U\rangle$ because M is set up in state $|M:U\rangle$. However, after the particle has had opportunity to interact with M the state evolves into,

$$|\Psi\rangle \rightarrow |\Psi'\rangle = \frac{1}{\sqrt{2}}(|U\rangle|M:U\rangle + |L\rangle|M:L\rangle) \qquad (3.2.2)$$

The intensity of the signal received on the screen is proportional to $\langle\Psi'|\Psi'\rangle$ which, from (3.2.2), is now simply,

$$\langle\Psi'|\Psi'\rangle = 1 \qquad (3.2.3)$$

i.e., the screen is now uniformly illuminated with no interference fringes. Why? Because the orthogonality of $|M:L\rangle$ and $|M:U\rangle$ causes the two terms on the RHS of (3.2.2) to be orthogonal. The cross-product between them is now zero, whereas previously it was the cross-product which gave rise to the $\cos\theta$ term in (3.1.3) and hence the interference.

It is clear, then, that the nature of the measurement, M, is irrelevant. The interference disappears by virtue of the perfect nature of the measurement as expressed by the orthogonality of the measurement states, (3.2.1). That is the only requirement. How the measurement is accomplished is irrelevant.

3.3 Imperfect "Which Path" Information

We can now see that the destruction of the interference pattern need not be an all-or-nothing affair. Suppose that M were an imperfect measuring device which has $\langle M: L | M: U \rangle \neq 0$. Then (3.2.2) gives,

$$\langle \Psi' | \Psi' \rangle = 1 + \Re\left(e^{i\theta(x)} \langle M: L | M: U \rangle\right) \tag{3.3.1}$$

Consequently, if $0 < |\langle M: L | M: U \rangle| < 1$ the interference pattern would be less marked but would not vanish completely. For example, if $|\langle M: L | M: U \rangle| = 0.5$ then $\langle \Psi' | \Psi' \rangle = 1 + 0.5 \cos\theta$ so the contrast between the darker and lighter fringes would be from 0.5 to 1.5 (as opposed to the full interference pattern for which the intensity varies from 0 to 2). An experimental example of an imperfect "which path" measurement resulting in a degraded interference pattern is given in Ref.[3.6].

3.4 Is Physical Interaction Necessary to Destroy the Interference?

In the early days of quantum mechanics great emphasis was placed on measurements requiring physical interaction with the measured system and the (supposed) impossibility of making measurements without disturbing what was being measured. Such physical disturbance was held to be responsible for the destruction of the interference in cases like that considered above. However, this is demonstrably incorrect. A graphic illustration of this is provided by the Elitzur-Vaidman bomb test (which is analysed in detail in chapter 6). This involves a bomb whose sensitivity is so exquisite that it is guaranteed to explode if exposed any physical interaction whatsoever, however slight. If this bomb is used as a measuring device it can successfully eliminate an interference pattern without exploding. So, elimination of the interference is clearly not caused by physical disturbance since any actual physical disturbance would have resulted in the bomb exploding. What *is* necessary is the *possibility* of physical interaction, not that the interaction actually takes place.

Another way of appreciating this point is to note that the evolution of the state into the form of Equ.(3.2.2) does not really constitute a measurement. It is merely the first part of a measurement process. The second part of a complete measurement is the "reduction of the state vector" in which only one or other of the two possibilities, either $|U\rangle|M:U\rangle$ or $|L\rangle|M:L\rangle$, is actually realised (see §2.6). But suppose the reduction of the state vector does not occur, so that what we are dealing with is not a completed measurement. The interference pattern is still predicted to disappear provided the combined system-plus-apparatus state is given by (3.2.2).

Interestingly, the state (3.2.2) can be obtained by a purely unitary evolution. In other words, $|\Psi\rangle \rightarrow |\Psi'\rangle$ is calculable from the Schrodinger equation, so that $|\Psi'\rangle = exp\{-i\hat{H}t/\hbar\}|\Psi\rangle$ where $\hat{H}$ is the appropriate Hamiltonian operator. Consequently, it is possible, by another unitary evolution, for (3.2.2) to be converted back to the initial product state, $|\Psi\rangle$, again. Clearly this is so because $|\Psi\rangle = exp\{i\hat{H}t/\hbar\}|\Psi'\rangle$. This is known as "quantum erasure" and will be considered further in chapter 19. The quantum erasure process re-establishes an interference pattern which would otherwise have been destroyed by the interaction with M. This would not be possible for a genuine measurement since the reduction of the state vector is irreversible. This is another way to see that it is not physical interaction *per se* which destroys interference, because disturbance by a physical interaction, once accomplished, could not be undone.

3.5 Three Slits and Partial Information

We have seen that an imperfect measurement on two slits causes degrading of the interference rather than complete elimination. What happens if we use three slits but obtain information from only one? Questions like this, which might seem rather abstruse, are easily answered using the formalism. In obvious notation, and with no measuring device, the state is,

$$|\psi\rangle = \tfrac{1}{\sqrt{3}}(|1\rangle + |2\rangle + |3\rangle) \tag{3.5.1}$$

Suppose the relative phase between the waves from slits j and k is $\theta_{jk}(x)$ at position x on the screen, e.g., $\langle 1|2\rangle = \langle 2|1\rangle^* = exp\{i\theta_{12}\}$, etc. Then there is an interference pattern on the screen given by,

$$\langle \psi|\psi\rangle = 1 + \tfrac{2}{3}[cos\,\theta_{12}(x) + cos\,\theta_{23}(x) + cos\,\theta_{13}(x)] \tag{3.5.2}$$

For example, if $\theta_{12} = \theta_{23} = \theta$ and $\theta_{13} = 2\theta$ then,

$$\langle \psi|\psi\rangle = 1 + \tfrac{2}{3}[2\,cos\,\theta\,(x) + cos\,2\,\theta(x)] \tag{3.5.3}$$

The average is unity, as it must be, and the intensity varies from 0 (when $\theta = 2.0944, 4.1888, etc$) to 3 (when θ is zero or a multiple of 2π).

Now, suppose we introduce a two-state device M into the path of slit 3 (only). This does not provide a measurement which discriminates between slits 1 and 2. So does a modified interference pattern remain, namely that due to slits 1 and 2 alone? The answer is "yes" as may be seen by considering the state algebra as follows. The combined state evolves thus,

$$|\Psi\rangle \rightarrow |\Psi'\rangle = \frac{1}{\sqrt{3}} (|1\rangle|M:U\rangle + |2\rangle|M:U\rangle + |3\rangle|M:L\rangle) \qquad (3.5.4)$$

Here $|M:U\rangle$ is the state in which M is initially setup, and hence it remains in this state if the particle takes paths 1 or 2. Only path 3 results in the modified M state, $|M:L\rangle$. Consequently, when we form the absolute square of the combined state, which provides the intensity seen at the screen, the third term in (3.5.4) is orthogonal to the first two (assuming perfect measurement), but the first two terms continue to provide an interference cross-product,

$$\langle\Psi'|\Psi'\rangle = 1 + \frac{2}{3}\cos\theta\,(x) \qquad (3.5.5)$$

Hence (3.5.5) continues to show an interference pattern, though with diminished contrast between the darker and lighter bands compared with that obtained with no measuring device. The minima are now $1/3$ and the maxima $5/3$, compared with 0 and 3 when all three slits contribute. The average is again unity, as it must always be. This is a further illustration of how partial information will degrade the interference pattern without removing it completely.

3.6 References

[3.1]   Davisson, C. J., and Germer, L. H. (1928). "Reflection of Electrons by a Crystal of Nickel". *Proceedings of the National Academy of Sciences of the United States of America*. **14** (4): 317–322. doi:10.1073/pnas.14.4.317

[3.2]   Jönsson, C. (1961). "Elektroneninterferenzen an mehreren künstlich hergestellten Feinspalten", *Zeitschrift für Physik*, **161**(4), 454–474. doi.org/10.1007/BF01342460.

[3.3]   P. G. Merli, G. F. Missiroli, and G. Pozzi, (1976). "On the statistical aspect of electron interference phenomena," *American Journal of Physics* **44**, 306–307. https://tinyurl.com/y5kgy5ly.

[3.4] Rosa, R., (2012). "The Merli–Missiroli–Pozzi Two-Slit Electron-Interference Experiment". *Phys Perspect.* **14**(2): 178–195. doi: 10.1007/s00016-011-0079-0.

[3.5] Taylor, G.I. (1909). "Interference Fringes with Feeble Light". *Proc. Camb. Phil. Soc.* **15**, 114.

[3.6] Mittelstaedt, P., Prieur, A., Schieder, R. (1987). "Unsharp particle-wave duality in a photon split-beam experiment". *Foundations of Physics.* **17** (9): 891–903. doi:10.1007/BF00734319.

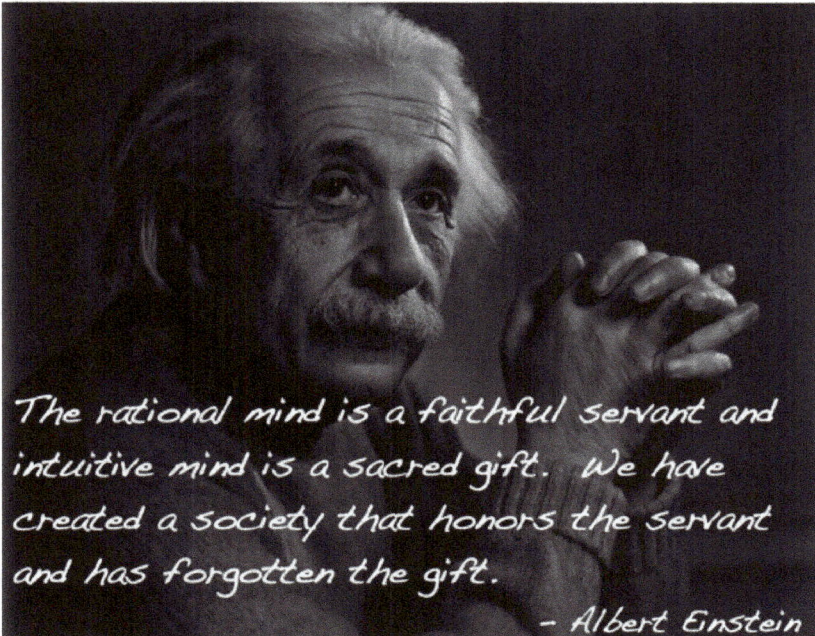

The rational mind is a faithful servant and intuitive mind is a sacred gift. We have created a society that honors the servant and has forgotten the gift.

– Albert Einstein

4

The No-Cloning Theorem

The impossibility of a general device which will faithfully produce a copy of an arbitrary pure quantum state is the basis of secure quantum communication.

4.1 Statement and Proof of the Theorem

The no-cloning theorem says that there is no device which will produce an exact copy of an arbitrary pure quantum state without altering the original. Any practical device can be applied only to systems of a specific physical make-up, so what exactly do we mean by "arbitrary"?. We can limit consideration to a particular class of systems. This is made precise by specifying a particular Hilbert space. The no-cloning theorem then states that there is no unitary operator which will copy an arbitrary quantum state of the Hilbert space in question without destroying the original. The difficulty in making a copy does not arise, therefore, simply because the definition of 'arbitrary' is too broad.

In classical computing it is taken for granted that digital data can be copied with (virtually) absolute fidelity an arbitrary number of times. The no-cloning theorem says that the equivalent for a quantum computer is not possible even in principle. The no-cloning theorem is therefore a substantial restriction on the facilities available to the programmer of a quantum computer. However, the no-cloning theorem is even more important in quantum communication, where the impossibility of copying an unknown quantum state is essential to the security of the information.

Given the huge importance of the theorem, it is remarkable for two things: firstly that it is so very simple to prove, and secondly, that it was not discovered until as late as 1982, by Dieks, Ref.[4.1] and by Wootters and Zurek, Ref.[4.2]. The proof is as follows.

Suppose that we are provided with a vector, $|s\rangle$, from the relevant Hilbert space, in some standard normalised initial state. This is the 'clay' from which our copy is to be moulded. We are required to transform this initial state into a copy of any arbitrary state, say $|\varphi\rangle$, with which we are provided, whilst also preserving the initial $|\varphi\rangle$ state. Thus, the initial pair of states, $|s\rangle$ and $|\varphi\rangle$, is to be transformed into two copies of $|\varphi\rangle$ via a unitary transformation, U. We can write this as,

$$\hat{U}(|\varphi\rangle|s\rangle) = |\varphi\rangle|\varphi\rangle \qquad (4.1.1)$$

It is understood that the left-hand "ket" state in the product refers to the original and the right-hand state to the copy. The unitary transformation refers, of course, to the solution of the corresponding Schrodinger equation, (2.4.1), so that it could be written in terms of a Hamiltonian operator, $\hat{U} = exp\{-i\hat{H}t/\hbar\}$, but this will not be required here.

(4.1.1) must work for an arbitrary state of the specified Hilbert space. So for any other state, $|\psi\rangle$, we must also have,

$$\hat{U}(|\psi\rangle|s\rangle) = |\psi\rangle|\psi\rangle \qquad (4.1.2)$$

Taking the scalar product of the LHSs of these two equations gives,

$$\langle\psi|\langle s|\hat{U}^{+}\hat{U}|\varphi\rangle|s\rangle = \langle\psi|\phi\rangle\langle s|s\rangle = \langle\psi|\phi\rangle \qquad (4.1.3)$$

where we have made use of the unitary nature of U. Similarly, taking the scalar product of the RHSs of the two equations gives,

$$\langle\psi|\langle\psi||\varphi\rangle|\varphi\rangle = (\langle\psi|\varphi\rangle)^{2} \qquad (4.1.4)$$

But (4.1.3) and (4.1.4) must be equal, so we conclude that $\langle\psi|\varphi\rangle = (\langle\psi|\varphi\rangle)^{2}$, and hence that $\langle\psi|\varphi\rangle$ is 0 or 1. Hence, the unitary copier, U, can only copy a set of orthonormal states, at best. It cannot copy an arbitrary state. The proof is simple but the result is profound.

4.2 Alternative Demonstration

To drive the message home, suppose we have a pair of orthogonal states, $|1\rangle$ and $|2\rangle$, which we *can* copy using the unitary transformation. Thus,

$$\hat{U}(|1\rangle|s\rangle) = |1\rangle|1\rangle \text{ and } \hat{U}(|2\rangle|s\rangle) = |2\rangle|2\rangle \text{ and } \langle 2|1\rangle = 0 \qquad (4.2.1)$$

Your intuition might lead you astray in imagining that an arbitrary superposition of the two states, $|\varphi\rangle = \alpha|1\rangle + \beta|2\rangle$, should also be copiable, after all U is a linear operator. But this is what happens,

$$\hat{U}|\varphi\rangle|s\rangle = \alpha\hat{U}|1\rangle|s\rangle + \beta\hat{U}|2\rangle|s\rangle = \alpha|1\rangle|1\rangle + \beta|2\rangle|2\rangle \qquad (4.2.2)$$

and the result is clearly not what is required since successful copying should have produced $|\varphi\rangle|\varphi\rangle = (\alpha|1\rangle + \beta|2\rangle)(\alpha|1\rangle + \beta|2\rangle)$. Instead, the "copier", U, produces, in (4.2.2), an entangled state, not a product state of two equal copies. The copier fails. (I have anticipated chapter 13 in referring to the state (4.2.2) as entangled).

4.3 Copying With Destruction

However, there is no objection to a device which will copy an arbitrary quantum state provided that the original is destroyed. Such a device would do this,

$$\hat{U}(|\varphi\rangle|s\rangle) = |t\rangle|\varphi\rangle \tag{4.3.1}$$

for an arbitrary quantum state φ, together with the "prepared state", s, and some arbitrary, but universal, state t (which may be the same as s). As an exercise the reader should attempt the no-go proofs of sections 4.1 and 4.2 to see how they fail in this case.

Such a device is hardly a copier. It is best described as a device which moves the state from one place to another. We will see in chapter 27 that this extends to teleporting the state to a remote place.

4.4 Exercises for the Reader

Attempting the no-go proof of section 4.1 we consider forming the inner product of (4.3.1) with $U(|\psi\rangle|s\rangle) = |t\rangle|\psi\rangle$. The inner product of the LHSs gives,

$$\langle\psi|\langle s|\hat{U}^+\hat{U}|\varphi\rangle|s\rangle = \langle\psi|\phi\rangle\langle s|s\rangle = \langle\psi|\varphi\rangle \tag{4.4.1}$$

Whilst the inner product of the RHSs gives,

$$\langle\psi|\langle t||t\rangle|\varphi\rangle = \langle\psi|\varphi\rangle \tag{4.4.2}$$

So, the two are consistent and there is no issue.

Alternatively, consider the argument of section 4.2, where we start from the assumption that $\hat{U}(|1\rangle|s\rangle) = |t\rangle|1\rangle$ and $\hat{U}(|2\rangle|s\rangle) = |t\rangle|2\rangle$ and consider the action of $\hat{U}$ on a superposition state $|\varphi\rangle = \alpha|1\rangle + \beta|2\rangle$,

$$\hat{U}|\varphi\rangle|s\rangle = \alpha\hat{U}|1\rangle|s\rangle + \beta\hat{U}|2\rangle|s\rangle = \alpha|t\rangle|1\rangle + \beta|t\rangle|2\rangle = |t\rangle|\varphi\rangle \tag{4.4.3}$$

So, again, there is no problem and no objection to the "copy with destruction" device.

4.5 References

[4.1] Dieks, D. (1982), "Communication by EPR devices", *Phys.Lett.* A, **92**, 271-271, 1982. https://doi.org/10.1016/0375-9601(82)90084-6

[4.2] Wootters, W.K., and Zurek, W.H. (1982), "A Single Quantum Cannot Be Cloned", *Nature*, **299**, 802-803, 1982. https://www.nature.com/articles/299802a0

5

Symmetry, Conservation Laws and Uncertainty

The relationship between symmetry and conservation laws is one of the most profound discoveries in all of physics. In fact, symmetry, conservation and uncertainty are best understood as a trio of intimately related phenomena.

5.1 Noether's Theorem

In this opening section I make reference to parts of physics with which the reader may not be familiar. Do not be deterred; this is only for the purpose of setting the rest of the chapter in its historical context and will not be required to understand the rest of the chapter.

So familiar has the relationship between symmetry and conserved quantities become to physicists now that there is a danger of forgetting that it is a relatively new understanding, being only a century old as I write. The discovery was made in the context of classical physics (which includes relativity theory). Curiously, though, this discovery came about only in the period that quantum mechanics was being born, and so was a very late development in classical physics. There were many intimations of the relationship between symmetry and conservation prior to the twentieth century, not least in the work of Hamilton. But its most general and most rigorous formulation in the classical context was presented by Noether in the period 1915 to 1918. This is the same period in which Einstein finally published his general relativistic field equations of gravity, and this is no coincidence. The historical account of Ref.[5.1] makes clear the relationship between Noether's work on the conservation problem and the development of relativity theory. Whilst relativity was the stimulus, it is not necessary that a theory be relativistically covariant for Noether's theorem to apply.

Noether's Theorem, in the classical context, relates to theories formulated in terms of Lagrangian dynamics. It shows that to every continuous symmetry, expressed as a differentiable transformation which leaves the Lagrangian invariant, there corresponds a conserved quantity (or "constant of motion"). Thus, we arrive at the profound insight that energy conservation is the result of the laws of physics being time-invariant, the same today as they were yesterday. Momentum conservation is the result of the homogeneity of space, and hence that the laws of physics are the same here as on Mars. And angular momentum is conserved because space is isotropic and hence there is no physically preferred spatial direction.

In field theories, this conservation is manifest by the existence of a tensor, specified explicitly in terms of the Lagrange density, whose divergence is zero – this being the expression of conservation in covariant form. This derivation of the relationship can seem somewhat opaque. In contrast, the derivation of the connection between symmetry and conservation in quantum mechanics is simple almost to the point of mathematical triviality. To this we now turn.

5.2 Invariance and Symmetry

Suppose we consider a transformation of every vector thus,

$$|\tilde{\varphi}\rangle = \hat{U}|\varphi\rangle \tag{5.2.1}$$

The transformation U must be unitary in order to preserve the norm of the vector. Now suppose that an operator, $\hat{Q}$, is such that every matrix element is left invariant by this transformation, i.e.,

$$\langle\tilde{\omega}|\hat{Q}|\tilde{\varphi}\rangle = \langle\omega|\hat{U}^+\hat{Q}\hat{U}|\varphi\rangle = \langle\omega|\hat{Q}|\varphi\rangle \tag{5.2.2}$$

For (5.2.2) to hold for every pair of vectors, it must be that $\hat{U}^+\hat{Q}\hat{U} = \hat{Q}$, which, due to $\hat{U}$ being unitary means that $\hat{Q}\hat{U} = \hat{U}\hat{Q}$. So, the condition that an operator be invariant under the transformation (5.2.1) is that it commutes with the transformation,

$$[\hat{Q}, \hat{U}] = 0 \tag{5.2.3}$$

The dynamics of quantum states is specified by Schrodinger's equation,

$$\hat{H}|\psi\rangle = i\hbar\partial_t|\psi\rangle \tag{2.4.1}$$

Consequently, a transformation $\hat{U}$ is said to be a symmetry of the system if it is not explicitly time dependent and it commutes with the Hamiltonian,

$$[\hat{H}, \hat{U}] = 0 \tag{5.2.4}$$

It follows that if $|\psi\rangle$ is a solution of Schrodinger's equation, and $\hat{U}$ is a symmetry of the system, then $\hat{U}|\psi\rangle$ is also a solution to Schrodinger's equation, as one would expect of a symmetry. This follows immediately from (2.4.1) since,

$$\hat{U}\hat{H}|\psi\rangle = \hat{H}\hat{U}|\psi\rangle = i\hbar\hat{U}\partial_t|\psi\rangle = i\hbar\partial_t\hat{U}|\psi\rangle \tag{5.2.5}$$

5.3 Conserved Operators

An operator is "conserved", i.e., does not vary over time, if all its matrix elements are constant over time. Thus, setting an arbitrary matrix element of operator $\hat{Q}$ at time t equal to its value at time zero gives,

$$\langle\hat{Q}\rangle = \langle\varphi(t)|\hat{Q}|\psi(t)\rangle = \langle\varphi(0)|e^{i\hat{H}t/\hbar}\hat{Q}e^{-i\hat{H}t/\hbar}|\psi(0)\rangle = \langle\varphi(0)|\hat{Q}|\psi(0)\rangle$$

where the Schrodinger equation, $|\psi(t)\rangle = exp\{-it\hat{H}/\hbar\}|\psi(0)\rangle$, has been used, for both states. Note that this expression assumes that $\hat{H}$ is not explicitly time-dependent. Because this must hold for all matrix elements it follows that $e^{i\hat{H}t/\hbar}\hat{Q}e^{-i\hat{H}t/\hbar} = \hat{Q}$. And because this must hold for all times it must be that $\hat{Q}$ commutes with $\hat{H}$ (if not obvious the proof is an exercise for the reader).

$$[\hat{Q},\hat{H}] = 0 \qquad (5.3.1)$$

In particular this means that the expectation value of the operator (which may be an observable) is constant,

$$\langle\hat{Q}\rangle = \langle\psi(t)|\hat{Q}|\psi(t)\rangle = \langle\psi(0)|\hat{Q}|\psi(0)\rangle = \langle\hat{Q}\rangle_0$$

Consequently, the condition for an operator to be conserved, (5.3.1), is the same as the condition for the operator to be a symmetry transformation, (5.2.4), in both cases the condition is to commute with the Hamiltonian. However, to be a symmetry transformation the operator must be unitary. In contrast, an observable must be Hermitian.

But we have already seen how we may convert an Hermitian operator (say $\hat{Q}$) into a unitary operator, namely by setting $\hat{U} = exp\{ia\hat{Q}\}$ for some real parameter a. Due to $\hat{Q}^+ = \hat{Q}$ we have $\hat{U}^+ = \hat{U}^{-1}$. Moreover, if $[\hat{H},\hat{U}] = 0$ then it follows that $[\hat{Q},\hat{H}] = 0$, and *vice-versa*.

Hence, if $\hat{U}$ is a symmetry transform then the observable $\hat{Q}$ which "generates" the transformation, in the sense that they are related by $\hat{U} = exp\{ia\hat{Q}\}$, is conserved. This is Noether's theorem as it is manifest in quantum mechanics. But the abstruse proof in classical Lagrangian field theory is replaced by a relationship in QM which is close to trivial.

The reader may worry that we have not proved that there necessarily exists a $\hat{Q}$ such that $\hat{U} = exp\{ia\hat{Q}\}$ for a given $\hat{U}$. We are saved for sufficiently small a parameter by $exp\{ia\hat{Q}\} \approx 1 + ia\hat{Q}$ and hence that $\hat{Q} = (\hat{U} - 1)/ia$. More generally what we are implicitly dealing with here are Lie groups ($\hat{U}$) and their corresponding Lie algebras ($\hat{Q}$), but we shall not be going further into those purely mathematical matters.

5.4 Can Non-Commuting Observables Both Be Conserved?

Yes.

Recall from §2.5.5 that non-commuting observables must have at least some different eigenvectors. Consequently we may consider a system to be in an eigenstate of one observable, and hence to have a definite value of that observable, whilst the value of the other observable will be indeterminate because the system is not in one of its eigenstates. A question which arises is whether both observables can be conserved – the issue being what meaning can be attached to the conservation of a quantity which (for the second observable) is indeterminate? If it is indeterminate, how can we speak of it being conserved?

Consider a Hamiltonian with no preferred direction in space, so that angular momentum is conserved. This means that all three components of angular momentum, $\hat{S}_x, \hat{S}_y, \hat{S}_z$, must commute with the Hamiltonian, despite the fact they do not commute with each other, see (2.9.7). The system can be in a definite (eigen)state of at most one of $\hat{S}_x, \hat{S}_y, \hat{S}_z$, so to what does the conservation of the others relate? The answer is that conservation relates, in general, to expectation values and hence does not require the system to be in an eigenstate of the conserved observable. So all three of $\hat{S}_x, \hat{S}_y, \hat{S}_z$ will be conserved in the sense that $\langle\psi(t)|\hat{Q}|\psi(t)\rangle = \langle\psi(0)|\hat{Q}|\psi(0)\rangle$ where $\hat{Q}$ may be any one of $\hat{S}_x, \hat{S}_y, \hat{S}_z$.

5.5 Conservation of Energy

The Hamiltonian, $\hat{H}$, clearly commutes with itself. So, providing it is not explicitly time dependent, it follows that $\hat{H}$ is conserved, and, of course, energy is simply the expectation value of $\hat{H}$, so energy is conserved.

We have already seen what symmetry corresponds to this conservation law, though we did not identify it as such. The solution to the Schrodinger equation, $|\psi(t)\rangle = exp\{-it\hat{H}/\hbar\}|\psi(0)\rangle$, can be regarded as a symmetry because it is of the same form as (5.2.1). It is a time-translation symmetry, where the unitary transform is $\hat{U} = exp\{-it\hat{H}/\hbar\}$. This transforms a state vector which obeys Schrodinger's equation into another such solution displaced in time by t.

That this all makes sense can be checked by the following observation. The Schrodinger equation, (2.4.1), means that $\hat{H}$ is equivalent in its action on a state vector to that of the differential operator $i\hbar\partial_t$. Substituting that for $\hat{H}$

in the formal solution presents it in the form $|\psi(t)\rangle = exp\{t\partial_t\}|\psi(0)\rangle$. Using the power series for the exponential, the RHS of this is seen to be,

$$exp\{t\partial_t\}|\psi(0)\rangle = \sum_k \frac{t^k}{k!}\partial_t^k|\psi(0)\rangle = |\psi(0)\rangle + t\partial_t|\psi(0)\rangle + \frac{t^2}{2!}\partial_t^2|\psi(0)\rangle + \cdots$$

(5.5.1)

Where $\partial_t^n|\psi(0)\rangle$ is the n'th time derivative of the state vector evaluated at time zero. We recognise (5.5.1) as the Taylor series for $|\psi(t)\rangle$, and hence $|\psi(t)\rangle = exp\{t\partial_t\}|\psi(0)\rangle$ is seen to be an algebraic identity. [There is a potential notational confusion here. The reader may think that terms like $\partial_t^n|\psi(0)\rangle$ should all be zero. That would, of course, be the case for a function, e.g., $\partial_t^n\psi(0) = 0$. But $|\psi(0)\rangle$ is the *state* at time zero, not merely a number, and has non-zero time derivatives – necessarily so or it could not change!].

5.6 Conservation of Momentum

Consider the x-component of momentum, represented by the Hermitian operator $\hat{P}_x$. We expect this to be conserved in many situations, and so we guess that we shall find $[\hat{P}_x, \hat{H}] = 0$ in such cases. Proceeding by analogy with the symmetry transform $exp\{-it\hat{H}/\hbar\}$ in the case of conservation of energy, we guess that the symmetry transform corresponding to conservation of momentum might be $exp\{ia\hat{P}_x/\hbar\}$, noting that the dimensions of $\hbar$ make the exponent dimensionless if a has the dimensions of length. The analogy with the Taylor series in time, (5.5.1), suggests that this transformation can be interpreted as a translation along the x-axis by a if we consider that the momentum operator can be represented by,

$$\hat{P}_x \rightarrow -i\hbar\partial_x \tag{5.6.1}$$

In which case $exp\{ia\hat{P}_x/\hbar\} \rightarrow exp\{a\partial_x\}$. Displaying explicitly the dependence of the state vector on position, x, we have,

$$exp\{a\partial_x\}|\psi(x)\rangle = \sum_k \frac{a^k}{k!}\partial_x^k|\psi(x)\rangle = |\psi(x)\rangle + a\partial_x|\psi(x)\rangle + \frac{a^2}{2!}\partial_x^2|\psi(x)\rangle + \cdots$$

(5.6.2)

which, being again a Taylor series, we can identify with $|\psi(x + a)\rangle$. Thus the effect of the operator $exp\{ia\hat{P}_x/\hbar\}$, subject to the interpretation (5.6.1), is to transform the state vector to its value displaced along the x-axis by a. If this is a symmetry of the system, i.e., if this unitary transform commutes with the

Hamiltonian, then so does $\hat{P}_x$ and hence this translational symmetry implies the conservation of momentum.

In this manner of developing the argument, the choice of representation (5.6.1) may be seen as motivated by the desire to link conservation of momentum to translational invariance, due to the prior knowledge that this is the case in classical physics (Noether's theorem). This is rather the reverse of the development in many texts. However, as a by-product we can now conclude, rather than assume, the commutation properties of the conjugate variables, namely,

$$[\hat{H}, t] = i\hbar \qquad \text{and} \quad [\hat{P}_x, x] = -i\hbar \qquad (5.6.3)$$

The term "conjugate variable" has been borrowed from classical mechanics, but has re-emerged here as a pair of variables consisting of a conserved quantity and the associated symmetry parameter, the former generating changes in the latter.

Note that the vectorial momentum operator which generalises (5.6.1) is,

$$\hat{\bar{P}} \rightarrow -i\hbar\bar{\nabla} \qquad (5.6.4)$$

The components of $\hat{\bar{P}}$ therefore commute and each component separately has a commutator with its conjugate like (5.6.3), i.e.,

$$[\hat{P}_i, r_k] = -i\hbar\delta_{ik} \qquad (5.6.5)$$

5.7 The Uncertainty Relation

Having deduced that conjugate variables do not commute, and knowing that non-commuting operators cannot have the same set of eigenvectors, it follows that a "particle" cannot have a well defined x-momentum and a well defined x-position at the same time. This is, of course, the essence of the famous Uncertainty Principle first enunciated by Heisenberg but first proved by Kennard, Ref.[5.2]. The quantitative statement of the Uncertainty Principle requires the "uncertainty" in an observable to be defined for an arbitrary state. We have already defined the expectation value (i.e., the average) of an observable, say $\hat{Q}$, in state $|\psi\rangle$ to be,

$$\langle Q \rangle = \langle \psi | \hat{Q} | \psi \rangle \qquad (5.7.1)$$

The "uncertainty" in Q may be defined in the usual statistical fashion as the standard deviation in the measurement of Q, i.e., as the square root of the mean of the square of the deviation from the mean, thus,

$$\sigma_Q = \sqrt{\langle\psi|(\hat{Q} - \langle Q\rangle)^2|\psi\rangle} \qquad (5.7.2)$$

This may be re-arranged as,

$$\sigma_Q = \sqrt{\langle\psi|(\hat{Q}^2 - \langle Q\rangle^2)|\psi\rangle} = \sqrt{(\langle\hat{Q}^2\rangle - \langle Q\rangle^2)} \qquad (5.7.3)$$

Using this definition, the Uncertainty Relation between conjugate variables may be shown, using the commutator (5.6.3), to be,

$$\sigma_x \sigma_{P_x} \geq \frac{\hbar}{2} \qquad (5.7.4)$$

An exercise for the reader is to prove that (5.7.4) follows from (5.6.3). I'd cheat if I were you and consult a standard text, or my answer in §5.9.

5.8 A Paradox and a Lesson Learned

Consider a system with momentum observable $\hat{P}_{sys}$ and position operator $\hat{x}$. We have,

$$[\hat{P}_{sys}, \hat{x}] = -i\hbar \qquad (5.8.1)$$

It suffices to consider just one spatial dimension. Suppose the system starts in a state of definite momentum and we then measure its position. Since (5.8.1) forbids simultaneous eigenstates of both momentum and position (see §2.5.5), the post-measurement state of defined position is a superposition of momentum states. Subsequent measurement of the system's momentum will select one of these momentum states (with probabilities given by the Born Rule, §2.6). Hence the final measured momentum will, in general, differ from the initial momentum. What has happened to the conservation of momentum?

The reader will readily appreciate that interaction of the system with measuring devices has been necessary to carry out the two measurements referred to in this scenario. No doubt you will guess that momentum has been exchanged with the apparatus so as to maintain overall conservation of total momentum. Indeed, that is the solution of the paradox. But how do we demonstrate that momentum is now conserved? We introduce the momentum operator for the apparatus, $\hat{P}_{app}$, so that the total momentum is $\hat{P}_{sys} + \hat{P}_{app}$. But we still have,

$$[\hat{P}_{sys} + \hat{P}_{app}, \hat{x}] = -i\hbar \qquad (5.8.2)$$

because the apparatus's momentum is independent of the position of the system, x, so $[\hat{P}_{app}, \hat{x}] = 0$.

Apparently we are no further forward. It seems that the total momentum will not survive measurement of $\hat{x}$ either. But wait: when we make a measurement of the system's position, it is actually a measurement of the system's position relative to that of the apparatus, not some absolute position. So we could say that $[\hat{P}_{app}, \hat{x}]$ is not actually zero if we interpret $\hat{x}$ as the relative position of the system wrt the apparatus. However, for clarity let us instead regard $\hat{x}$ as measured with respect to some other coordinate frame, independent of both the system and the apparatus. But let's also introduce the position operator of the apparatus with respect to this same coordinate system, call it $\hat{X}$.

The position of the system wrt the apparatus is thus $\hat{x} - \hat{X}$ where $[\hat{P}_{app}, \hat{x}] = 0$ continues to hold. However we now also have,

$$[\hat{P}_{app}, \hat{X}] = -i\hbar \tag{5.8.3}$$

so that,

$$[\hat{P}_{sys} + \hat{P}_{app}, \hat{x} - \hat{X}] = 0 \tag{5.8.4}$$

We now see that the total momentum can be known whilst also measuring the *relative* position of the system wrt the apparatus. The total momentum of system plus apparatus can consistently be regarded as invariant whilst the (relative) position measurement is made, because (5.8.4) allows both the total momentum and the relative position to be simultaneously known.

Moreover, the total momentum is conserved because the symmetry condition is a displacement of both the system and the apparatus by the same amount, say a, which leaves the relative positive, $x - X$, unchanged. Hence, measurements of the relative position (which is what the apparatus inevitably measures) does not prejudice the conservation of total momentum.

That total momentum is conserved is particularly transparent in this example. Since space is homogeneous, the Hamiltonian cannot depend on absolute positions, x or X, but, due to the physical interaction between the system and the apparatus it can depend upon their distance apart, $x - X$. Hence, the most general Hamiltonian depends upon the difference $\hat{x} - \hat{X}$ only. Assuming that the Hamiltonian can therefore be expressed as a sum over powers of $\hat{x} - \hat{X}$, perhaps an infinite series, then (5.8.4) immediately tells us that,

$$\left[\hat{P}_{sys} + \hat{P}_{app}, \hat{H}\right] = 0 \tag{5.8.5}$$

and hence that the total momentum is conserved. Hence, whilst we initially deduced the commutation relations (5.6.3) by enforcing the connection between symmetry and conservation, we have now shown how this can be done in reverse, deriving essentially Noether's theorem, (5.8.5), from the assumed commutation relations, (5.8.1) and (5.8.3).

5.9 Exercises for the Reader

(1) Show that if $e^{i\hat{H}t/\hbar}\hat{Q}e^{-i\hat{H}t/\hbar} = \hat{Q}$ for all times then $\left[\hat{Q}, \hat{H}\right] = 0$. It suffices to consider very small times so that,

$$e^{\frac{i\hat{H}t}{\hbar}}\hat{Q}e^{-\frac{i\hat{H}t}{\hbar}} \approx \left(1 + \frac{i\hat{H}t}{\hbar}\right)\hat{Q}\left(1 - \frac{i\hat{H}t}{\hbar}\right) = \hat{Q} + \frac{it}{\hbar}\left(\hat{H}\hat{Q} - \hat{Q}\hat{H}\right)$$

Hence, for this to equal $\hat{Q}$ we must have $\left[\hat{Q}, \hat{H}\right] = 0$. QED.

(2) Derive the Uncertainty Relation, $\sigma_x\sigma_{P_x} \geq \hbar/2$

$$\Delta\hat{x} = \hat{x} - x_0 \quad \text{and} \quad \Delta\hat{P} = \hat{P} - P_0 \tag{5.9.1}$$

where x_0 and P_0 are just numerical values. Hence, the mean square deviation of the observables $\hat{x}$ and $\hat{P}$ from these constant values are given by,

$$\langle\Delta x^2\rangle = \langle\psi|\Delta\hat{x}^2|\psi\rangle \quad \text{and} \quad \langle\Delta P^2\rangle = \langle\psi|\Delta\hat{P}^2|\psi\rangle \tag{5.9.2}$$

where the arbitrary state of the system is labelled ψ. This state can be expanded in some arbitrary orthonormal basis as $|\psi\rangle = c_i|\varphi_i\rangle$, where summation over repeated subscripts will be assumed hereafter. Note that the basis states $\{|\varphi_i\rangle\}$ need not be eigenstates of either $\hat{x}$ or $\hat{P}$. The action of these operators on the basis can be written,

$$\Delta\hat{x}|\varphi_i\rangle = a_{ij}|\varphi_j\rangle \quad \text{and} \quad \Delta\hat{P}|\varphi_i\rangle = b_{ij}|\varphi_j\rangle \tag{5.9.3}$$

Substitution into (5.9.2) yields,

$$\langle\Delta x^2\rangle = |\bar{u}|^2 \quad \text{and} \quad \langle\Delta P^2\rangle = |\bar{v}|^2 \tag{5.9.4}$$

where the complex-valued vectors $\bar{u}$ and $\bar{v}$ are defined by,

$$u_j \equiv c_i a_{ij} \quad \text{and} \quad v_j \equiv c_i b_{ij} \tag{5.9.5}$$

But Schwartz's inequality tells us that, for any complex valued vectors in any number of dimensions,

$$|\bar{u}|^2|\bar{v}|^2 \geq |\bar{u}^* \cdot \bar{v}|^2 \tag{5.9.6}$$

[This inequality is *almost* obvious. It says that the product of the squared-lengths of two vectors must be greater than or equal to the square of their dot product. Since the dot product is the product of the lengths of the vectors times the cosine of the included angle, this is clearly true – the equality holding only when the vectors are parallel and the included angle is zero. The only thing that complicates this is the complex nature of the vectors. However, this has no effect on the LHS, which involves only the moduli of the vectors. On the RHS the effect of the phase difference between the two vectors can only lead to partial cancellations, and hence further reduces the magnitude of the RHS. QED].

Schwartz's inequality suggests we consider the expectation value of the product of $\Delta \hat{x}$ and $\Delta \hat{P}$, i.e. $\langle \psi | \Delta \hat{x} \Delta \hat{P} | \psi \rangle$, which, on substitution of (5.9.3,5) does indeed turn out to equal $|\bar{v}^* \cdot \bar{u}| = |\bar{u}^* \cdot \bar{v}|$. Hence, Schwartz's inequality gives,

$$\langle \Delta x^2 \rangle \langle \Delta P^2 \rangle \geq \left| \langle \psi | \Delta \hat{x} \Delta \hat{P} | \psi \rangle \right|^2 \tag{5.9.7}$$

The operator appearing on the RHS is not Hermitian, because $\hat{x}$ and $\hat{P}$ do not commute. It may be re-written, $\tag{5.9.8}$

$$\Delta \hat{x} \Delta \hat{P} = \frac{1}{2}[\Delta \hat{x}, \Delta \hat{P}] + \frac{1}{2}\left(\Delta \hat{x} \Delta \hat{P} + \Delta \hat{P} \Delta \hat{x}\right) = i\frac{\hbar}{2} + \frac{1}{2}\left(\Delta \hat{x} \Delta \hat{P} + \Delta \hat{P} \Delta \hat{x}\right)$$

The last operator on the RHS is now Hermitian, by virtue of its symmetry and the Hermitian nature of $\hat{x}$ and $\hat{P}$. Consequently it has real eigenvalues and hence a real expectation value wrt any state. In contrast, the first (purely numerical) term on the RHS is imaginary. It follows that, when taking the absolute square on the RHS of (5.9.7) there is no cross-product between the two terms, and we have,

$$\langle \Delta x^2 \rangle \langle \Delta P^2 \rangle \geq \frac{\hbar^2}{4} + \frac{1}{4}\left| \langle \Delta \hat{x} \Delta \hat{P} + \Delta \hat{P} \Delta \hat{x} \rangle \right|^2 \tag{5.9.9}$$

But the second term on the RHS is positive, so we get,

$$\langle \Delta x^2 \rangle \langle \Delta P^2 \rangle \geq \frac{\hbar^2}{4} \tag{5.9.10}$$

which is the usual expression of the uncertainty relation. The term we have dropped from the RHS of (5.9.9) can be zero, and hence (5.9.10) is the strongest general form of the expression. **QED**

Note that (5.9.10) has been derived for arbitrary values of the constants x_0 and P_0. It is usual to interpret these constants as the expectation values of the operators $\hat{x}$ and $\hat{P}$, i.e.,

$$x_0 = \langle \hat{x} \rangle = \langle \psi | \hat{x} | \psi \rangle \quad \text{and} \quad P_0 = \langle \hat{P} \rangle = \langle \psi | \hat{P} | \psi \rangle \tag{5.9.11}$$

so that $\langle \Delta \hat{x} \rangle = \langle \psi | \Delta \hat{x} | \psi \rangle = 0$ and $\langle \Delta \hat{P} \rangle = \langle \psi | \Delta \hat{P} | \psi \rangle = 0$. It is easily seen that adopting $x_0 = \langle \hat{x} \rangle$ and $P_0 = \langle \hat{P} \rangle$ leads to the smallest mean square deviations, $\langle \Delta x^2 \rangle$ and $\langle \Delta P^2 \rangle$, and hence provides the strongest form of uncertainty relation.

5.10 References

[5.1]   Byers, N. (1998). "E. Noether's Discovery of the Deep Connection Between Symmetries and Conservation Laws", arXiv: hep-th/9807044

[5.2]   Kennard, E. H. (1927). "Zur Quantenmechanik einfacher Bewegungstypen", *Zeitschrift für Physik* **44** (4–5): 326–352, doi:10.1007/BF01391200.

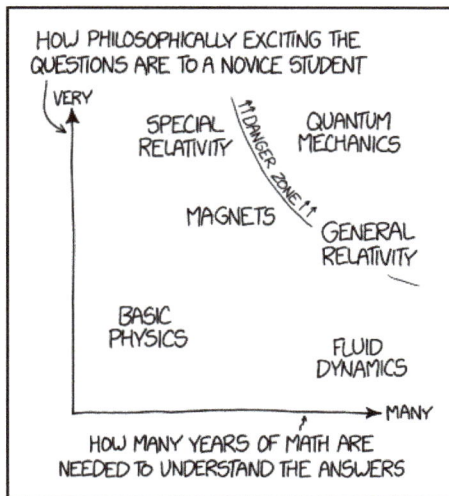

6

The Elitzur-Vaidman Bomb Test

The Observability of Counterfactuals: In classical physics the only things which can influence the future are things which actually happened. Not so in quantum mechanics.

6.1 Counterfactuals

Suppose something could have happened, but actually did not happen. In classical physics the fact that an event could have happened makes no difference to any future outcome if, in actual fact, it did not happen. Only those things which actually happen can influence the future evolution of the world. But in quantum mechanics it is otherwise. The potential for an event to happen can influence future outcomes even if the event does not happen. Something that could have happened but actually did not is called a "counterfactual". In quantum mechanics, counterfactuals are observable: they have measurable consequences. The gedanken experiment known as the Elitzur-Vaidman bomb test provides a striking illustration of this. But first we must revise how interferometers work, specifically the Mach-Zehnder interferometer.

Referring to Figure 6.1, the incoming beam of particles (perhaps photons perhaps something else) hits a beam splitter which divides the beam into a lower beam (LB), which is simply transmitted, and an upper beam (UB) which is reflected. Ignoring any optional objects in the beam path for the moment, the two beams are reflected off mirrors at the top left and bottom right. The sole function of these mirrors is to focus the two beams onto the same spot on the other beam splitter at the top right. From there any particle travelling through the interferometer may be detected in one or other of the two detectors.

A convenient simplification is to assume that both beam splitters split the beam 50/50, and that the mirrors are 100% efficient. This is merely to simplify the explanation and is not crucial to the principles of the device. I'll use terms like "beam" and "wave" and "particle" interchangeably here. The word used indicates nothing of significance.

Figure 6.1 The Mach-Zehnder Interferometer
LB = Lower Beam; UB = Upper Beam

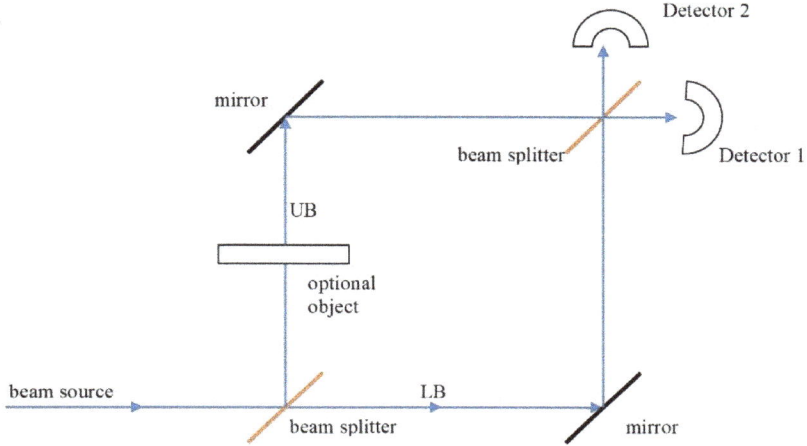

The key to the behaviour of the interferometer is the phase change suffered by the reflected and transmitted waves at the mirrors/beam splitters. There is potential confusion over this since the phase change depends upon the type of object used as a "mirror". For simplicity we shall assume plane glass plates are used as both mirrors and beam splitters. This makes a small simplification in the phase change rules compared with real mirrors, though the outcome would be the same. Real mirrors would be used experimentally, of course, and half-silvered mirrors for the beam splitters.

Any transparent plate with a refractive index, n, greater than air will generally cause an incident beam to be partly transmitted and partly reflected. The phase change rule is simply that the reflected wave acquires a phase factor of i compared to the transmitted wave. An exercise for the reader is to research why this relative phase factor occurs (my answer in §6.6 below).

6.2 The Mach-Zehnder Interferometer: No Specimen

Initially we analyse what signals are expected at the detectors 1 and 2 shown in Figure 6.1 if there is no optional object ("specimen") put in the way of the upper beam (nor the lower beam, of course). Consider the phase changes of the upper beam (UB) with respect to the lower beam (LB) for exit into detector 2. Beam LB undergoes just one reflection (at the second plate) in order to enter detector 2, whereas beam UB requires three reflections, one at each of the three plates. Their relative phase entering detector 2 is therefore

$i^3/i = -1$. Consequently the waves travelling via paths LB and UB interfere destructively and hence no particles are recorded by detector 2.

Repeating this analysis for detector 1 we see that both the LB and UB paths involve two reflections. The two beams therefore enter detector 1 with the same relative phase and so there is constructive interference. Hence, all the particles are recorded by detector 1 and none by detector 2. This is the signature of interference in this device. From the classical perspective, a particle would travel by one or other of the two paths and have a 50/50 chance of ending up in either detector regardless of the path followed. Without admitting the possibility of interference (i.e., superposition) it would be incomprehensible why all the particles end up in detector 1.

6.3 Mach-Zehnder with "Which Path" Measurement

We have already seen in chapter 2 that, in the context of a double slit interference pattern, obtaining "which path" information will destroy the interference. Exactly the same is true for a Mach-Zehnder Interferometer, and for identical algebraic reasons. However the matter is so fundamental that it bears repeating in this context.

Suppose the state of a particle which follows path LB into detector j is written $|LB\rangle_j$. Then the state of a particle following path UB into detector j is $|UB\rangle_j = e^{i\delta_j}|LB\rangle_j$. The preceding analysis has shown that $\delta_1 = 0$ hence $e^{i\delta_1} = 1$ and $\delta_2 = \pi$ hence $e^{i\delta_2} = -1$. The state at the detectors is thus,

$$|\psi\rangle = \frac{1}{\sqrt{2}}\left(|LB\rangle_j + e^{i\delta_j}|LB\rangle_j\right) \tag{6.3.1}$$

Hence the flux of particles into the detectors is proportional to,

$$\langle\psi|\psi\rangle = \frac{1}{\sqrt{2}}\left(1 + e^{-i\delta_j}\right)\frac{1}{\sqrt{2}}\left(1 + e^{i\delta_j}\right) = 1 + \cos\delta_j \tag{6.3.2}$$

thus giving constructive interference, $\langle\psi|\psi\rangle = 2$, at detector 1 and destructive interference, $\langle\psi|\psi\rangle = 0$, at detector 2.

How does this behaviour change if a measuring device is inserted into (say) the upper beam path (the "optional object" in Figure 6.1)? A device to detect a particle in path UB must have two states, one of which corresponds to "a particle was detected in path UB" and the other corresponding to "a particle was not detected in path UB". These states of the measuring device will be written $|M:UB\rangle$ and $|M:LB\rangle$ respectively. The notation reflects the assumption of perfectly efficient detection, and no particle losses, for which it follows that not detecting the particle in path UB implies that it followed path LB.

Now a perfect measurement requires the two states of the measuring device to be absolutely distinguishable. This means that the two measurement states must be orthogonal, $\langle M: UB | M: LB \rangle = 0$. However, it is possible to envisage a poor measurement which might indicate a probability of the particle being in path LB, but with some residual possibility of being in path UB. For such an imperfect measurement $\langle M: UB | M: LB \rangle \neq 0$.

The combined state of the particle and the measuring device, at either detector, is thus,

$$|\psi\rangle = \frac{1}{\sqrt{2}} \left(|LB\rangle_j |M: LB\rangle + e^{i\delta_j} |LB\rangle_j |M: UB\rangle \right) \qquad (6.3.3)$$

The flux of particles into the detectors is thus,

$$\langle \psi | \psi \rangle = 1 + \Re\{ e^{i\delta_j} \langle M: UB | M: LB \rangle \} \qquad (6.3.4)$$

Consequently, if we have performed a perfect measurement which definitely detects which path the particle took, then, because $\langle M: UB | M: LB \rangle = 0$, we have from (6.3.4) that the particle flux is unity, $\langle \psi | \psi \rangle = 1$, into *both* detectors. All interference has been lost, both detectors register particles equally often.

This establishes quite generally that any means of detecting which path the particle takes around the interferometer will destroy the interference (in the sense that both detectors will detect equal numbers of particles, as they would if the situation were classical with each particle taking a definite path around the interferometer).

However, the occurrence or not of interference is not an all-or-nothing affair. An imperfect measurement, which has $0 < |\langle M: UB | M: LB \rangle| < 1$, will still leave residual interference and hence more particles into detector 1 than into detector 2. All this is just as we found previously in chapter 3 for the double-slit experiment. The algebra is identical.

6.4 The Elitzur-Vaidman Bomb Test

Everything is now set up for this lovely example of the observability of counterfactuals. Imagine that you have manufactured a large number of bombs. These bombs are so exquisitely sensitive that the slightest physical interaction, however tenuous - say with a single very low energy neutrino - would make the bomb explode. The trouble is that you know that some bombs are duds, but you need to identify one which is definitely not a dud without it exploding.

How on earth can you do this? The problem defines the bombs to be so sensitive that the slightest physical interaction with any given bomb will make it explode. So you need to determine if a bomb is not a dud without interacting with it at all! Of course you can easily determine which are the duds by poking all the bombs. The ones that don't explode are the duds. But unfortunately that leaves you with no live bombs.

In classical physics the problem is insoluble. But in quantum physics, amazingly, it can be solved.

A bomb is placed[1] within the UB beam path of the interferometer in such a way that a passing particle may or may not interact with it. For example, this might be accomplished by attaching a rod to the mirror at the top left such that there was the tiniest gap between the rod and the bomb. If the mirror is free to move when struck by a particle, the rod would then strike the bomb and set if off. (The mechanism is impractical, of course, but the principle is what matters – a truly practical device is perfectly feasible).

An active bomb has thus been made into a measuring device regarding which of the paths, UB or LB, the particle takes. If it takes path UB, the bomb explodes. If the bomb does not explode then either the particle took path LB or the bomb is a dud. The very sensitivity of an active bomb makes it a perfect measuring device.

Now if the bomb is a dud, then the bomb does not constitute a measuring device. A dud bomb might as well not be there. So, with a dud bomb, the particles will always emerge into detector 1, never into detector 2.

But if the bomb is *not* a dud, and assuming it does not explode, then the measurement implemented by the bomb destroys the interference and the particle could be detected in either detector 1 or 2. But if it is detected in detector 2 the bomb cannot be a dud! So we have successfully identified a bomb which is definitely not a dud but without it exploding. Miraculous! Note that in terms of the state algebra, and assuming a live bomb, the absence

---

[1] Of course, you can't actually pick up the bomb to move it, or it would go off. So this is really shorthand for "a Mach-Zehnder interferometer is constructed around a given bomb". You might also be wondering what use these bombs could possibly be, since you could never move them to where you might want to destroy something. Hey, don't take this example so literally!

of an explosion means that the measurement has selected (i.e., the state vector has reduced to) the first term in the state,

$$|\psi\rangle = \frac{1}{\sqrt{2}}(|LB\rangle|M:LB\rangle no\ bang + |UB\rangle|M:UB\rangle bang!) \qquad (6.4.1)$$

So the particle state hitting the final beam splitter is just $|LB\rangle$, which thus has a 50/50 chance of ending up at detector 1 or 2.

The fact that the bomb *might* have gone off, despite the fact that it did not, is crucial to the bomb constituting a measuring device and hence to the identification of the unexploded bomb as not being dud. The counterfactual has had an observable consequence, namely that the bomb is now known with certainty to be live.

This curious phenomenon, and the bomb scenario described above, was originally described by Elitzur and Vaidman, Ref.[6.1]. Do not think that it is too theoretical to be demonstrated experimentally (though not literally with exquisitely sensitive bombs). On the contrary, this was done almost as soon as the effect was predicted, by Anton Zeilinger's group in Vienna, Ref.[6.2].

As described, the bomb test is terribly inefficient. Of the live bombs, half are exploded, and of the remaining 50% only half of these result in a particle at detector 2 and hence are identified definitely as live bombs. Thus, a single application of the bomb tester identifies ¼ of the live bombs. But another ¼ still remain unexploded and unidentified. Running these (with all the duds, of course) through the bomb tester again results in a further ¼ of this ¼ being identified as live. Hence, repeated applications identify,

$$\frac{1}{4} + \frac{1}{4} \times \frac{1}{4} + \frac{1}{4} \times \frac{1}{4} \times \frac{1}{4} + \frac{1}{4} \times \frac{1}{4} \times \frac{1}{4} \times \frac{1}{4} + \cdots = \frac{1}{3}$$

of the live bombs. But the remaining two-thirds are destroyed – terribly inefficient. However, this is avoidable. In principle virtually all the live bombs can successfully be identified with a sufficiently cunning approach, see for example Ref.[6.3].

6.5 Forward to Magic - but not really

The startling realisation that information can be obtained without interacting with the object opens up many possibilities. One is to replace the bomb with a quantum computer. The claim is sometimes loosely made that this can provide a means of obtaining the result of a quantum computation without even switching the computer on. This is rather misleading because we are apt to interpret the "switching on" as a classical process, i.e., involving a real macroscopic switch and the initiation of large currents or voltages. In contrast what is really envisaged is that the quantum computer includes an "activation" qubit (chapter 13), such that the computer does nothing if the qubit is set to a certain state $|0\rangle$ but runs its computation if the qubit is switched to $|1\rangle$. The difference between the classical and quantum computer scenarios is that the qubit is a pure quantum state and will in general be in a superposition of states, something not available to a classical switch.

This is crucial to the claim, which is more correctly stated as being that it is possible to extract the result of the computation whilst also confirming that the activation qubit remains in state $|0\rangle$, the "off position". This is referred to as "counterfactual computation", see for example Ref.[6.4]. It is directly analogous to obtaining the information that a bomb is live whilst not exploding it (i.e., whilst not "switching it on").

The great curse of quantum computers is decoherence. The possibility of obtaining the result of a quantum computation without the computer actually having to run might appear to be a very neat way of avoiding the problem of decoherence. Unfortunately this is not so. Obtaining counterfactual information depends crucially on unitary evolution (which is another way of talking about maintaining coherence). Just as the *possibility* of the computer running is essential in obtaining its result counterfactually, so the *possibility* of its decoherence if it did run is sufficient to undermine a successful outcome. The magic does not extend to getting something for nothing.

6.6 Exercises for the Reader

(1) Derive the reflected beam relative phase of i wrt the transmitted beam for a plane glass plate.

Consider a glass plate whose thickness lies at $x \in [0, a]$, so the glass has thickness, a. The incident plus reflected wave in the region $x < 0$ is $e^{ikx} + Be^{-ikx}$. Within the plate material, $0 < x < a$, the right plus left going waves are $Ee^{ik'x} + Fe^{-ik'x}$ where $k' = nk$ is the wavenumber in glass and n is the refractive index. In the region $x > a$ the transmitted wave is Ce^{ikx}. By applying the boundary conditions (continuity of the wavefunction and its x-derivative) at both boundaries $x = 0$ and $x = a$, the coefficients B, E, F, C can be found. In particular the ratio of the reflection (B) and transmission (C) coefficients is,

$$\frac{B}{C} = i \cdot \left(\frac{n^2 - 1}{2n}\right) \sin k'a \cdot e^{ika} \qquad (6.6.1)$$

Now the factor of e^{ika} just accounts for the fact that the phase of the reflected wave has been reference to position $x = 0$ whereas the transmitted wave phase is referenced to $x = a$. Referencing them both to the same point, as required, leaves the phase factor from (6.6.1) as just i (because the other terms in (6.6.1) are real and determine the relative amplitudes of the two beams, not their relative phase). **QED**.

6.7 References

[6.1] Elitzur, A. C. and Vaidman, L. (1993). "Quantum mechanical interaction-free measurements". *Found. Phys.* **23**, 987-97. arxiv:hep-th/9305002.

[6.2] Kwiat, P.G., Weinfurter, H., Herzog, T., Zeilinger, A., and Kasevich, M.A. (1995). "Interaction-free Measurement". *Phys. Rev. Lett.* **74** (24): 4763. doi: https://doi.org/10.1103/PhysRevLett.74.4763

[6.3] Kwiat, P.G., White, A.G., Mitchell, J.R., Nairz, O., Weihs, G., Weinfurter, H., and Zeilinger, A. (1999), "High-efficiency Quantum Interrogation Measurements via the Quantum Zeno effect", *Phys. Rev. Lett.* **83** (23), 4725-4728. Doi: https://doi.org/10.1103/PhysRevLett.83.4725. See also Paul Kwiat, http://physics.illinois.edu/people/kwiat/interaction-free-measurements.asp

[6.4] G.Mitchison and R.Jozsa (2001), "Counterfactual Computation", *Proc.Roy.Soc.Lond.* **A457**, 1175-1194. Doi: https://doi.org/10.1098/rspa.2000.0714

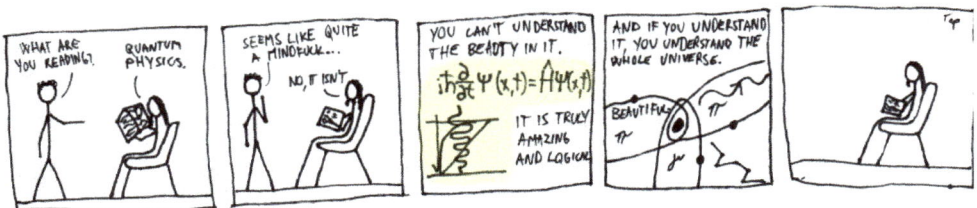

7

State Vector Reduction Demonstrated

What happens if we decide whether or not to bring together a pair of particle beams (obtained by splitting an original beam) only after the beams have been split? We thus permit interference to occur only after the time when, classically, the path taken by the particle would have been determined. Do we find interference? If so, what does it imply about the case where we choose instead to detect without permitting interference? This is Wheeler's delayed choice interference experiment realised.

7.1 Wheeler's Delayed Choice Experiment

Consider a Mach-Zehnder interferometer (see Figure 6.1 in the last chapter, but there will be no "optional object" this time). The manifestation of interference in this device is that all output photons appear in detector 1, and none in detector 2. Why this happens was explained in chapter 6. We know that if we measure which path the photon takes then interference is destroyed, and photons will be found by both detectors. Similarly, in the case of a double slit experiment, obliteration of the interference fringes occurs if a measurement is made to determine the path of the photon between the slits and the screen. The reason why this is inevitable follows from the same state algebra, as demonstrated in chapter 3.

Wheeler, Ref.[7.1], posed the following gedanken experiment. Consider a double-slit experiment and suppose that at some time after the photon has passed through the slits we choose at random whether to leave the screen or to snatch it away revealing a pair of telescopes, one focussed on each slit, thus observing whether the photon originated from the left or the right slit. Considering only those cases when the screen is left in place, the interference pattern will (one presumes) be found. But the remaining cases (when the screen is whipped away) have no interference pattern, because the slit which is illuminated betrays the path taken. But the photon is already committed regarding its behaviour when the decision to leave or remove the screen is taken, in the sense that it has already passed beyond the slits. This purely theoretical experimental arrangement suggests that whether one gets an interference pattern or a deterministic indication of which slit the photon went through is not determined until the point of detection (on a screen or in a telescope). The experiment has not been done in quite this form to confirm the expected behaviour.

However, the same quantum behaviour can be displayed by a Mach-Zehnder interferometer. Consider the final beam splitter (top right in Figure 6.1). If the final beam splitter were omitted, detector 1 would measure only photons taking the path UB, and detector 2 would measure only photons taking path LB. Both would occur with equal frequency (assuming the first beam splitter is 50/50). There is no interference (rather trivially in this case since, in the absence of the final beam splitter, the two paths are never brought together in order to interfere).

But we can now consider leaving or removing the final beam splitter at random, and at a time after the photon has passed through the first beam splitter. To be more precise, we can ensure that the event defined by the photon passing through the first beam splitter cannot be causally influenced by the decision to retain or remove the final beam splitter. If quantum mechanics is correct, and if we confine attention to those randomly selected cases when the final beam splitter is retained, we shall find all the photons enter detector 1 and none into detector 2 (which is the manifestation of interference in this device).

This version of Wheeler's experiment, employing a Mach-Zehnder interferometer, has actually been performed by Jacques et al, Ref.[7.2]. The retention or removal of the final beam splitter was achieved electronically and the decision was made by a random number generator sufficiently quickly to be spacelike separated from the photon passing through the first beam splitter. Interference occurs unambiguously when the final beam splitter is activated, thus confirming the quantum mechanical expectation, but does not occur in those cases where the final beam splitter is removed.

What does this result imply? The interpretation would seem to be that,

(i)   For randomly selected cases which retain the final beam splitter, the photon must be regarded as "going both ways at once" in order to generate the observed interference pattern.

(ii) But the photon must therefore also "go both ways at once" for the remaining cases, when the final beam splitter is removed, since there is no causal link between this decision and the photons prior to the slits.

(iii) This suggests that the "reduction of the state vector", which leads to detection in one or other of the two detectors, does not occur until the photon encounters the detectors. Hence, when the final beam splitter is absent, detection of a photon in detector 1 does not quite mean that the photon took path UB. Prior to detection, either of detectors 1 or 2 might register the photon since the photon would have been in a superposition of both states. The counterfactual, that it might have taken path LB, only becomes a counterfactual after the detectors have detected a photon. Before that, both options were open.

When the final beam splitter is omitted, the state vector does appear to reduce at the point the photon is detected. The outcome of the detection is truly indeterminate – there are no hidden variables which determine the 'actual path taken'. Or, as the authors of Ref.[7.2] concluded,

"Our realization of Wheeler's delayed choice gedanken experiment demonstrates beyond any doubt that the behavior of the photon in the interferometer depends on the choice of the observable which is measured, even when that choice is made at a position and a time such that it is separated from the entrance of the photon in the interferometer by a space-like interval….. Once more, we find that Nature behaves in agreement with the predictions of Quantum Mechanics even in surprising situations where a tension with Relativity seems to appear."

7.2 References

[7.1]   Wheeler, J.A., and Zurek, W.H. (editors), Quantum Theory and Measurement (Princeton Series in Physics). https://press.princeton.edu/books/hardcover/9780691641027/quantum-theory-and-measurement. For a discussion of Wheeler's Delayed Choice experiment see https://www.anthonypeake.com/627/

[7.2]   Jacques, V, Wu, E., Grosshans, F., Treussart, F., Grangier, P., Aspect, A., and Roch, J-F. (2007) "Experimental Realization of Wheeler's Delayed Choice Gedanken Experiment", *Science* **315** (5814) 966–968. https://science.sciencemag.org/content/315/5814/966.abstract. See also arXiv: quant/ph-0610241.

8

A Watched Pot Never Boils

The Quantum Zeno Effect: continually observing something prevents it evolving.

8.1 What is the Quantum Zeno Effect?

Most of the phenomena discussed in this book can be analysed using the state algebra alone without having to solve the Schrodinger equation, (2.4.1). However, the subject matter of this chapter is an exception. Here we are explicitly interested in the unitary time evolution of a system – and specifically how this can be subverted by making repeated measurements. I shall make use of first order perturbation theory in this chapter. (The required result is derived in §8.4).

The phenomenon known as the quantum Zeno effect has, in recent years, been associated with Misra and Sudarshan (1977), Ref.[8.1]. I first became aware of the effect via Wolsky (1976), Ref.[8.2]. It is noteworthy that Wolsky also refers to Zeno in his text. However, many authors had noted the phenomenon earlier. It was reputedly known to Alan Turing in 1954, see Ref.[8.3].

Zeno's arrow paradox seems to deny the possibility of motion (or change generally) on the grounds that at every instant of time no motion (or change) is apparent. The quantum Zeno effect is that the natural unitary evolution of a system, in accord with the Schrodinger equation, is suppressed if the system is "measured" sufficiently frequently. Whilst Zeno's paradox is merely a philosophical conundrum, the quantum Zeno phenomenon is a real physical effect which has been observed in many experiments, e.g., Refs.[8.4,5].

To be more accurate, the quantum Zeno effect requires the initial state of the system to be an eigenstate of the measurement. The measurement process will then have the effect of continually knocking the system back to its initial state due to the repeated reduction of the state vector – providing that the measurements are sufficiently frequent compared to the timescale of the system's free evolution. It is not hard to establish this behaviour, as demonstrated below. The key to it is that whilst probability amplitudes (essentially state vectors) evolve linearly in time for short periods, the associated probabilities (essentially expectation values) vary quadratically over short periods. Consequently, if N measurements are made at equal intervals over a total time t, the probability of change will be proportional to $N(t/N)^2 = t^2/N$, which therefore tends to zero for large N. The rest of this chapter simply fills in the details.

8.2 Derivation of the Quantum Zeno Effect

Suppose a system starts in a state $|1\rangle$, which is perhaps an eigenstate of some undisturbed Hamiltonian, $\hat{H}_0$. Now suppose we apply some influence to the system, described by an interaction Hamiltonian, $\hat{H}_I$ ("heating the pot to make it boil"). This has the potential to change the state of the system. Indeed, if we left the combined effects of $\hat{H}_0 + \hat{H}_I$ alone to do their work, the state of the system *would* change (the pot *would* boil). But what happens if we repeatedly measure the system's energy by applying $\hat{H}_0$ over and over again? What happens if we continually 'watch the pot'?

In classical physics this would make no difference, of course. A classical observation may be made, in principle, without disturbing the system: but not so in quantum mechanics if the system departs from being an eigenstate, because measurement involves reduction of the state vector[2]. It turns out that if we measure the system's energy state often enough, then its state will never change: a watched quantum pot never boils.

The most general time-dependent pure quantum state of the system can be written in terms of the eigenstates, $|j\rangle$, of the free Hamiltonian as,

$$|t\rangle = \sum_j a_j(t)\, exp\left\{-i\frac{E_j t}{\hbar}\right\} |j\rangle \tag{8.1}$$

We shall be concerned only with changes over small periods of time, so we can use first order perturbation theory secure in the knowledge that this will be exact in the limit. Suppose at $t = 0$ the system is in eigenstate $j = 1$, i.e., $a_1(0) = 1$, $a_j(0) = 0$ for $j > 1$. First order perturbation theory tells us that the time dependent expansion coefficients are given by,

For $j > 1$ $$\frac{\partial a_j}{\partial t} = \frac{\langle j|\hat{H}_I|1\rangle}{i\hbar} exp\left\{i\frac{(E_j - E_1)}{\hbar}t\right\} \tag{8.2}$$

An exercise for the reader is to consult a standard QM text for the derivation of this perturbation theory result (my derivation is in §8.4). In passing I note

---

[2] The reader may be puzzled on the grounds that, in chapter 6, we showed that Elitzur-Vaidman bombs could be identified as live without physically interacting with them. But this crucially did not involve a completed measurement on such bombs; their state vector was never reduced.

that this result motivates referring to matrix elements like $\langle j|\hat{H}_I|1\rangle/i\hbar$ as "transition amplitudes" as they control the rate at which the amplitude of other eigenstates, $a_j, j \neq 1$, appear in the perturbed state.

Because we are concerned only with the limit of very short times, it suffices to consider $\hat{H}_I$ to be time-independent, without loss of generality, apart from the fact that it is turned on at $t = 0$. We can then integrate (8.2) explicitly, giving,

$$\text{For } j > 1 \qquad a_j(t) = \frac{\langle j|\hat{H}_I|1\rangle}{(E_j-E_1)}\left(1 - exp\left\{i\frac{(E_j-E_1)}{\hbar}t\right\}\right) \tag{8.3}$$

Hence,

$$|a_j(t)|^2 = 2\left|\frac{\langle j|\hat{H}_I|1\rangle}{E_j-E_1}\right|^2\left(1 - cos\left\{\frac{(E_j-E_1)}{\hbar}t\right\}\right) \tag{8.4}$$

We are only claiming that this is universally correct in the limit of short times, so more properly we write (8.3) and (8.4) as,

$$\text{For } j > 1 \text{ and } LIM(t \to 0): \qquad a_j(t) = \frac{\langle j|\hat{H}_I|1\rangle}{i\hbar}t \tag{8.5}$$

$$\text{For } j > 1 \text{ and } LIM(t \to 0): \qquad |a_j(t)|^2 = \left|\frac{\langle j|\hat{H}_I|1\rangle}{\hbar}t\right|^2 \tag{8.6}$$

Hence, whenever a measurement is made, the probability that the system will be found to be in the original state, $|1\rangle$, is $P = |a_1(t)|^2$, which is given by,

$$P = |a_1(t)|^2 = 1 - \sum_{j>1}|a_j(t)|^2 = 1 - \left(\sum_{j>1}\left|\frac{\langle j|\hat{H}_I|1\rangle}{\hbar}\right|^2\right)t^2 \equiv 1 - \chi t^2$$

Here χ is defined as the constant,

$$\chi \equiv \sum_{j>1}\left|\frac{\langle j|\hat{H}_I|1\rangle}{\hbar}\right|^2 \tag{8.7}$$

Consequently, if we leave it long enough before measuring the state of the system, we may find that it is no longer in the initial state (because $P < 1$). If you don't watch the pot, it will eventually boil. (In fact, if it's milk, I can guarantee that it will boil over).

But see what happens if the system is measured N times at equal intervals during the period t, i.e., at intervals of t/N. The probability that the system remains in state $|1\rangle$ after the first measurement is,

$$P_1 = 1 - \chi\left(\frac{t}{N}\right)^2 \tag{8.8}$$

Assuming that the system is in state $|1\rangle$ after the first measurement, the probability that it is still in state $|1\rangle$ after the second measurement is also given by (8.8). So the overall probability of being in state $|1\rangle$ after two measurements is,

$$P_2 = P_1 \times P_1 = \left[1 - \chi\left(\tfrac{t}{N}\right)^2\right]^2 \qquad (8.9)$$

Similarly, after N measurements, the probability that the state remains $|1\rangle$ is,

$$P_N = P_1{}^N = \left[1 - \chi\left(\tfrac{t}{N}\right)^2\right]^N \qquad (8.10)$$

Now as we let N become very large, so we are making a very large number of measurements in the fixed time period, t, this probability tends to unity:

$$LIM(N \rightarrow \infty): P_N \rightarrow 1 \qquad (8.11)$$

which means that the system remains in its initial state: evolution has been suppressed by the repeated measurements of its energy.

If the limit (8.11) is not immediately obvious to you, note that (8.10) can be expanded thus,

$$1 - N\chi\left(\tfrac{t}{N}\right)^2 + \tfrac{N(N-1)}{2!}\left[\chi\left(\tfrac{t}{N}\right)^2\right]^2 - \tfrac{N(N-1)(N-2)}{3!}\left[\chi\left(\tfrac{t}{N}\right)^2\right]^3 + \ldots \qquad (8.12)$$

and successive terms after unity are of order $1/N, 1/N^2, 1/N^3, \ldots$ and hence all tend to zero. The reason that the probability limit is unity can be traced to the quadratic dependence on time in (8.6), and hence to the quadratic dependence of the denominator on N in (8.10).

Physically, the repeated measurements continually knock the system back into its initial state, neutralising the influence of the disturbance ($\hat{H}_I$) which would otherwise cause its state to change. The "knocking back" consists of a continual reduction of the state vector ("collapse of the wavefunction"). The quantum Zeno effect has been demonstrated experimentally many times, for examples Refs.[8.4-5]. These experiments are therefore an indirect means of confirming that measurement involves state vector reduction.

8.3 OK, the Boiling Pot Thing Was a Joke

Perhaps I should warn the reader not to take seriously the whimsical analogy of the quantum Zeno effect with "a watched pot never boils". A pan of water at a finite temperature is not in a pure quantum state — very far from it, in fact. Consequently, the quantum Zeno effect is not applicable to literal pans of water. At this point in the book, however, I have not yet explained why "large" (i.e., classical) systems are not pure quantum states – and, if not, how they can be described in QM. That will come in chapter 10 when we discuss mixed states.

8.4 Reader Exercises

Derive the first-order perturbation theory result, equ.(8.2): For the reader's delight and edification, I first derive a more general result. Time dependent perturbation theory addresses the time dependent solution to,

$$\hat{H}|\psi\rangle = i\hbar \frac{\partial}{\partial t}|\psi\rangle \qquad (8.4.1)$$

As in the main text we assume that the Hamiltonian consists of an unperturbed part plus a relatively small perturbation, $\hat{H} = \hat{H}_0 + \hat{H}_I$. Also consistent with the main text we assume that the perturbation is switched on only at time zero, so more exactly we should write,

$$\hat{H} = \hat{H}_0 + \Theta(t)\hat{H}_I \qquad (8.4.2)$$

where Θ is the step function. At any time, t, the wavefunction is given in terms of the unperturbed energy eigenstates, $\hat{H}_0|u_n\rangle = E_n|u_n\rangle$, by,

$$|\psi\rangle = \sum_n a_n(t)|u_n\rangle \, exp\{-iE_n t/\hbar\} \qquad (8.4.3)$$

Which is just the same as equ.(8.1) in slightly different notation. However, we allow for the possibility that the state starts, not in an eigenstate of the unperturbed Hamiltonian, but any state at all. Nevertheless, the coefficients, $a_n(0)$, are defined at time zero by the initial state. If there were no perturbation, the time dependent state would be given by,

$$|\psi\rangle_{unperturbed} = \sum_n a_n(0)|u_n\rangle \, exp\{-iE_n t/\hbar\} \qquad (8.4.4)$$

Note that this means, in general, that the unperturbed state would have been non-trivially time dependent. Substituting (8.4.2,3) into (8.4.1) for $t > 0$,

$$(\hat{H}_0 + \hat{H}_I) \sum_n a_n(t)|u_n\rangle \, exp\left\{-\frac{iE_n t}{\hbar}\right\}$$

$$= \sum_n a_n(t)[E_n + \hat{H}_I]|u_n\rangle \, exp\left\{-\frac{iE_n t}{\hbar}\right\}$$

$$= i\hbar \frac{\partial}{\partial t}\left(\sum_n a_n(t)|u_n\rangle \, exp\left\{-\frac{iE_n t}{\hbar}\right\}\right)$$

$$= \sum_n \left[a_n(t)E_n + i\hbar \frac{\partial a_n}{\partial t}\right]|u_n\rangle \, exp\{-iE_n t/\hbar\} \qquad (8.4.5)$$

The first terms cancel, leaving,

$$\sum_n a_n(t)\hat{H}_I|u_n\rangle\, exp\{-iE_nt/\hbar\} = \sum_n i\hbar\frac{\partial a_n}{\partial t}|u_n\rangle\, exp\{-iE_nt/\hbar\} \qquad (8.4.6)$$

Taking the scalar product with $\langle u_k|$ and appealing to the orthonormality of the eigenstates gives,

$$\sum_n a_n(t)\langle u_k|\hat{H}_I|u_n\rangle\, exp\{-iE_nt/\hbar\} = i\hbar\frac{\partial a_k}{\partial t}exp\{-iE_kt/\hbar\} \qquad (8.4.7)$$

Hence,

$$\frac{\partial a_k}{\partial t} = \frac{1}{i\hbar}\sum_n\langle u_k|\hat{H}_I|u_n\rangle\, exp\{i(E_k - E_n)t/\hbar\}\, a_n(t) \qquad (8.4.8)$$

Equ.(8.4.8) is exact. So far we have introduced no approximations. It gives a coupled set of first order differential equations for the expansion coefficients in terms of the matrix elements of the perturbing interaction with respect to the unperturbed eigenstates, and a phase factor dependent upon the unperturbed energy levels. Numerical integration of (8.4.8) would provide a means of solving the time dependent problem including the perturbing interaction, if its matrix elements can be computed.

First order perturbation theory consists of approximating the coefficients on the RHS of (8.4.8) by their initial values, i.e.,

$$\frac{\partial a_k}{\partial t} \approx \frac{1}{i\hbar}\sum_n\langle u_k|\hat{H}_I|u_n\rangle\, exp\{i\omega_{kn}t\}\, a_n(0) \qquad (8.4.9)$$

where we have written $\omega_{kn} = (E_k - E_n)/\hbar$. If the initial state is an eigenstate of the unperturbed Hamiltonian, say the 1st energy eigenstate of the unperturbed Hamiltonian, (8.4.9) becomes,

$$\frac{\partial a_k}{\partial t} \approx \frac{exp\{i\omega_{k1}t\}}{i\hbar}\langle u_k|\hat{H}_I|u_1\rangle \qquad (8.4.10)$$

which is just equ.(8.2) in slightly different notation.   **QED**

8.5 References

[8.1]   Misra, B. and Sudarshan, E.C.G.; (1977). "The Zeno's paradox in quantum theory". Journal of Mathematical Physics **18**, 756–763. https://doi.org/10.1063/1.523304

[8.2]   A.M.Wolsky (1976), "Why A Watched Pot Never Boils", Foundations of Physics, **6** (3) 367-369. https://link.springer.com/article/10.1007/BF00708809

[8.3]   C.Teuscher, editor, (2003). "Alan Turing: Life and Legacy of a Great Thinker", Springer.

[8.4] Zheng, W., Xu, D.Z.2,Peng, X., Zhou X., Du, J., and Sun, C.P. (2013) "Experimental demonstration of the quantum Zeno effect in NMR with entanglement-based measurements", Phys.Rev. **A87**, 032112. https://doi.org/10.1103/PhysRevA.87.032112

[8.5] Kalb, N., Cramer, J., Twitchen, D.J., Markham, M., Hanson, R., and Taminiau, T.H. (2016). "Experimental creation of quantum Zeno subspaces by repeated multi-spin projections in diamond", Nature Communications **7**, 13111. https://www.nature.com/articles/ncomms13111

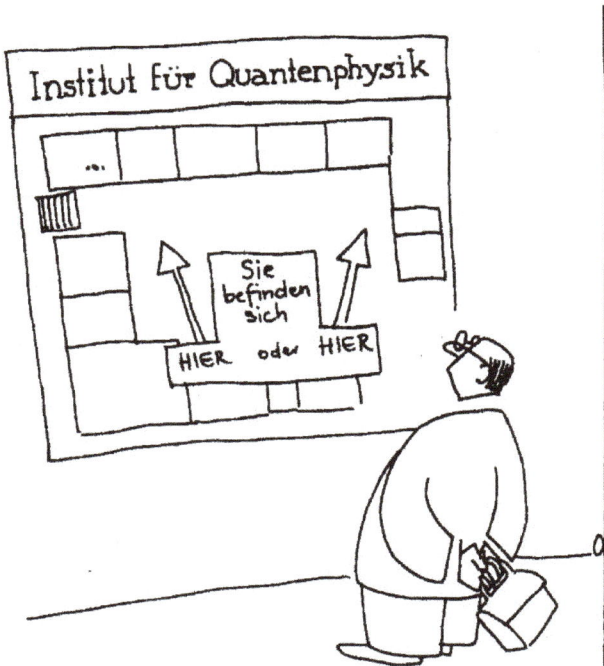

9

There Are No Hidden Variables: Part 1

There are powerful reasons for believing that the indeterminacy of quantum measurements is inescapable: that there are no "hidden variables" which would make the outcome of every measurement uniquely determined. This belief is founded on two pillars: the theorem of Kochen and Specker, and the experimental violations of Bell's Inequality. Here we prove the theorem of Kochen and Specker which rules out a broad class of hidden variables, after first reviewing von Neumann's false argument. We shall complete the hidden variable no-go story when we return to Bell's Inequality in chapter 14.

9.1 von Neumann's False Start

All physicists are initially (and perhaps permanently) uncomfortable with the indeterminism of quantum measurement. Raised as they are on classical physics, where strict determinism holds sway, physicists in the early days of quantum mechanics sought ways to smuggle determinism back into the theory. This endeavour is generically known as appealing to "hidden variables", which we may symbolically represent as Λ. One need not know of what Λ consists. It may be many variables, vectors, tensors, or whatever. But the idea is that if we only knew the values taken by all the quantities in Λ then the outcome of any measurement, in any quantum state, would be uniquely determined.

I must add a qualification. It is usual also to assume that Λ does not depend upon the details of the measurement process but is "innate" within the system itself. This is referred to as "non-contextuality", i.e., that the hidden variables do not vary with context (by which is meant the details of any measurement process). One might refer to non-contextual hidden variables as representing the hard-line determinist view. It is this hard-line, non-contextual, version of hidden variables which is ruled out by the Kochen and Specker theorem to be described here. There are, however, theories which permit contextuality, such as that of Bohm, Ref.[9.1], which are not ruled out by the theorem. Bohm's theory is essentially a rediscovery of de Broglie's pilot wave theory, presented at the 1927 Solvay Conference. The de Broglie-Bohm theory is also explicitly non-local, as it must be to agree with QM predictions as we shall see later.

For several decades after the dawn of quantum theory, the Copenhagen interpretation was dominant. In part this was due to the dominant influence of Niels Bohr. But the Copenhagen interpretation was also buttressed by von Neumann's purported proof that hidden variables could not exist. This false proof was published in 1932, Ref.[9.2]. Interestingly it was three years later, in 1935, that Einstein, Podolsky and Rosen published their famous paper arguing that QM was incomplete, Ref.[9.3]. This paper, universally known as EPR, will be discussed in more detail in chapters 13 and 14.

EPR's claim that QM is incomplete is essentially the same as claiming there are hidden variables. So, these authors appear not to have been impressed by von Neumann's purported no-go theorem of three years earlier. Either EPR thought it incorrect, or else they were unaware of it. The latter seems unlikely, although Einstein had other things on his mind in the period 1932-35, such as fleeing for his life from Nazi Germany and making a life elsewhere.

Von Neumann's "proof" is fatally flawed, as pointed out by Bell in 1966, Ref.[9.4], as well as by other people much earlier. It is worth examining what went wrong with it. Rather than reproducing von Neumann's original 'proof' we shall use the much simpler argument of Bell, Ref.[9.4]. Von Neumann's false premise is that the expectation value of a linear combination of observables equals the same linear combination of their individual expectation values. Expressed algebraically,

$$\langle aP + bQ \rangle = a\langle P \rangle + b\langle Q \rangle \tag{9.1.1}$$

In other words, "expectation value" is a homomorphism. Quantum mechanics has this property, of course, because it is simply a re-writing of the linearity of the Hilbert space operators. But von Neumann imposed this condition on the hypothetical hidden variable theories as well. To be more precise, von Neumann supposed that, given some hidden variables, Λ, then any observable, say $\hat{P}$, would have a uniquely determined outcome when measured on state ψ. We can represent this deterministic value $v(\hat{P}; \psi, \Lambda)$. Von Neumann claimed that the additivity property would apply to this deterministic value, i.e.,

$$v(a\hat{P} + b\hat{Q}; \psi, \Lambda) = av(\hat{P}; \psi, \Lambda) + bv(\hat{Q}; \psi, \Lambda) \tag{9.1.2}$$

Now it is easy to show via a counter-example that this eliminates deterministic (i.e., hidden variable) theories. Consider a spin ½ particle. In units of $\hbar/2$, on whatever axis we choose to measure the spin, the answer will be either +1 or -1, nothing in between. Suppose we interpret $\hat{P}$ and $\hat{Q}$ as spin measurements in the x and y directions respectively, and let us now consider $\hat{R} =$

$(\hat{P} + \hat{Q})/\sqrt{2}$ which represents a spin observable at 45-degrees to x and y. Measurement of the spin in this direction can also only return ± 1. But according to von Neumann's assumption, the measured value of this observable would have to be $v(\hat{R}) = \left(v(\hat{P}) + v(\hat{Q})\right)/\sqrt{2}$ and hence could only be zero or $\pm\sqrt{2}$. That's all von Neumann's "proof" amounts to.

But actually (9.1.2) is far too strong a condition. The QM condition, (9.1.1), refers to expectation values, not individual measurements. Hence, to be consistent with QM predictions, it is only necessary that (9.1.2) be true when averaged over possible hidden variables. In other words we really only need our hidden variable theory to respect,

$$\int v\left(a\hat{P} + b\hat{Q}; \psi, \Lambda\right)d[\Lambda] = a \int v\left(\hat{P}; \psi, \Lambda\right)d[\Lambda] + b \int v\left(\hat{Q}; \psi, \Lambda\right)d[\Lambda]$$

$$(9.1.3)$$

where $d[\Lambda]$ is some suitable measure on Λ-space. In fact, in Ref.[9.4], Bell gave an explicit prescription for deterministic outcomes of the measurement of spin ½ in an arbitrary direction in terms of a specific hidden variable formulation which obeyed condition (9.1.3). Do not get the wrong idea. This is not a viable hidden variable theory, nor was Bell presenting it as such. Its purpose was only to refute von Neumann's purported "proof". We shall see below that there is no local, non-contextual hidden variable theory which can reproduce all the predictions of quantum mechanics.

In the same 1966 paper, Ref.[9.4], Bell also anticipated the result that became generally recognised only in the following year when rediscovered by Kochen and Specker. Bell proved the result as a corollary of Gleason's Theorem, Ref.[9.5]. Consequently the Kochen and Specker theorem is more properly called the Bell, Kochen and Specker theorem. To that we now turn.

9.2 The Theorem of Bell, Kochen and Specker

There are many different ways of stating the conditions which lead to the Bell, Kochen and Specker (BKS) no-go theorem. Let's express it in terms appropriate for someone looking to develop a hidden variable theory. The desire is to be able, for any given state, to assign a uniquely determined value to every observable. This is what we have previously written $v(\hat{P}; \psi, \Lambda)$. If the "overt" state, ψ, and the "covert" hidden variables, Λ, are understood then we can contract this simply to $v(\hat{P})$, the value taken by the observable $\hat{P}$ in the understood state with understood hidden variables.

But we also want the theory to agree with QM when QM gives a deterministic outcome: in other words when the state is an eigenstate of the observable. In fact, if $\{p_i\}$ are the eigenvalues of $\hat{P}$ then we require that the values taken by this observable in our putative hidden variable theory are confined to these eigenvalues, as this is the position in QM, i.e., $v(\hat{P}) \in \{p_i\}$.

So far this is surely the absolute minimum one would want fulfilled by a nice deterministic hidden variable theory. But there is one more condition which is needed to make the BKS theorem provable. It is akin to von Neumann's additivity condition, (9.1.2), namely that,

$$v(\hat{P} + \hat{Q}) = v(\hat{P}) + v(\hat{Q}) \qquad (9.2.1)$$

but enormously weakened by requiring only that (9.2.1) hold when $\hat{P}$ and $\hat{Q}$ are "compatible", i.e., they commute. Note that it follows immediately that additivity extends to any number of commuting observables,

$$v(\hat{P} + \hat{Q} + \cdots \hat{R}) = v(\hat{P}) + v(\hat{Q}) + \cdots v(\hat{R}) \qquad (9.2.2)$$

In QM (9.2.1) is unavoidable when evaluated for a common eigenstate (§2.5.6) of commuting observables, which is the only case where there is a deterministic value function. In this case we must have $v(\hat{P}) \in \{p_i\}$ and $v(\hat{Q}) \in \{q_i\}$, where the subscript i denotes their common eigenstate. It follows that the RHS of (9.2.1) is just one of $\{p_i + q_i\}$. But this is also the set of eigenvalues of $\hat{P} + \hat{Q}$, and hence in standard QM the LHS of (9.2.1) must take values from this set, when the outcome is determinate. The imposition of (9.2.1) as a pre-condition for the BKS theorem therefore corresponds to the value-structure of commuting observables in standard QM, but, in the spirit of hidden variables, extends its applicability to all states because all states are to have deterministic measurement outcomes in the hidden variable theory.

In this sense the preconditions of the BKS theorem are the minimal requirements of a deterministic hidden variable theory. Actually there is one assumption I have sneaked past you: non-contextuality. A theory is non-contextual if, whenever an observable can be said to "have a value" it does so independent of the measurement context. In non-contextual theories, values of observables are innate properties of the system, independent of the means of measurement.

Everything is now in place to prove the BKS theorem. The theorem states, simply, that the minimal conditions described above for a non-contextual hidden variable theory cannot be fulfilled in any Hilbert space of dimension 3 or larger: they are mathematically inconsistent. This is so remarkable on first acquaintance that it is hard to believe. The conditions stated above seem so innocuous that it is hard to see how any problem could arise. The origin of the inconsistency lies in the arbitrarily large number of observables, all of which need to be assigned deterministic values taken from a discrete set, $v(\hat{P}) \in \{p_i\}$. In particular, we may consider a set of observables with far more elements than the dimension of the Hilbert space.

The original proof of Kochen and Specker, Ref.[9.6], is notoriously abstruse. A far simpler proof, in full generality, has been provided by Peres, Ref.[9.7]. However we opt here for a simpler proof still, due to Cabello et al, Ref.[9.8]. This proof applies in a Hilbert space of dimension 4 (and hence higher), but does not prove the result for Hilbert spaces of dimension 3, for which see Ref.[9.6] or Ref.[9.7]. (The reader should have no fear that the door has been left open for two-dimensional Hilbert spaces. That door will be firmly shut by other means, see chapter 14).

In a four-dimensional Hilbert space, consider a set of mutually orthogonal vectors, say $\{|u_i\rangle, i = 1 - 4\}$ where $\langle u_k|u_i\rangle = \delta_{ki}$. From these we can construct four projection operators, $\hat{P}_i = |u_i\rangle\langle u_i|$ (see §2.3.4), which are clearly Hermitian. As with any projection operators we have $(\hat{P}_i)^2 = \hat{P}_i$ (projecting again does nothing more) and hence the eigenvalues of $\hat{P}_i$ are 0 or 1 (see §2.5.3). Moreover, because $\sum_i \hat{P}_i = \sum_i |u_i\rangle\langle u_i| = 1$, the sum of all four projection operators is the unit operator which leaves all states invariant. We also note that $\hat{P}_i\hat{P}_k = 0$ when $i \neq k$ because $|u_i\rangle\langle u_i||u_k\rangle\langle u_k| = 0$ due to orthogonality (e.g., the projection in the x-direction of the y-component is zero). This means that the projection operators commute, i.e., $\hat{P}_i\hat{P}_k = \hat{P}_k\hat{P}_i$, rather trivially in this case as both are zero.

We may choose the orthogonal vectors, $\{|u_i\rangle\}$, to be the eigenvectors of some observable, say $\hat{Q}$, with eigenvalues $\{|q_i\rangle\}$. In this case we can interpret the projection operator $\hat{P}_i$ as an observable in its own right. It asks the question "does the observable $\hat{Q}$ take the value q_i?" and the result of "observing" $\hat{P}_i$ is 1 or 0 according to whether the answer is yes or no. Alternatively, we may simply define an observable $\hat{Q}$ by stating that its eigenvectors are the arbitrarily chosen orthogonal set $\{|u_i\rangle\}$. However, it is upon the projection operators we concentrate as these are mutually commuting observables.

According to the pre-condition of the BKS theorem that $v(\hat{P}) \in \{p_i\}$ it follows that the only values that can be taken by the projection operators are 0 and 1. But because $\sum_i \hat{P}_i = 1$ the other BKS pre-condition, the additivity rule, (9.2.2), tells us that,

$$v(\hat{P}_1) + v(\hat{P}_2) + v(\hat{P}_3) + v(\hat{P}_4) = 1 \qquad (9.2.3)$$

But each of the terms on the LHS of (9.2.3) can only be 0 or 1, from which it follows that exactly one of them is 1 and the other three are 0. This is only true in standard QM if, (i) the state is an eigenstate of the $\hat{P}_i$, or, (ii) we interpret (9.2.3) in the sense of expectation values, because $\sum_i \langle \varphi | u_i \rangle \langle u_i | \varphi \rangle \equiv 1$. But (9.2.3) will not hold for individual measurement outcomes in standard QM because, other than for eigenstates, the outcomes are indeterminate. Depending on the state, and upon random chance, in QM (9.2.3) may evaluate to 0, 1, 2, 3 or 4. Hence, the imposition of (9.2.3) on the type of hidden variable theories covered by the BKS theorem marks their distinction from standard QM.

Recall that we can associate a particular ray, or direction, $|u_i\rangle$, in Hilbert space with each projection operator, $\hat{P}_i$. Hence we require, for an arbitrarily chosen set of mutually orthogonal rays in Hilbert space, $\{|u_i\rangle\}$, that we can assign the value 1 to one of them whilst assigning 0 to the rest. For a single orthogonal set of directions, this is clearly completely trivial. The key to the BKS no-go theorem is the recognition that it is impossible to consistently assign unique values to every direction so that every orthogonal set of directions has $N - 1$ zeros and just one value 1. The reason is that a given direction is orthogonal to many (actually an infinite number of) directions in the $N - 1$ dimensional orthogonal sub-space.

All versions of the proof proceed by explicit construction of a number of sets of orthogonal vectors for which it can be shown that it is impossible to assign unique values to every vector which respects the condition that, for every orthogonal set of four vectors, one is assigned a value of 1 and the remaining $N - 1$ are assigned a value of zero. Following Cabello et al, Ref.[9.8], for $N = 4$, I simply write down, in Table 9.1, their 9 sets of four mutually orthogonal vectors. These sets of orthogonal vectors have been cleverly chosen so that every vector occurs precisely twice, as can very quickly be checked with the aid of the colour scheme.

Of the four vectors in every column in Table 9.1, one must be assigned the value 1 and the other three the value zero. The total value of all 4 x 9 vectors in the Table must therefore be 9 as there are 9 columns. But every vector appears exactly twice. So whichever vectors are assigned the value 1, their total must be an even number. As 9 is not even, this provides a contradiction. We conclude that it is impossible to assign values 0 or 1 to the 18 different vectors in Table 9.1 such that every column contains exactly one vector with value 1. This establishes the theorem in four dimensions. The theorem in more than four dimensions follows immediately by consideration of any four-dimensional sub-space.

The BKS theorem establishes that the minimum conditions one would require of a deterministic and non-contextual hidden variable theory, including the requirement (9.2.2), are contradictory, and hence that a deterministic, non-contextual hidden variable theory which reproduces QM measurements is not possible. Deeper discussions of the underlying presumptions of the proof and their interrelationship can be found in Refs.[9.9-10].

The key diagram from Kochen & Specker's original proof (not needed here)

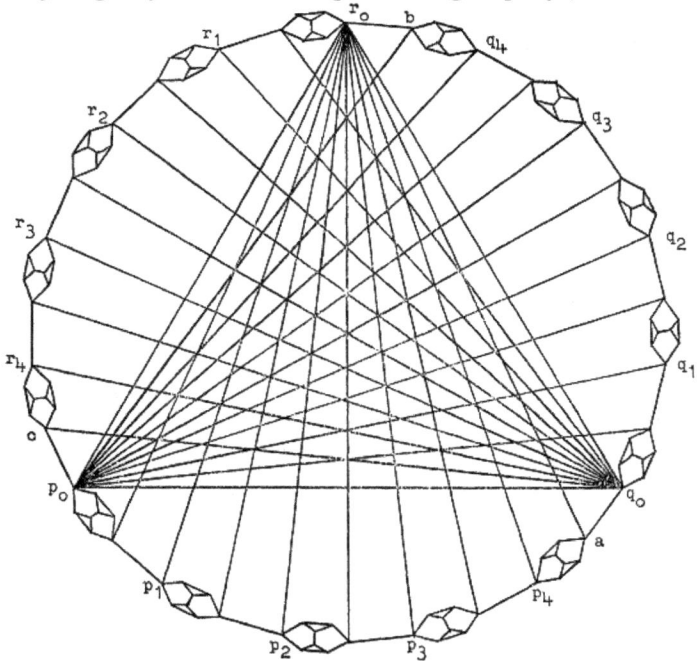

Table 9.1: An explicit set of 9 orthogonal sets of four vectors in 4D space (after Cabello, et al, Ref.[9.8])

u_1	(0, 0, 0, 1)	(0, 0, 0, 1)	(1, -1, 1, -1)	(1, -1, 1, -1)	(0, 0, 1, 0)	(1, -1, -1, 1)	(1, 1, -1, 1)	(1, 1, -1, 1)	(1, 1, 1, -1)
u_2	(0, 0, 1, 0)	(0, 1, 0, 0)	(1, -1, -1, 1)	(1, 1, 1, 1)	(0, 1, 0, 0)	(1, 1, 1, 1)	(1, 1, 1, -1)	(-1, 1, 1, 1)	(-1, 1, 1, 1)
u_3	(1, 1, 0, 0)	(1, 0, 1, 0)	(1, 1, 0, 0)	(1, 0, -1, 0)	(1, 0, 0, 1)	(1, 0, 0, -1)	(1, -1, 0, 0)	(1, 0, 1, 0)	(1, 0, 0, 1)
u_4	(1, -1, 0, 0)	(1, 0, -1, 0)	(0, 0, 1, 1)	(0, 1, 0, -1)	(1, 0, 0, -1)	(0, 1, -1, 0)	(0, 0, 1, 1)	(0, 1, 0, -1)	(0, 1, -1, 0)

The proof of the BKS theorem in four dimensions: Of the four vectors in each column, exactly one is to be assigned the value 1, whilst the other three are assigned the value 0. The total value of all 36 vectors in the Table would then be 9, as there are 9 columns. But every vector appears exactly twice in the Table (as indicated by the colour scheme, equal colours meaning equal vectors). So, whichever vectors are assigned the value 1, their total must be an even number. As 9 is not even, this is a contradiction. Hence it is not possible to assign unique values to all the 18 different vectors so that each column has exactly one vector of value 1 and three of value 0.

9.3 References

[9.1]   David Bohm (1952), 'A Suggested Interpretation of the Quantum Theory in Terms of "Hidden Variables" I and II', *Phys Rev* **85**(2), 166 and 182, available at https://cqi.inf.usi.ch/qic/bohm1.pdf and https://cqi.inf.usi.ch/qic/bohm2.pdf

[9.2]   John von Neumann, *"The Mathematical Foundations of Quantum Mechanics"*, English translation 1955 Princeton University Press (first published in German in 1932)

[9.3]   Einstein, A; B Podolsky, N Rosen (1935). "*Can Quantum-Mechanical Description of Physical Reality be Considered Complete?*". Physical Review **47**, 777–780. http://msekce.karlin.mff.cuni.cz/~holub/soubory/EPRoriginal.pdf

[9.4]   Bell, J S, (1966), *"On the problem of hidden variables in quantum mechanics"*, Rev.Mod.Phys. **38**, 447-452. http://www-hep.phys.unm.edu/~gold/phys491/Bell-Hidden-Variables-RevModPhys.38.447.pdf

[9.5]   Gleason, A M, (1957), J.Maths. & Mech. **6**, 885

[9.6]   Kochen, S, and Specker, E.P., (1967), "The Problem of Hidden Variables in Quantum Mechanics", J.Maths. & Mech. **17**, 59-87. https://www.jstor.org/stable/24902153?seq=1#page_scan_tab_contents

[9.7]   Peres, A., 1991, "Two Simple Proofs of the Kochen-Specker Theorem", *Journal of Physics*, A 24: L175–8. https://pdfs.semanticscholar.org/94ab/dca6caae890e351114f420aa65a81b15fe76.pdf

[9.8]   Cabello, A., Estebaranz, J. and Garcìa-Alcaine, G., 1996, "Bell-Kochen-Specker Theorem: A Proof with 18 vectors", *Physics Letters*, A 212: 183–87. https://arxiv.org/abs/quant-ph/9706009

[9.9]   The Sandford Encyclopedia of Philosophy, "The Kochen-Specker Theorem", last revision Feb 7, 2018. https://plato.stanford.edu/entries/kochen-specker/

[9.10] Redhead, Michael (1987), "Incompleteness, Nonlocality, and Realism", Clarendon Press, Oxford.

10

The General Quantum State is a Mixed State

So far in this book we have been dealing only with pure quantum states. But pure quantum states are rather special states. The macroscopic world does not consist of pure quantum states. Yet if QM is a fundamental theory of physics, it should be possible – at least in principle – to describe any system in QM terms. In this chapter I describe how this is done using the concept of mixed states and the density matrix.

10.1 Epistemic and Aleatory Uncertainty

When I was learning QM as a student, I used to skip over the bits in texts referring to the density matrix. I thought it was something I could do without. This was a serious error. I was left with the erroneous belief that all physical systems should be describable as pure quantum states. Not so. Such thinking leads to nonsense like Schrodinger's Cat (next chapter). Pure quantum states, fully describable as Hilbert space vectors, are exceptional. More general physical states require a different description.

Even in classical physics it is possible that the state of a system is unknown, or only partially known, simply because of our ignorance. In fact, the exact microstate of a macroscopic system is invariably unknown. In thermodynamics, knowledge of the temperature, pressure and density of a gas does not determine the specific state of all the constituent molecules. In classical physics this information would be regarded as knowable in principle, but just not known in practice. Every molecule, at every instant of time, has a definite momentum (according to the classical view of things). But because we do not have such detailed knowledge of every molecule, we are content that there is a certain probability that a randomly chosen molecule possesses a kinetic energy in the range, say, E to $E + dE$. Thus, the gas is a mixture of molecules of various energies, moving in various directions, and these quantities can be assigned probability densities. In this situation the uncertainty in the energy of a given molecule is due only to our ignorance. It is said to be an "epistemic" uncertainty; the uncertainty lies in our lack of knowledge, rather than being a fundamental uncertainty in the thing itself. Such situations are successfully addressed in statistical thermodynamics by accepting that the appropriate description must deploy probability density distributions.

This is very different from the uncertainty, i.e., the lack of determinism, in the outcome of a measurement on a pure quantum state. In that case the Born Rule appears to indicate an irreducible uncertainty that cannot, even in principle, be relieved by acquiring more information; there are no hidden variables. This insurmountable QM uncertainty, arising from lack of determinism, is referred to as aleatory uncertainty.

However, it is possible for our lack of knowledge also to contribute to the uncertainty in quantum mechanics, in which case these "ordinary" uncertainties must be codified in the formalism as well as the quantum mechanical probabilities associated with indeterminism and the Born Rule. In other words, the more general formulation must allow for both epistemic and aleatory uncertainties. How is this done?

It cannot be done simply by adding two quantum state vectors. This can be demonstrated with a simple example. Consider two spin ½ particles. One is prepared in the "spin up" state wrt the z-axis, and the other in the "spin up" state wrt the x-axis. Let us prepare a great many (N) of these two-particle mixtures, and let us measure the spin of just one of the pair of particles chosen at random from each of the N mixtures. Suppose we choose to measure the spin in the z-direction. It is clear what the distribution of our measurement outcomes will be. About half the particles chosen will be in the z-spin-up state, and these will all give the result +1 (measuring in units of $\hbar/2$). The other half will be in the x-spin-up state. These particles will give a z-spin value of +1 half the time, and -1 the other half (see §2.9). So, overall, we would observe a z-spin of +1 in 75% of occasions and a z-spin of -1 in 25% of cases.

But see what happens if we naïvely assumed that the two particle mixture could be described by adding their wavefunctions. If $|\uparrow\rangle$ and $|\downarrow\rangle$ represent spin up & down wrt the z axis, then the x-spin-up state is $(|\uparrow\rangle + |\downarrow\rangle)/\sqrt{2}$ (see §2.9). As there is one of each in every two-particle mixture, then if adding state vectors is the way to go, they can only be added with the same weighting, giving,

$$|\psi\rangle = A\left[|\uparrow\rangle + \frac{|\uparrow\rangle+|\downarrow\rangle}{\sqrt{2}}\right] \tag{10.1.1}$$

where A is chosen to normalise the resulting two-particle state, and hence takes the value $A = 1/\left(\sqrt{2+\sqrt{2}}\right)$. Hence (10.1.1) becomes,

$$|\psi\rangle = 0.92388|\uparrow\rangle + 0.38268|\downarrow\rangle \tag{10.1.2}$$

This state would produce measurements with z-spin-up 85.36% of the time, and z-spin-down 14.64% of the time. This is the wrong result, and hence the mixture of the two particles cannot be represented by simply adding the Hilbert space states as in (10.1.1). (If the reader is worried that (10.1.2) should represent two particles, and hence should not be normalised to unity, just multiply the whole state by $\sqrt{2}$. It makes no difference to the argument).

A more fundamental physical objection to (10.1.1) is that it represents a pure quantum state. But a pure quantum state cannot be made by throwing two particles together willy-nilly. Equ.(10.1.1) is a coherent quantum superposition, whereas in a classical mixture the two particles are incoherent. This is the crucial distinction.

10.2 The Density Matrix

So how can a mixture be codified correctly in quantum mechanics? The answer is in terms of the density matrix. A mixed state is not represented by a vector in Hilbert space. By definition, a vector in Hilbert space refers to a pure quantum state. A mixed state is represented instead by an operator in Hilbert space. This operator is called the "density matrix". The terminology is slightly unfortunate because, whilst the density matrix can indeed be represented as a matrix, more generally it is simply an operator, i.e., a mapping from the Hilbert space to itself.

For a pure state, $|\psi\rangle$, the density matrix is identified with the projection operator which projects onto that state, i.e.,

$$\hat{\rho} = |\psi\rangle\langle\psi| \tag{10.2.1}$$

Clearly, in the case of a pure state, there is no loss in moving from a description of the state as $|\psi\rangle$ to the density matrix description as $|\psi\rangle\langle\psi|$. Suppose now that we have a mixture in which state $|\psi_1\rangle$ occurs with probability p_1, and state $|\psi_2\rangle$ occurs with probability p_2, etc. The density matrix of the mixture is defined by,

$$\hat{\rho} = \sum_{i=1}^{n} p_i |\psi_i\rangle\langle\psi_i| \tag{10.2.2}$$

where p_i are probabilities, and hence are real numbers in the range [0, 1] such that $\sum_{i=1}^{n} p_i = 1$, where n is the (arbitrary) number of terms in the mixture, (10.2.2). Note that the mixture may be considered as literally a mixture of particles, and hence a multiparticle state, or a single particle whose state is subject to "ordinary" non-quantum, epistemic, uncertainty. For example, in the limit of very large n, the probabilities p_i approximate to a probability density function which represents only our ignorance of the precise microstate of the particle.

Note that the density matrix, (10.2.2), is an operator in Hilbert space (it maps states to states) and hence is a well defined quantity in our formalism. The reason why the density matrix is the correct vehicle for formulating a mixture in quantum mechanics is that it incorporates both types of uncertainty, the epistemic and also the indeterminate, quantum, aleatory uncertainty. The density matrix represents probabilities, as contrasted with the state vector which represents probability amplitudes.

The reformulation of QM in terms of the density matrix requires that the outcome of measurements be re-expressed using the density matrix rather than the state vector. This is equivalent to defining the expectation value of observables in terms of the density matrix. To do this we need to define the "trace" of an operator. With respect to some orthonormal basis, $\{|\varphi_i\rangle\}$, the trace of an operator, $\hat{Q}$, is defined as,

$$Tr(\hat{Q}) = \Sigma_i \langle \varphi_i | \hat{Q} | \varphi_i \rangle \tag{10.2.3}$$

i.e., the trace of an operator is the trace of the matrix of its components. Despite appearances, the trace of an operator depends upon the operator alone; it does not depend upon the basis chosen. Consider bases $\{|\varphi_i\rangle\}$ and $\{|\psi_i\rangle\}$ connected by the unitary transformation: $|\varphi_i\rangle = U|\psi_i\rangle$, for all i. Then,

$$\Sigma_i \langle \varphi_i | \hat{Q} | \varphi_i \rangle = \Sigma_i \langle \psi_i | U^+ \hat{Q} U | \psi_i \rangle$$
$$= \Sigma_{i,k,n} \langle \psi_i | U^+ | \psi_k \rangle \langle \psi_k | \hat{Q} | \psi_n \rangle \langle \psi_n | U | \psi_i \rangle$$
$$= \Sigma_{i,k,n} \langle \psi_n | U | \psi_i \rangle \langle \psi_i | U^+ | \psi_k \rangle \langle \psi_k | \hat{Q} | \psi_n \rangle$$
$$= \Sigma_{k,n} \langle \psi_n | U U^+ | \psi_k \rangle \langle \psi_k | \hat{Q} | \psi_n \rangle$$
$$= \Sigma_{k,n} \langle \psi_n || \psi_k \rangle \langle \psi_k | \hat{Q} | \psi_n \rangle$$
$$= \Sigma_{k,n} \delta_{nk} \langle \psi_k | \hat{Q} | \psi_n \rangle = \Sigma_k \langle \psi_k | \hat{Q} | \psi_k \rangle \tag{10.2.4}$$

thus establishing that the trace is the same in all orthonormal bases.

The trace can be used to express the expectation value of an arbitrary observable in terms of the density matrix. In the case that the state is pure, if we choose to represent the state of the system, not by the vector $|\psi\rangle$ directly, but by the corresponding projection operator, $\hat{\rho} = |\psi\rangle\langle\psi|$, the expectation value of observable Q is given by,

$$Tr(\hat{\rho}\hat{Q}) = \Sigma_i\langle\varphi_i||\psi\rangle\langle\psi|\hat{Q}|\varphi_i\rangle = \Sigma_i\langle\psi|\hat{Q}|\varphi_i\rangle\langle\varphi_i||\psi\rangle = \langle\psi|\hat{Q}|\psi\rangle$$

$$(10.2.5)$$

where we have used $\Sigma_i|\varphi_i\rangle\langle\varphi_i| = 1$ (see §2.3.4). Hence, (10.2.5) shows that the expectation value of an observable is given by $Tr(\hat{\rho}\hat{Q})$ in the case of a pure state. But the utility of the density matrix formulation is that this works also for mixed states. Hence, in full generality we have,

$$\langle\hat{Q}\rangle = Tr(\hat{\rho}\hat{Q})$$

$$(10.2.6)$$

To see this, substitute (10.2.2) into (10.2.6)

$$Tr(\hat{\rho}\hat{Q}) = \Sigma_{i,k}\langle\varphi_k|p_i|\psi_i\rangle\langle\psi_i|\hat{Q}|\varphi_k\rangle$$

$$(10.2.7)$$

$$= \Sigma_{i,k}\,p_i\langle\psi_i|\hat{Q}|\varphi_k\rangle\langle\varphi_k||\psi_i\rangle = \Sigma_i\,p_i\langle\psi_i|\hat{Q}|\psi_i\rangle$$

We see that (10.2.7) involves an "ordinary" averaging $\Sigma_i\,p_iQ_i$ over the mixture whose component states occur in the relative frequencies p_i together with each state contributing its quantum mechanical expectation value to this averaging process, $Q_i = \langle\psi_i|\hat{Q}|\psi_i\rangle$.

Note that if we expand the states $|\psi_i\rangle$ in terms of the basis formed by the eigenvectors of the measured observable, $\hat{Q}$, i.e., $|\psi_i\rangle = \Sigma_k\,a_{ik}|q_k\rangle$, then the probability that the component $|\psi_i\rangle$ in the mixed state leads, after measurement-state-vector-reduction, to the state being $|q_k\rangle$ is $|a_{ik}|^2$, as per the usual Born Rule. Hence, for the mixed state as a whole, the probability that measurement leaves it in state $|q_k\rangle$ is $\Sigma_i\,p_i\,|a_{ik}|^2$.

It is important to note that there is no requirement for the states $\{|\psi_i\rangle\}$ in (10.2.2) or (10.2.7) to be orthogonal. In general they will not be.

10.3 Example: Resolution of the Mixture Problem

The example mixture introduced above consisted of one particle in a z-spin-up state and a second particle in an x-spin-up state. Note that these are not orthogonal states, i.e., $(|\uparrow\rangle + |\downarrow\rangle)/\sqrt{2}$ is not orthogonal to $|\uparrow\rangle$. The density matrix for this example is, using (10.2.2),

$$\hat{\rho} = 0.5|\uparrow\rangle\langle\uparrow| + 0.5 \times \frac{1}{\sqrt{2}}[|\uparrow\rangle + |\downarrow\rangle]\frac{1}{\sqrt{2}}[\langle\uparrow| + \langle\downarrow|]$$

$$= 0.75|\uparrow\rangle\langle\uparrow| + 0.25|\downarrow\rangle\langle\downarrow| + 0.25|\uparrow\rangle\langle\downarrow| + 0.25|\downarrow\rangle\langle\uparrow|$$

$$= (|\uparrow\rangle \quad |\downarrow\rangle)\begin{pmatrix}0.75 & 0.25\\0.25 & 0.25\end{pmatrix}\begin{pmatrix}\langle\uparrow|\\\langle\downarrow|\end{pmatrix}$$

$$(10.3.1)$$

where the z-basis has been used throughout. Where this basis is understood, the density matrix can literally be represented simply as the matrix in the last line of (10.3.1). When the components of the density matrix are expressed wrt an orthonormal basis, its diagonal elements are the probabilities of the corresponding measurement outcomes. So the 75%/25% up/down z-spin result is reproduced correctly in this formalism.

Since a z-spin of +1 (in units of $\hbar/2$) occurs 75% of the time, and a z-spin of -1 occurs 25% of the time, the average (or expectation value) of the z-spin is 0.5 (or $\hbar/4$ if you prefer). To see how the general formula (10.2.6) for the expectation value works in this case, note that the operator for the z-spin may be represented in the z-basis as the corresponding Pauli matrix (see §2.9), i.e.,

$$\hat{\sigma}_z = \begin{pmatrix} 1 & 0 \\ 0 & -1 \end{pmatrix} \tag{10.3.2}$$

Thus the expectation value of the z-spin according to (10.2.6) is,

$$\langle \sigma_z \rangle = Tr \left[\begin{pmatrix} 0.75 & 0.25 \\ 0.25 & 0.25 \end{pmatrix} \begin{pmatrix} 1 & 0 \\ 0 & -1 \end{pmatrix} \right] = Tr \begin{pmatrix} 0.75 & -.25 \\ 0.25 & -.25 \end{pmatrix} = 0.5$$

So the correct result is reproduced.

10.4 Diagonalising the Density Matrix

It follows from (10.2.2) that the density matrix is Hermitian, $\hat{\rho}^+ = \hat{\rho}$, and hence that it has real eigenvalues and a complete set of orthogonal eigenvectors. The latter provide a unitary transformation which diagonalises the density matrix. That is, if we change to the basis $\{|\varphi_j\rangle\}$ which consists of the eigenvectors of $\hat{\rho}$, then it becomes,

$$\hat{\rho} = \sum_{i=1}^{N_s} \lambda_i |\varphi_i\rangle\langle\varphi_i| \tag{10.4.1}$$

where N_s is the dimension of the Hilbert space. (This is the spectral representation of the density matrix, §2.5.4). A key distinction between (10.2.2) and (10.4.1) is that the latter involves a sum over the number of basis states, N_s, i.e. the dimension of the Hilbert space, as opposed to (10.2.2) which is a sum over an arbitrary number of terms in the mixture. Moreover, the basis states occurring in (10.4.1) are, of course, orthogonal, whereas the states occurring in (10.2.2) need not be. The diagonal components, λ_i, of the diagonalized density matrix are its eigenvalues. We shall see below that $Tr(\hat{\rho}) = 1$, from which it follows that the eigenvalues of a density matrix always sum to unity.

The density matrix is only diagonal in this specific representation. Transforming to an alternative orthonormal basis it will become a full matrix, i.e., putting $|\varphi_i\rangle = U_{ij}|\varphi_j'\rangle$ gives,

$$\hat{\rho} = \sum_{i,j,k=1}^{N_s} \lambda_i U_{ij} U_{ik}^* |\varphi_j'\rangle\langle\varphi_k'| = \sum_{j,k=1}^{N_s} C_{jk} |\varphi_j'\rangle\langle\varphi_k'| \qquad (10.4.2)$$

Hence, the matrix $C_{jk} = \sum_{i=1}^{N_s} \lambda_i U_{ij} U_{ik}^*$ is the density matrix wrt the transformed basis. For example, even if we were dealing with a pure state, the density matrix would appear to be a full matrix when expressed in a general basis.

How, then, does one tell whether a given density matrix actually represents a pure state? One way is to find its eigenvalues. If a density matrix actually represents a pure state, on diagonalization only one term (one eigenvalue) will be non-zero (and hence necessarily unity). However, there is a computationally simpler way which avoids having to solve the eigenvalue problem.

10.5 Testing for a Pure State

An important property of the density matrix is that $Tr(\hat{\rho}) = \sum_i p_i = 1$. This follows directly from (10.2.2). Consider an orthonormal basis $\{|\varphi_j\rangle\}$. Then,

$$Tr(\hat{\rho}) = \sum_{j=1}^{N_s} \sum_{i=1}^{n} p_i \langle\varphi_j|\psi_i\rangle\langle\psi_i|\varphi_j\rangle = \sum_{j=1}^{N_s} \sum_{i=1}^{n} p_i \langle\psi_i|\varphi_j\rangle\langle\varphi_j|\psi_i\rangle$$
$$= \sum_{i=1}^{n} p_i \langle\psi_i|\psi_i\rangle = \sum_i p_i = 1 \qquad (10.5.1)$$

where N_s is the dimension of the Hilbert space and n is the (arbitrary) number of terms in the sum defining the density matrix in (10.2.2). We have used the fact that $\sum_{j=1}^{N_s}|\varphi_j\rangle\langle\varphi_j| \equiv 1$, and the states $|\psi_i\rangle$ in (10.2.2) must be normalised, though not necessarily orthogonal.

If $\hat{\rho}$ is simply a pure state, $|\psi\rangle\langle\psi|$, then it is clear that $\hat{\rho}^2 = \hat{\rho}$ and hence $Tr(\hat{\rho}^2) = 1$. In fact this is *only* true if $\hat{\rho}$ is a pure state. Otherwise we have $Tr(\hat{\rho}^2) < 1$. This follows in a similar manner to (10.2.2), i.e.,

$$Tr(\hat{\rho}^2) = \sum_{j=1}^{N_s} \sum_{i,k=1}^{n} p_i p_k \langle\varphi_j|\psi_i\rangle\langle\psi_i|\psi_k\rangle\langle\psi_k|\varphi_j\rangle$$

$$= \sum_{j=1}^{N_s} \sum_{i,k=1}^{n} p_i p_k \langle\psi_k|\varphi_j\rangle\langle\varphi_j|\psi_i\rangle\langle\psi_i|\psi_k\rangle$$
$$= \sum_{i,k=1}^{n} p_i p_k \langle\psi_k|\psi_i\rangle\langle\psi_i|\psi_k\rangle = \sum_{i,k=1}^{n} p_i p_k |\langle\psi_i|\psi_k\rangle|^2 < 1 \qquad (10.5.2)$$

where the strict inequality holds provided that $\hat{\rho}$ is not a pure state. The last step follows because each of the $|\langle\psi_i|\psi_k\rangle|^2$ is in the range $[0, 1]$, and strictly less than 1 when $i \neq k$, and hence $\sum_{k=1}^{n} p_k |\langle\psi_i|\psi_k\rangle|^2 < 1$ and therefore $\sum_{i,k=1}^{n} p_i p_k |\langle\psi_i|\psi_k\rangle|^2 < \sum_{i=1}^{n} p_i = 1$. Note again that this holds even though the states $|\psi_i\rangle$ may not be orthogonal (though they must be normalised) and there may be any arbitrary number of them, possibly exceeding the dimensionality of the Hilbert space, i.e., $n > N_s$.

In the particular case that the $|\psi_i\rangle$ *are* orthogonal, then (10.5.2) gives $Tr(\hat{\rho}^2) = \sum_{i=1}^{n} p_i^2$, but this is not true in general.

The fact that $\hat{\rho}^2 = \hat{\rho}$ and $Tr(\hat{\rho}^2) = 1$ hold only for a pure state provides a convenient test of a pure state. It will generally be easier to check if $\hat{\rho}^2 = \hat{\rho}$ than to solve for the eigenvalues. As an illustration, consider the density matrix from our example above, (10.3.1). Squaring the matrix readily shows that $\hat{\rho}^2 \neq \hat{\rho}$, rather,

$$\hat{\rho}^2 = \frac{1}{16}\begin{pmatrix} 10 & 4 \\ 4 & 1 \end{pmatrix}$$

consistent with this being a mixed state. On the other hand if we consider the density matrix,

$$(\hat{\rho}) = \begin{pmatrix} 0.75 & 0.433 \\ 0.433 & 0.25 \end{pmatrix}$$

whilst not immediately obvious that it represents a pure state, this is found to be the case by showing that $\hat{\rho}^2 = \hat{\rho}$ and one can check that its eigenvalues are indeed 0 and 1.

10.5 Ontology

At one level the shift from the description of states as Hilbert space vectors to the description of states as Hilbert space operators is merely a bit of formalism. But there is a significant ontological change involved. A Hilbert space vector, however abstract, is an "object", a "thing", a noun. Thus, a physical "thing", a physical system, is represented by another "thing", albeit an abstract one, when states are defined as vectors. But a Hilbert space operator is a mapping; it is an instruction to replace one vector with another. Yes, yes, I know one can also regard mappings as "things". But this is to deny the significant shift in the nature of what is being used to describe a physical state. A mapping is naturally perceived intuitively as a verb rather than a noun. It implies a change induced in the object on which it acts. And, yes, I know that vectors can be interpreted as actions also, such as a displacement vector being an instruction to move a certain distance in a certain direction. Nevertheless, there is food for thought here, though such philosophical musings lie beyond the scope of this book. It is worth pointing out, though, that quantum fields are also operator-valued. They are not descriptions of the state of a system but instructions to create or destroy particles, thus changing the physical state.

10.6 Exercises for The Reader

Are the following density matrices pure or mixed states? If pure, what is the state, expressed as a column vector in the same basis? If mixed, what are the two mixed states, and with what probabilities?

(a) $\dfrac{1}{25}\begin{pmatrix} 9 & 12 \\ 12 & 16 \end{pmatrix}$ (b) $\dfrac{1}{25}\begin{pmatrix} 9 & 10 \\ 10 & 16 \end{pmatrix}$

Answers

(a) Pure state $\dfrac{1}{5}\begin{pmatrix} 3 \\ 4 \end{pmatrix}$

(a) Mixed state: 7.62% of $\begin{pmatrix} 0.8156 \\ -0.5787 \end{pmatrix}$ and 92.38% of $\begin{pmatrix} 0.5787 \\ 0.8156 \end{pmatrix}$

"Actually I started out in quantum mechanics, but somewhere along the way I took a wrong turn."

(this one is personal)

11

Schrodinger's Cat

'When I hear of Schrodinger's cat, I reach for my gun' (attributed to Steven Hawking, Ref.[11.1])

I concur with Hawking's sentiment. Whole books have been written on Schrodinger's cat, and innumerable articles. The apparent paradox arises merely from a misunderstanding of quantum mechanics. Specifically, the confusion arises from the misunderstanding I have alluded to previously: the mistaken belief that if something is quantum mechanical (and everything is) then it can be represented by a ray in Hilbert space. But this is to assume that everything is in a pure quantum state. But chapter 10 has disabused us of that naïve notion. At best – if QM is right – a general physical system can be described by a density matrix. The general physical system is in a mixed state. This is not a concept you can do without.

No blame attaches to Schrodinger or his contemporaries. They were groping towards an understanding of quantum mechanics with the help of few signposts. What that generation of physicists did in the first three decades of the twentieth century was amazingly impressive. In fact, the resolution of the paradox had been implicit within standard quantum mechanics almost since the problem was first posed. In this chapter I use Schrodinger's Cat to illustrate the crucial importance of appreciating that pure quantum states are special and not to be expected in things of macroscopic size or things at non-zero temperature. It also provides a vehicle for delivering a message about the nature of measurement.

11.1 The Precarious Cat's False Paradox

We are to imagine placing a poor cat in a box together with a contrivance which activates a poison gas canister if a single radioactive nucleus decays. Here is Schrodinger's difficulty. Quantum mechanics tells us that the state of the radioactive nucleus is described by a vector (strictly, a ray) in Hilbert space. We will have no quibble with this: it is reasonable to assume a single isolated nucleus to be in a pure quantum state. Simplifying somewhat, we can regard this nucleus as having two states available to it, undecayed and decayed, which we write as $|u\rangle$ and $|d\rangle$. If we have not interfered with the nucleus, and given that the nucleus is unstable, it will be in some superposition of these states, $\alpha|u\rangle + \beta|d\rangle$.

Now here is the essence of the supposed paradox. Quantum mechanics holds that the nucleus is neither decayed nor undecayed until its state has been measured. In common presentations of the problem, "measurement" is taken to be accomplished by a human observing whether the decay has taken place. If it has, the cat will also be found dead. If not, the cat will be alive.

The puzzle presented by this setup is that the cat apparently shares the indeterminacy of the nucleus. The cat will be dead if and only if the nucleus has decayed. Since the state of the nucleus is indeterminate, prior to observation, it is also indeterminate (we are told) whether the cat is alive or dead. The full force of this conundrum is felt only if one has already absorbed the lesson that quantum indeterminacy is absolute, that there are no hidden variables. So we seem to be forced to believe that the cat is neither alive nor dead. It is not merely that we don't know if it is alive or dead though it really is one or the other – no, its state is strictly indeterminate: the cat is neither dead nor alive but in a superposition of both states – so we are told.

Note that the proposed experimental arrangement is perfectly feasible (though unlikely to pass the ethics board).

To drive home the nature of the concern, suppose we represent the states of the cat available after we observe it as either $|alive\rangle$ or $|dead\rangle$. If we peeked inside the box and discovered the nucleus to still be intact and the cat alive and well, then the state of the nucleus-plus-cat bipartite system would appear to be $|u\rangle|alive\rangle$. Conversely, if we peeked inside the box and discovered the nucleus to have decayed and the cat to be dead, then the state of the nucleus-plus-cat system would appear to be $|d\rangle|dead\rangle$. It would seem, therefore, that prior to our observation the nucleus-plus-cat system must have been in a superposed quantum state of the form $\alpha|u\rangle|alive\rangle + \beta|d\rangle|dead\rangle$. Hence, the cat was neither alive nor dead, but in a quantum superposition of aliveness and deadness, a situation which runs counter to our intuition.

11.2 Absolute Smallness and Pure Quantum States

There are two key observations, either sufficient to resolve the paradox. The first hinges on the fact that a cat is very different from an atom: a cat is a great deal bigger. The resolution lies in the absolute smallness of things like atoms and nuclei, and the absolute largeness of cats. As we will see in chapters 21-24, the issue here is not, strictly speaking, size as measured in metres (or Angstroms or fermis) although that is most often a pretty good surrogate for a more accurate arbiter of quantum pureness.

The lesson of quantum mechanics is that the states of small quantum systems like atoms can be completely and exactly defined by a small number of integers. Ignoring excited states of the nucleus, the quantum states of an atom can be defined by the quantum numbers n, j, m, s_z for each of its electrons. In principle, though it is less well understood, the state of our unstable nucleus can also be expressed in terms of quantum numbers analogous to n, j, m, s_z for each of its component protons and neutrons. Thus, improving somewhat on our previous description, the state of our radioactive nucleus can be written $|\{n_i, j_i, m_i, s_{zi}\}\rangle$. In practice our previous description may be quite adequate since there will be some particular state, $|\{n_i, j_i, m_i, s_{zi}\}_u\rangle$, which defines the radionuclide in question, and perhaps just one final state (assuming there is only a single decay mode), $|\{n_i, j_i, m_i, s_{zi}\}_d\rangle$. So we might as well write these states using the shorthand $|u\rangle$ and $|d\rangle$. But it is important that the correct, completely specified, quantum states $|\{n_i, j_i, m_i, s_{zi}\}_u\rangle$ and $|\{n_i, j_i, m_i, s_{zi}\}_d\rangle$ lie behind this shorthand. This is because these really are genuine, pure quantum states.

Now tell me, what is the quantum state of the cat? It is entirely fallacious to suppose that a pure quantum state has been defined simply by enclosing the word "alive" inside a ket, thus: "$|alive\rangle$". This is **not** a pure quantum state. It has just been written to look like one textually. But actually a living cat cannot be in a pure quantum state.

It is worth pausing awhile to consider how many integers would be required to specify a pure quantum state of a cat (ignoring for the moment that the endeavour involves a logical contradiction). A lower bound will suffice. A cat contains of the order of 10^{26} atoms. Let us assume that all these atoms' nuclei are in their ground state, and not liable to be excited. Let us further assume that all electrons in the atoms will also remain unexcited, except for a small number of the outer valence electrons. These will be involved (or not) in forming chemical bonds with other atoms. How many different states might these valence electrons be in? I do not know – but it suffices to assume just two states are available. You may imagine this to be choice between being bonded to another atom or not, or between being in an excited state or not, as you wish. In case you are of the view that giving every atom two available states is going too far, let's say only one in a million atoms can be in either of two states, and all the rest are in fixed states. That is still 10^{20} atoms with two available states. All this is a huge simplification and will certainly cause our estimate of the number of possible quantum states of the cat to be a gross underestimate. However, this convincingly demonstrates that a lower bound to the number of electron states in the cat must be at least 2 raised to the power 10^{20}. In other words, at least 10^{20} binary digits are required to specify the electrons' state – and really far more than this. And I have not even counted the number of possible states of the atoms considered as a whole, such as vibrational modes.

Now 2 raised to the power 10^{20} is a colossal number. It makes the number of protons in the observable universe ($\sim 10^{80}$) look trivially small, as it does the number of photons ($\sim 10^{89}$). The maximum number of elementary binary computations that could have been carried out using all the resources of the entire observable universe in its whole 13.8 Byr history is only $\sim 10^{123}$ and so is also trivial in comparison.

And yet a pure quantum state of a cat must specify which of these 2-to-the-power-10^{20} states prevails, or specify a coherent superposition of them, i.e., 2-to-the-power-10^{20} coefficients must be defined specifying the particular pure superposed state.

11.3 Finite Temperatures Are Mixed States

Hopefully we are now agreed that specifying a particular pure quantum state of a cat is not a practical possibility. But let us suppose that by some stupendous miracle of experimental technique we were able to place a cat into a pure quantum state, whether we knew what that state was or not. I have bad news for you. Your cat would already be dead.

Why? Well, what constitutes being alive? Here there is a slight problem: there is no universal agreement on what constitutes being alive. You can reasonably argue either way for a virus. However we sidestep this problem for now by confining attention to warm blooded animals about which there is no dispute. Amongst other things, a living organism is a seething mass of complex, ceaseless, physical and chemical reactions. These continual interactions and reactions are essential to life. They **are** life. This means that the electron states of your cat must be in continual flux if it is to have any chance of being alive – in other words metabolism, plus a myriad of vital functions. The cat's micro-state must be constantly changing via a mess of incoherent (thermally driven) reactions. So I am afraid that your miraculous experimental technique was merely a sophisticated way of killing your cat.

Actually there is an even simpler way of seeing that your pure-quantum-state-cat is a dead cat. If the cat is in a pure quantum state then it is not in a thermal state – because a thermal state is an incoherent mixture of energy states. But living organisms survive only within quite a restricted range of temperature. They rather depend upon being in a thermal state. By 'freezing' your cat into a pure quantum state you bring about its death.

Finally, here is yet another way of understanding the pure-quantum-state-cat's demise. Suppose this pure quantum state is an energy eigenstate. These are stationary states. Their only time dependence is through a phase factor, which makes no physical change to the cat. Such a cat is frozen in time. Nothing about it changes. Since life is a dynamic condition, this is synonymous with death.

11.4 The Cat's Actual QM State

I trust I have hammered home the message: a cat cannot be in a pure quantum state, and pretending that it is by writing something like $|alive\rangle$ or $|dead\rangle$ is a contradiction, a subterfuge, a form of misdirection. The notation implies a pure quantum state for a "system", the cat, which cannot be in one. And since the cat cannot be described by a pure quantum state, a superposed state like $\alpha|u\rangle|alive\rangle + \beta|d\rangle|dead\rangle$ does not arise. Hence there is no reason to attribute weird quantum indeterminacy to the cat. The cat is not simultaneously both alive and dead. It is not in a quantum superposition of states of aliveness and deadness. There is no such thing as a quantum superposition of states of aliveness and deadness, because the state of being alive is not a pure quantum state.

Yes, but…is the cat alive or dead? Of course we do not know unless we look inside the box. But our lack of knowledge on this subject is just the usual, classical, deterministic, every-day lack of knowledge that arises when anything happens beyond the reach of our senses or instruments. There is nothing quantum mechanical or mysterious about this unknown status of the cat. At any time the cat is in a definite state, either alive or dead. We just do not know which. Our uncertainty over the matter is entirely epistemic. The radioactive nucleus, on the other hand, does indeed start off in a quantum superposition. However, whilst the nucleus has the power to deal death on the cat, it does not have the power to contaminate the cat with its quantum indeterminacy.

The truth is quite the reverse. The cat contaminates the nucleus with its classical determinacy. What is going on here is that the cat is a measuring device, measuring the state of the nucleus. Like all measuring devices it projects out the corresponding eigenstate of the thing measured, i.e., the cat causes reduction of the indeterminate quantum state of the nucleus and renders it either decayed, or not, definitely.

The measurement is not made when you happen to look in the box, which would be horribly subjective. A measurement is a physical interaction between the system being measured and the measuring apparatus – in this case a cat. The measurement is made when the interaction takes place. It is a physical event, not some mystical process. Let us hear no talk of your consciousness being responsible for the collapse of the wavefunction, or other similar nonsense.

You may, however, still be feeling dissatisfied. There is something too glib about this explanation, perhaps? This feeling of unease probably springs from the fact that we have not as yet provided the cat with a description in quantum mechanical terms. The essence of our explanation is to veto the possibility of a pure-quantum-state-cat. But if quantum mechanics is the correct description of the world, everything – cats included – must be expressible within its lexicon. Yes indeed. And it is. And we know what that is: the density matrix. The cat is in a mixed state; an extremely mixed state, in fact. Hence, the description of the combined cat-plus-nucleus state must be of the form $\hat{\rho} = \sum_j p_j |\psi_j\rangle\langle\psi_j|$.

A brief aside on the nature of states at finite temperature. A thermal state is described by a density matrix in which the probabilities p_i take values characteristic of a system in thermal equilibrium, as controlled by the familiar Boltzmann factor, $p_i \propto exp\{-E_i/kT\}$. A thermal state is an ordinary (deterministic) mixture of pure quantum states. This is another way of saying that the various pure quantum states, of various energies, combine randomly, i.e., incoherently. This incoherent mixing of a large number of pure states is the essence of the thermal state.

Suppose we partition the density matrix for the cat into parts which represent a living cat and the remainder which represent a dead cat. The density matrix for the cat can thus be written,

$$\hat{\rho} = \sum_j p_j^{alive} |\psi_j\rangle\langle\psi_j| + \sum_k p_k^{dead} |\psi_k\rangle\langle\psi_k| \qquad (11.4.1)$$

The undecayed state of the nucleus can be represented just as well by the density matrix $|u\rangle\langle u|$ as by the Hilbert space vector $|u\rangle$ itself, and similarly $|d\rangle\langle d|$ is a complete representation of the decayed nuclear state. The first of these is associated with the first term on the RHS of (11.4.1), i.e. with a living cat, whereas $|d\rangle\langle d|$ is associated with the second term in (11.4.1), i.e. the dead cat. If the nucleus had been in a state $\alpha|u\rangle + \beta|d\rangle$ prior to the introduction of the cat into the box, then the state $|u\rangle\langle u|$ is associated with a probability of $|\alpha|^2$, and $|d\rangle\langle d|$ is associated with a probability of $|\beta|^2$ by the usual Born rule. Thus, the combined cat-plus-nucleus system is represented by the sum of the corresponding product states,

$$\hat{\rho} = |\alpha|^2 \left[\sum_j p_j^{alive} \left(|\psi_j\rangle\langle\psi_j|\right)\right] \otimes (|u\rangle\langle u|) + $$
$$|\beta|^2 \left[\sum_k p_k^{dead} \left(|\psi_k\rangle\langle\psi_k|\right)\right] \otimes (|d\rangle\langle d|) \qquad (11.4.2)$$

What this reveals is that, not only has the cat no weird quantum superposition properties, but the nucleus is no longer in a quantum superposition either. The state vector of the nucleus has reduced, as it does in any measurement, leaving it in a definite state. We do not know which state, of course , until we look inside the box. This purely deterministic (epistemic) uncertainty is represented in (11.4.2) by the sum of the two terms. The probability of the cat being alive is $|\alpha|^2$ and the probability of it being dead is $|\beta|^2$. These are, of course, exactly the probabilities that would be expected from the original quantum state of the nucleus, $\alpha|u\rangle + \beta|d\rangle$.

The crucial difference from the "paradoxical" interpretations is that, having interacted with the cat, the measurement of the nucleus's state has already been made. The wavefunction has already collapsed before you look in the box. Your consciousness is not responsible for the state vector reduction: the cat is, because the cat is a measurement device.

You may demur on the grounds that this interpretational position should not merely be asserted but demonstrated experimentally. And you could be forgiven for thinking that such an experimental demonstration is impossible, since it would seem to require the observer to know the state of the cat before being conscious of it! Be confounded, then, to be told that such experiments have been performed and can be very simple (though the literal use of a cat is frowned upon), see for example Carpenter and Anderson, Ref.[11.2]. The trick is to involve two observers each of whom acquires a piece of classical information. Neither piece of information is sufficient to conclude whether the 'cat' remains alive. Some time later the two pieces of information are brought together, i.e., some single observer acquires both pieces of data, and this proves to be a perfectly reliable indicator of the state of the 'cat'. Hence all the information required to reliably and deterministically deduce the state of the 'cat' existed before any human was aware of it consciously.

In summary, rather than the nucleus contaminating the cat with its quantum indeterminacy, the opposite happens. The cat contaminates the nucleus with its classical determinacy, thus reducing the state vector of the nucleus as usually happens in a measurement. This measurement of the state of the nucleus is performed by an instrument known as a cat. The outcome of the measurement is recorded in terms of the cat's life, which is a classical state.

11.5 References

[11.1] Barrow, J.D., and Tipler, F.J., (1986) "The Anthropic Cosmological Principle", Oxford University Press.

[11.2] R.H.S.Carpenter and A.J.Anderson (2006), "The death of Schrödinger's cat and of consciousness based quantum wave-function collapse", Annales de la Fondation Louis de Broglie, Volume 31, no 1, 2006, 45. http://aflb.ensmp.fr/AFLB-311/aflb311m387.pdf

DR YUKALOT PROVES THAT CATS DON'T HAVE WAVE PROPERTIES, THEREBY LAYING TO REST, ONCE AND FOR ALL, THE PROBLEM OF SCHRÖDINGER'S CAT.

Please don't try this at home

12

Quantum Entropy

The entropy of a quantum state, the von Neumann entropy, is defined and contrasted with the classical Boltzmann entropy, or Shannon information. The increase in a system's entropy upon measurement is demonstrated.

12.1 Von Neumann Entropy Versus Classical Entropy

In classical statistical thermodynamics the entropy is defined as,

$$S_{Bolzmann} = k_B \, log \, W \qquad (12.1.1)$$

where k_B is Boltzmann's constant and W is the number of accessible microstates consistent with the known macrostate of the system. It applies if all microstates are equally probable. If so then each has probability $p = 1/W$. The Boltzmann entropy can then be written $S = -k_B \, log \, p$. Alternatively, if the different microstates have differing probabilities, say p_i for the ith state, then the contribution to the entropy of this state would be $-k_B \, log \, p_i$ and this occurs with a relative frequency of p_i. So the ensemble average entropy would be,

$$S_{Boltzmann} = -k_B \sum_i p_i \, log \, p_i \qquad (12.1.2)$$

This is the more general expression for classical entropy. Dimensionless entropy is defined by dropping the Boltzmann constant factor. We shall assume dimensionless entropy from here on.

In information theory, the Shannon 'entropy' (or information) is defined in the same way. Let's say a message is transmitted using a set of distinguishable symbols x_i and that each of these are known to occur with some probability p_i. This may be known, for example, because past messages have shown that x_1 occurs with a relative frequency p_1, etc. How much information is there in a message N symbols long? Well, it is N times the average information per symbol transmitted, and the latter is defined as,

$$S_{Shannon} = -\sum_i p_i \, log_2 \, p_i \qquad (12.1.3)$$

Note that whereas entropy in classical physics, i.e., (12.1.2) above, is defined using the natural logarithm, in information theory log_2 is used. This is natural because it means that one 50/50 binary choice corresponds to one unit of information (one bit). Some authors employ entropy defined using logarithms to some other integer base, for example the dimension of the relevant Hilbert space.

What is the entropy of a system in a pure quantum state? The answer, of course, is zero. A classical state of many particles, or a mixed quantum state, could be in any one of many microstates. But a pure quantum state is the definition of what is meant by "a microstate". So the number of microstates which a system in a pure quantum state could be in is, trivially, $W = 1$ and hence its entropy is zero (because $log(1) = 0$).

Consequently, if we want to extend the concept of entropy to quantum states, it is mixed quantum states that are of interest. Mixed quantum states are specified by a density matrix (chapter 10) which, being Hermitian, can always be put in diagonal form with respect to some orthonormal basis, $\{|\varphi_i\rangle\}$, i.e., written as $\hat{\rho} = \sum_i p_i |\varphi_i\rangle\langle\varphi_i|$ where the sum is over the dimension of the Hilbert space, though some of the p_i terms may be zero. The particular property of density matrices is that their eigenvalues, i.e., p_i, lie in the range $[0,1]$ and sum to unity, $\sum_i p_i = 1$.

Since each of the contributing $|\varphi_i\rangle\langle\varphi_i|$ is a pure state, it is reasonable to expect that (12.1.3) will again be the appropriate entropy. However, it is desirable to have an expression for the entropy of a mixed state when the density matrix is expressed wrt an arbitrary basis and hence may not be in diagonal form. This general expression defines the von Neumann entropy,

$$S_{vN} = -Tr(\hat{\rho}\ log_2(\hat{\rho})) \tag{12.1.4}$$

where Tr denotes the trace (see §10.2). Recall that a function of an operator is defined via the corresponding power series. This means that if an operator is represented by a diagonal matrix, with diagonal elements p_i, a function f of the operator is also diagonal with elements $f(p_i)$. Firstly we show that (12.1.4) gives the correct result when the density matrix is in diagonal form,

$$S_{vN} = -Tr(\hat{\rho}\ log_2(\hat{\rho}))$$
$$= -Tr(\sum_j p_j |\varphi_j\rangle\langle\varphi_j| \sum_i log_2[p_i|\varphi_i\rangle\langle\varphi_i|])$$
$$= -\sum_k\langle\varphi_k| \sum_j p_j |\varphi_j\rangle\langle\varphi_j| \sum_i log_2[p_i|\varphi_i\rangle\langle\varphi_i|] |\varphi_k\rangle$$
$$= -\sum_{k,j,i} \delta_{kj}\ p_j\delta_{ji}\ log_2 p_i\ \delta_{ik}$$
$$= -\sum_i p_i\ log_2 p_i \tag{12.1.5}$$

It follows immediately that (12.1.4) produces the same result in whatever basis the density matrix is expressed. This is because the density matrix in any basis may be written $\hat{\rho} = \hat{U}\hat{\rho}_{diag}\hat{U}^+$ for some unitary transformation, $\hat{U}$. But we proved in §10.2 that the trace of an operator is invariant under a unitary transformation (i.e., a change of basis), hence the general expression (12.1.4) is valid in any basis. [Note that we have appealed here to the fact that, for a function understood to be defined by a power series, we have $f(\widehat{U}\hat{\rho}_{diag}\widehat{U}^+) = \widehat{U}f(\hat{\rho}_{diag})\widehat{U}^+$. The reader should convince himself that this is correct.]

Note that not only is the entropy zero for a pure state, but it is only zero for a pure state. Consequently, a mixed quantum state may be recognised as such by $S_{vN} \neq 0$.

12.2 Coherent (Unitary) Evolution Preserves Entropy

Provided we do not contaminate a quantum system with our classical impurity, for example by performing a measurement or by interaction with the environment, it will evolve unitarily, or coherently, according to the Schrodinger equation. The Schrodinger equation (see §2.4) shows that a pure quantum state evolves in time via a unitary transformation, $|\psi_t\rangle = \widehat{U}|\psi_0\rangle$, where $\widehat{U}$ is the unitary operator $exp\{-i\widehat{H}t/\hbar\}$. This applies to all the states which contribute to the density matrix and so the density matrix evolves to $\hat{\rho}_t = \widehat{U}\hat{\rho}\widehat{U}^+$. Hence, because the trace is invariant under such a unitary transformation, the entropy is constant over time so long as unitary evolution applies.

This may seem inconsistent with the second law of thermodynamics, but the situation is no different from classical dynamics in this respect. Even in classical physics, if evolution is reversible, then entropy is constant. Unitary evolution in QM is reversible (as implemented by the inverse operator $\widehat{U}^{-1}$) and so entropy is invariant for the same reason.

12.3 Perfect Distinguishability and von Neumann Entropy

Care is required when $\hat{\rho} = \sum_i p_i |\tilde{\varphi}_i\rangle\langle\tilde{\varphi}_i|$ but the states $\{|\tilde{\varphi}_i\rangle\}$ are not orthogonal. These states are therefore not distinguishable with certainty. If we ignore this fact we would be tempted to assume the entropy is given by $-\sum_i p_i \log_2 p_i$. But in truth the amount of information conveyed by a sequence of $\{|\tilde{\varphi}_i\rangle\}$ states must be rather less than this, because it is possible to confuse, say, $|\tilde{\varphi}_1\rangle$ with $|\tilde{\varphi}_2\rangle$ if $\langle\tilde{\varphi}_1|\tilde{\varphi}_2\rangle \neq 0$. The von Neumann entropy, (12.1.4), properly accounts for this. An example makes this clear.

Consider a mixture of two spin ½ particles, one particle in the z-up spin state, and the other in the x-up state. Now these states are not orthogonal. (Do not confuse the orthogonality of the spins in physical space with the (lack of) orthogonality of the state vectors in Hilbert space). From the classical point of view, if we falsely regarded the two electrons as perfectly distinguishable, we would have a mixture with $p_1 = p_2 = 0.5$, giving an entropy of $-(0.5\,log_2\,0.5) \times 2 = 1$. But now let us find the correct (von Neumann) entropy. Using (2.9.8a), the density matrix in the z-basis is,

$$\hat{\rho} = 0.5|\uparrow\rangle\langle\uparrow| + 0.5 \times \frac{1}{\sqrt{2}}[|\uparrow\rangle + |\downarrow\rangle]\frac{1}{\sqrt{2}}[\langle\uparrow| + \langle\downarrow|]$$

$$= (|\uparrow\rangle \quad |\downarrow\rangle) \begin{pmatrix} 0.75 & 0.25 \\ 0.25 & 0.25 \end{pmatrix} \begin{pmatrix} \langle\uparrow| \\ \langle\downarrow| \end{pmatrix} \tag{12.3.1}$$

To find the von Neumann entropy we need to diagonalise the density matrix, i.e., find its eigenvalues. This is done by solving the secular equation (see §2.5.7),

$$\left\| \begin{pmatrix} 0.75 - \lambda & 0.25 \\ 0.25 & 0.25 - \lambda \end{pmatrix} \right\| = 0 \tag{12.3.2}$$

This yields $\lambda = 0.1464$ or 0.8536. The von Neumann entropy is thus,

$$S_{vN} = -(0.1464\,log_2\,0.1464 + 0.8536\,log_2\,0.8536) = 0.6008$$

So, the correct entropy is only ~0.6, compared to the entropy of 1 which would apply had the states been orthogonal, i.e., perfectly distinguishable. Quite generally we find,

$$S_{vN} \leq -\textstyle\sum_i p_i\,log_2\,p_i \tag{12.3.3}$$

when the density matrix is of the form $\hat{\rho} = \sum_i p_i\,|\tilde{\varphi}_i\rangle\langle\tilde{\varphi}_i|$ but the states $\{|\tilde{\varphi}_i\rangle\}$ in the sum may not all be orthogonal.

12.4 Maximum Entropy

The maximum entropy of a single particle in a Hilbert space of dimension N is $S_{vN}^{max} = log_2\,N$. The same inequality, $S_{vN} \leq log_2\,N$, applies for classical entropy or Shannon information if N is interpreted as the number of distinct classical symbols or microstates. If p_i are the eigenvalues of a density matrix in a Hilbert space of dimension N, the von Neumann entropy is $-\sum_i p_i\,log_2\,p_i$, and an algebraically identical expression applies for Shannon information. The maximum of $-\sum_i p_i\,log_2\,p_i$ subject to the constraint $\sum_i p_i = 1$ is found from the turning point of $-\sum_i p_i\,log_2\,p_i - \lambda\sum_i p_i$ where λ is a Lagrange multiplier. Taking the derivative wrt p_i gives $\frac{1}{ln(2)} + \lambda + log_2\,p_i = 0$ which means that all the p_i must be equal, and hence equal to $1/N$. The entropy is then $log_2\,N$ and this must be its maximum value (the minimum being zero).

12.5 The Ensemble Entropy Inequality

There are a great many inequalities satisfied by the von Neumann entropy. We shall meet a few more when we discuss the entropy of multipartite states in chapter 16. Here is one which generalises (12.3.3). Suppose we have an ensemble of states, each described by a density matrix $\hat{\rho}_i$, such that the ith state occurs with relative frequency p_i in the ensemble. The density matrix of the ensemble is therefore $\hat{\rho} = \Sigma_i p_i \hat{\rho}_i$. Denote the von Neumann entropy of a density matrix $\hat{\rho}$ as $S_{vN}(\hat{\rho})$. It can be shown (e.g., Ref.[12.1] §4.3.4), that,

$$S_{vN}(\Sigma_i p_i \hat{\rho}_i) \leq -\Sigma_i p_i \log_2 p_i + \Sigma_i p_i S_{vN}(\hat{\rho}_i) \tag{12.5.1}$$

The equality is achieved only if all the contributing states in one component density matrix are orthogonal to all the contributing states in every other component density matrix (in maths speak, if the component density matrices "have support on orthogonal subspaces"). In the case that each of the individual states $\hat{\rho}_i$ in the ensemble is a pure state, (12.5.1) reduces to (12.3.3) because $S_{vN}(\hat{\rho}_i) = 0$.

12.6 The Entropy of Measurement

Physical evolution in accord with the Schrodinger equation is unitary. But those curious processes called "measurements" involve the reduction of the state vector, which is not a unitary (or reversible) effect. We have seen that unitary evolution leaves the von Neumann entropy unchanged. But what does a measurement do to the entropy?

Suppose we have a mixed state with density matrix $\hat{\rho} = \Sigma_i p_i |\varphi_i\rangle\langle\varphi_i|$ wrt some orthonormal basis $\{|\varphi_i\rangle\}$. Let us carry out a measurement of an observable $\hat{Q}$ on this mixed system. Any given $|\varphi_i\rangle$ will give rise to the measurement outcome q_j with a probability, according to the Born Rule, of $|C_{ij}|^2$, where, $C_{ij} = \langle\varphi_i|q_j\rangle$, where the $|q_j\rangle$ are the eigenvectors of $\hat{Q}$. But each state $|\varphi_i\rangle$ occurs in the original mixture with relative frequency p_i, so the overall probability of the measurement outcome q_j is,

$$\tilde{p}_j = \Sigma_i p_i |C_{ij}|^2 \tag{12.6.1}$$

When the measurement outcome is q_j the system will be left in the state $|q_j\rangle$ so the density matrix after measurement will be $\hat{\tilde{\rho}} = \Sigma_j \tilde{p}_j |q_j\rangle\langle q_j|$. Of course, this density matrix applies only so long as we have not looked to see what measurement outcome has actually been realised. There is nothing strange about this as long as you recall that the density matrix accounts for our ordinary lack of knowledge about purely deterministic systems, as well as quantum mechanical effects. If we did look we would necessarily find just one of the states, $|q_j\rangle$, with probability $\tilde{p}_j$.

After the measurement the entropy is $S_{vN}(\hat{\tilde{\rho}})$ and it can be shown, as a consequence of (12.6.1), that,

$$S_{vN}(\hat{\tilde{\rho}}) = -\sum_j \tilde{p}_j log_2(\tilde{p}_j) \geq S_{vN}(\hat{\rho}) = -\sum_j p_j \, log_2 \, p_j \qquad (12.6.2)$$

Hence, the entropy of the system after the measurement is greater than that before the measurement. For a proof see Ref.[12.1], section 4.3.3.

As an example consider a 2D density matrix in diagonal form wrt an orthonormal basis and with diagonal components p and $1 - p$ before measurement. The matrix, $|C_{ij}|^2$, in (12.6.1) which transforms the probabilities before measurement to those after measurement consists of real, positive numbers such that every row and every column add to unity. Hence the most general form of this matrix in 2D is,

$$\begin{pmatrix} c & 1-c \\ 1-c & c \end{pmatrix} \qquad (12.6.3)$$

where $0 \leq c \leq 1$. Hence, the probabilities after measurement are,

$$\tilde{p}_1 = 1 + 2cp - c - p \quad \text{and} \quad \tilde{p}_2 = 1 - \tilde{p}_1 = c + p - 2cp \qquad (12.6.4)$$

The entropy before measurement is,

$$S_{vN}^{before} = -(p \, log_2 \, p + (1 - p) \, log_2(1 - p)) \qquad (12.6.5)$$

whereas the entropy after measurement is,

$$S_{vN}^{after} = -\left\{ \begin{matrix} (1 + 2cp - c - p) \, log_2(1 + 2cp - c - p) \\ +(c + p - 2cp) \, log_2(c + p - 2cp) \end{matrix} \right\} \qquad (12.6.6)$$

Figure 12.1 plots the ratio of the entropy before to that after measurement against p for a number of different values of c. It is readily seen that the entropy after measurement is always greater except when equality holds, i.e., when $p = 0.5$.

What about the entropy after a measurement on a pure state? The entropy starts as zero. Usually by "measurement" is implied that we have (a) used some apparatus to interact with the system, and, (b) looked at the outcome. If this is what we mean by "measurement" then state vector reduction means that the system is once again in a pure state and hence also has zero entropy after measurement. However, what if we have carried out only step (a) but not step (b): we have not yet looked at the outcome? Suppose the initial pure state is written in terms of the eigenstates of the observable to be measured as $|\psi\rangle = \sum_i a_i |q_i\rangle$. After step (a), but before step (b), the system has been rendered into a mixed state. We do not know which of the various outcome possibilities has been realised only because we haven't looked yet. This is described by a density matrix which is diagonal with elements $p_i = |a_i|^2$. So the entropy after step (a) but before step (b) is non-zero in general, namely $-\sum_i |a_i|^2 \log_2 |a_i|^2$, so inequality (12.6.2) is again respected.

Why does measurement increase entropy? Taking the case of an initially pure state, it is because measurement changes the number of states accessible to the system from one to many. This seems paradoxical because a measurement is supposed to increase our information, not reduce it. But the entropy associated with the post-measurement mixed state applies only before we look at the result of the measurement.

Figure 12.1: *Ratio of entropies before and after measurement*

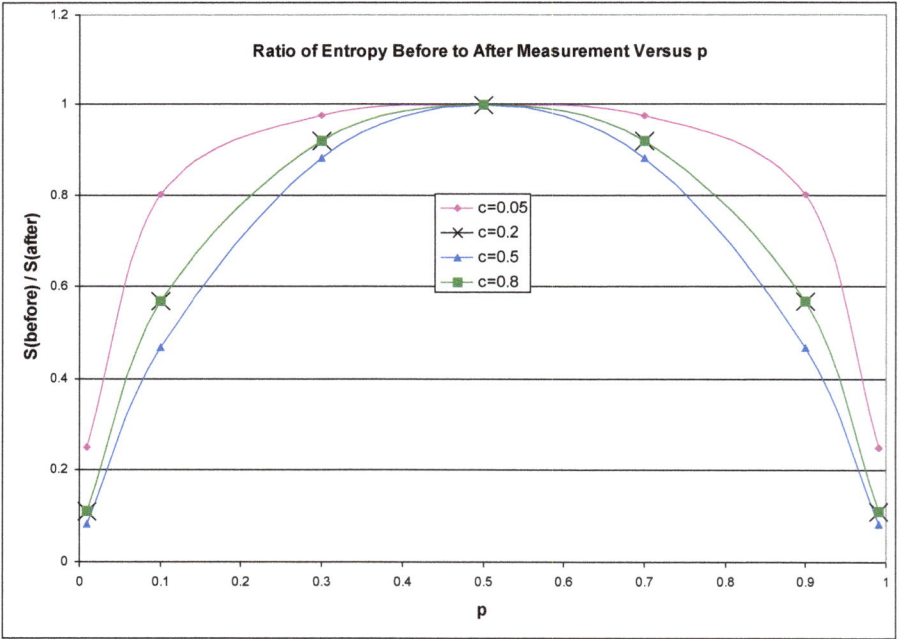

This change, when we observe the outcome of a measurement, from a multiplicity of possibilities (entropy) to our knowledge of the unique outcome (zero entropy) appears to present another paradox, namely a reduction of entropy. This might be thought to violate the second law of thermodynamics. The resolution of this conundrum lies in the *physical* nature of "our knowledge". Taking this into account reveals a hidden increase of entropy elsewhere. Rather than bringing consciousness or our brain into the matter, the acquired knowledge (information) can be considered as recorded in a computer register. But such recording is inevitably also associated with erasure of the register's previous contents, and erasure, being irreversible, involves an increase of entropy. This is discussed in more detail in Plenio and Vitelli, Ref.[12.2], and is essentially identical to the resolution of the famous Maxwell Sorting Demon paradox. These insights are due primarily to Rolf Landauer and Charles Bennett. For further details see Refs.[12,3-4].

12.7 References

[12.1] M.A.Nielsen (1998), "Quantum Information Theory", Doctoral dissertation, The University of New Mexico, Albuquerque. Available as arXiv: quant-ph/0011036.

[12.2] M.B.Plenio and V.Vitelli (2001) "The physics of forgetting: Landauer's erasure principle and information theory", Contemporary Physics **42**, 25 – 60. arXiv:quant-ph/0103108v1

[12.3] R.P.Feynman, (1999) "Feynman Lectures on Computation", eds. J.G.Hey and R.W.Allen, Penguin, GB.

[12.4] H.S.Leff and A.F.Rex, eds, (2002) "Maxwell's Demon 2: Entropy, Information, Computing", (revised edition, CRC Press)

13

Entanglement of Bipartite Pure States

The reader is reminded of how states which consist of several distinct parts are formulated in quantum mechanics, namely in terms of the tensor product of the individual Hilbert spaces. This algebraic formulation appears innocuous, even sterile. Yet it gives rise to the most bizarre physical phenomena involving the indeterminate yet correlated outcomes of measurements on spatially separate parts which can have no causal connection. In this chapter the notion of entanglement is given a precise algebraic definition and quantification for pure states. It is shown that entanglement is the norm for pure bipartite states. The definition and quantification of entanglement in mixed states will be addressed in chapter 17.

13.1 The EPR Paradox

Einstein famously did not like the indeterminacy of quantum mechanics. He felt the theory was incomplete. This was the motivation for a paper by Einstein, Podolski and Rosen in 1935, Ref.[13.1], which became so famous that it is known to all physicists simply as "EPR". This paper purported to show that either quantum mechanics was incomplete or that relativistic causality was violated. The EPR paper centred on a gedanken experiment (of which Einstein was so fond). However, discussions of EPR rarely now use the original gedanken experiment, preferring to use instead a variant due to David Bohm, Refs.[13.2,3], which is even more compelling and easier to appreciate. The essence of the argument is as follows.

Suppose an atom or particle with zero angular momentum decays into a pair of particles which have two spin states, such as spin ½ particles with mass (e.g., an electron and a positron), or alternatively two massless spin 1 particles (photons). These particles then move away from each other at great speed. Suppose also that due to some selection rule, or by some other means, we know that the pair of particles are created in an S-wave state, i.e., without orbital angular momentum. It follows from the conservation of angular momentum that the particles' spin state must be the singlet state of spin zero, which we can write as $(|\uparrow\rangle_1|\downarrow\rangle_2 - |\downarrow\rangle_1|\uparrow\rangle_2)/\sqrt{2}$ (see §2.10). This means that, as we would have expected, if we measure particle 1 to be "spin up" then we will certainly measure particle 2 to be "spin down" wrt the same axis, and vice-versa. But we have no way to tell in advance of an individual measurement which of these two outcomes (i.e., which particle will be spin-up) will be found.

So far no problem is apparent. The puzzlement arises when we note that the measurements on the two particles can be arranged to be at a spacelike separation, so that no causal connection between the measurements is possible. But quantum mechanics would have us believe that before the measurement on particle 1, the spin state of particle 2 is a superposition of both spin up and spin down, i.e., $(|\uparrow\rangle_1|\downarrow\rangle_2 - |\downarrow\rangle_1|\uparrow\rangle_2)/\sqrt{2}$. And yet, after a measurement of the spin of particle 1 has been carried out, which could result in either an up or down spin, the outcome of a spin measurement on particle 2 becomes determined despite no causal connection between them being possible. Thus if particle 1 is spin-up, then particle 2 will always be found to be spin-down, and vice-versa (assuming perfect experiments).

EPR argued that either a faster-than-light interaction must have occurred ("spooky action at a distance") or quantum mechanics must be incomplete. Ruling out the former, the implication appears to be that there are some "hidden variables" carried by each particle which determine the spin that will be registered for each particle separately. Otherwise how could the second particle contrive to always have the opposite spin to the first, if the spin of the first were not decided until it is measured and that measurement is at a space-like separation from the measurement on the second particle? In other words, EPR were arguing that measurements must really be deterministic and determined by some hidden variables, rather than being irreducibly indeterminate as quantum mechanics claims.

On the face of it the argument seems sound – an Einstein special. And yet we already have a powerful reason to believe that there cannot be any hidden variables (see chapter 9). Ironically, it turns out that experimental arrangements of the EPR type can be used to consolidate the case *against* hidden variables. As we shall see in chapter 14, Einstein's argument can be turned against him in a most convincing manner. For now, though, we are concerned with clarifying, and quantifying, the property of a quantum system which allows such phenomena as the EPR paradox to occur: namely, entanglement.

13.2 Notation for Bipartite States

States composed of multiple parts can be formulated as vectors within the Hilbert space formed from the tensor product of the Hilbert spaces of the individual parts (see §2.7). The product state of two sub-systems A and B can thus be written simply as a juxtaposition, $|\varphi\rangle_A|\psi\rangle_B$, which the purist would write as $|\varphi\rangle_A \otimes |\psi\rangle_B$. No actual operation or computation is implied, but the tensor product space is crucially different from a mere set of ordered pairs of vectors in spaces A and B, as explained in §2.7 and repeated below.

The notation for the bra state formed by conjugation from the ket state $|\varphi\rangle_A|\psi\rangle_B$ will be written here as $\langle\varphi|_A\langle\psi|_B$, i.e., the left-to-right ordering of A-then-B is maintained between bra and ket. Where the subscripts A and B are retained there would be no objection or ambiguity in writing this instead as $\langle\psi|_B\langle\varphi|_A$. However, we will generally drop the subscripts which denote to which part, or particle, the states relate, simply for compactness. It is then essential to adopt a consistent convention regarding the ordering. The convention adopted here is that the leftmost state relates to the first particle, and that this applies for both bra and ket vectors. This means we can contract the notation further without ambiguity thus, $|\varphi\psi\rangle \equiv |\varphi\rangle|\psi\rangle \equiv |\varphi\rangle_A|\psi\rangle_B$ and the conjugate of the same state is written $\langle\varphi\psi| \equiv \langle\varphi|\langle\psi| \equiv \langle\varphi|_A\langle\psi|_B$.

13.3 Recognising Entangled Pure States

The most general bipartite state formed from parts whose individual states are within Hilbert spaces $\mathcal{H}_A$ and $\mathcal{H}_B$ is any vector within the tensor product space, $\mathcal{H}_A \otimes \mathcal{H}_B$. However, this should not be misunderstood to consist only of states of the form $|\varphi_A\rangle|\psi_B\rangle$. The product space can be spanned by states of this form, but the general product space vector is not a product of vectors. This is the crucial fact that creates the possibility of entanglement. If the individual spaces are spanned by the complete orthonormal bases $\{|\varphi_i\rangle_A\}$ and $\{|\psi_k\rangle_B\}$, then the most general vector in the product space is,

$$\sum_{i,k} a_{ik}|\varphi_i\rangle_A|\psi_k\rangle_B \tag{13.3.1}$$

where a_{ik} are complex numbers, and normalisation requires $\sum_{i,k}|a_{ik}|^2 = 1$. For example,

$$[|\varphi_1\rangle_A|\psi_1\rangle_B + |\varphi_2\rangle_A|\psi_2\rangle_B]/\sqrt{2} \tag{13.3.2}$$

cannot be expressed as the product of an A-state and a B-state. This contrasts with, for example,

$$[|\varphi_1\rangle_A|\psi_1\rangle_B + |\varphi_2\rangle_A|\psi_2\rangle_B - |\varphi_1\rangle_A|\psi_2\rangle_B - |\varphi_2\rangle_A|\psi_1\rangle_B]/2 \tag{13.3.3}$$

which *is* a product state, namely it is identical to,

$$[(|\varphi_1\rangle_A - |\varphi_2\rangle_A)(|\psi_1\rangle_B - |\psi_2\rangle_B)]/2 \tag{13.3.4}$$

Quite generally then, we have the following definition of entanglement in pure bipartite states,

> A pure bipartite state is said to be entangled if it cannot be expressed as the product of single-particle states. If it can be so expressed it is said to be separable.

Note that "separable" has exactly the same meaning as it does in ordinary algebra. That is, a function of two variables, $f(x, y)$, is separable if it can be expressed as a product of single variable functions, $f(x, y) \equiv g(x)h(y)$.

Hence (13.3.2) is an entangled state whereas (13.3.3) is not because it can be rewritten as (13.3.4). It is quite obvious from inspection of (13.3.3) that it factorises into the product state (13.3.4). But this is not always so obvious. For example (dropping the unnecessary A, B subscripts), is the state (13.3.5) entangled?

$$|\theta\rangle = 0.5913|\varphi_1\rangle|\psi_1\rangle - 0.4008|\varphi_2\rangle|\psi_2\rangle - 0.4281|\varphi_1\rangle|\psi_2\rangle + 0.5536|\varphi_2\rangle|\psi_1\rangle$$

$$(13.3.5)$$

What about this one,

$$0.5|\varphi_1\rangle|\psi_1\rangle - 0.4|\varphi_2\rangle|\psi_2\rangle + 0.5|\varphi_1\rangle|\psi_2\rangle - 0.5831|\varphi_2\rangle|\psi_1\rangle \qquad (13.3.6)$$

Both are normalised. But (13.3.5) is separable, whereas (13.3.6) is entangled. The separability of the first is demonstrated by writing it as,

$$|\theta\rangle = (0.73|\varphi_1\rangle + 0.6835|\varphi_2\rangle)(0.81|\psi_1\rangle - 0.5864|\psi_2\rangle) \qquad (13.3.7)$$

But how can we be sure that (13.3.6) cannot be factorised in this way? More generally, how can we decide whether the state in (13.3.8), below, is separable?

$$A|\varphi_1\rangle|\psi_1\rangle + D|\varphi_2\rangle|\psi_2\rangle + B|\varphi_1\rangle|\psi_2\rangle + C|\varphi_2\rangle|\psi_1\rangle \qquad (13.3.8)$$

The answer is that (13.3.8) is separable if, and only if,

$$AD - BC = 0 \qquad (13.3.9)$$

The proof is an exercise for the reader (see §13.9.1 for an answer).

13.4 The Reduced Density Matrix: Quantifying Entanglement

Note something a little strange about entangled states like $|\omega\rangle = [|\varphi_1\rangle_A|\psi_1\rangle_B + |\varphi_2\rangle_A|\psi_2\rangle_B]/\sqrt{2}$. We are assuming that $|\varphi_1\rangle_A$ and $|\varphi_2\rangle_A$ are distinct states, and that the same is true of $|\psi_1\rangle_B$ and $|\psi_2\rangle_B$, otherwise $|\omega\rangle$ would be separable. Suppose A and B are particles travelling in different directions away from each other, and that you can access only particle A because the other is now miles away. Further suppose that a large number of these bipartite states are being produced by some mechanism, all in the identical initial bipartite state. And finally suppose that, unbeknown to us, and at some spacelike distance out of causal connection with our laboratory, someone is measuring the state of particle B. What arrives in our lab is a stream of A particles in a mixed state; some particles in state $|\varphi_1\rangle_A$ and some in state $|\varphi_2\rangle_A$. And yet the initial state was a pure state, and there can have been no physical interaction with the A particles as they are causally disconnected from the particles B which have been "measured".

This is exactly the EPR paradox which led those authors to posit (in effect) that there must be some hidden variables; that QM is incomplete. But there are very strong reasons now for rejecting hidden variables. Hence, if we discount faster-than-light influences, we are forced to concede that state vector collapse is not "physical" in the usual sense, i.e., that it can occur due to spacelike separated events. This is the non-local implication of entanglement.

Let's consider how this appears in terms of the density matrix. A useful concept is the "reduced density matrix". This is what you get if you form the sum of the expectation values of the density matrix over all states of one of the sub-spaces. It is natural to do this if we have no way of knowing the state of one of the particles, in this case particle B. Hence, the effective or reduced density matrix for particle A is,

$$\hat{\rho}_A = \Sigma_i \langle \psi_{iB} | \hat{\rho} | \psi_{iB} \rangle = Tr_B(\hat{\rho}) \qquad (13.4.1)$$

where, in general, $\{\psi_{iB}\}$ would be a basis of the B-subspace. (13.4.1) is also referred to as "tracing out" the B-states. For example, the product state $[(|\varphi_1\rangle_A - |\varphi_2\rangle_A)(|\psi_1\rangle_B - |\psi_2\rangle_B)]/2$ results in a reduced density matrix for the A subspace of,

$$(|\varphi_1\rangle_A - |\varphi_2\rangle_A)(\langle\varphi_1|_A - \langle\varphi_2|_A)/2 = |\varphi_A'\rangle\langle\varphi_A'| \qquad (13.4.2)$$

The original (pure) product state corresponds to a pure state when reduced to a one-particle state by "tracing out" (i.e. averaging over) the other particle. But this is not the case for an entangled state. The state,

$$[|\varphi_1\rangle_A|\psi_1\rangle_B + |\varphi_2\rangle_A|\psi_2\rangle_B]/\sqrt{2} \qquad (13.4.3)$$

gives a reduced density matrix,

$$[|\varphi_{1A}\rangle\langle\varphi_{1A}| + |\varphi_{2A}\rangle\langle\varphi_{2A}|]/2 \qquad (13.4.4)$$

This is a mixed state, with the maximum 2D entropy, i.e. 1. This confirms that, if we interpret these as spin states, no spin measurement, wrt any axis, will produce deterministic outcomes. And yet this arises from a pure state simply by averaging over the other particle state. In contrast to product states, entangled states become mixed states when one part is traced out.

Since pure product states give rise to pure states when one part is traced out, it follows that the von Neuman (vN) entropy of the composite state and the reduced state are both zero. In contrast, an entangled pure state gives rise to a mixture when one part is traced out. Hence, whilst the vN entropy of the composite state is zero, the vN entropy of the reduced state is non-zero. The signature of entanglement in a pure state is therefore that,

$$S_{vN}(\hat{\rho}_A) \neq 0 \qquad (13.4.5)$$

This suggests that we can use the vN entropy of the reduced state, $S_{vN}(\hat{\rho}_A)$, to quantify the degree of entanglement of the original bipartite state. Hence, writing the composite, pure bipartite density matrix as $\hat{\rho}_{AB}$, its entanglement is defined as,

$$E(\hat{\rho}_{AB}) = S_{vN}(\hat{\rho}_A) \qquad (13.4.6)$$

where, $\hat{\rho}_A = Tr_B(\hat{\rho}_{AB})$. The reader should be concerned that this definition apparently treats the two sub-systems, A and B, differently. Since the entanglement is a property of the combined, bipartite state, why should we trace out B and then define the entanglement as the vN entropy of the reduced density matrix for A? Why not vice versa? But (13.4.6) is a sensible definition of entanglement because one gets the same result whichever way round the calculation is done. Thus, the reduced density matrix for B, obtained by tracing-out A, is written $\hat{\rho}_B = Tr_A(\hat{\rho}_{AB})$ and we find that

$$E(\hat{\rho}_{AB}) = S_{vN}(\hat{\rho}_A) = S_{vN}(\hat{\rho}_B) \qquad (13.4.7)$$

That $S_{vN}(\hat{\rho}_A) = S_{vN}(\hat{\rho}_B)$ is identically true is by no means obvious. The proof is left as an exercise for the reader (see §13.9.2).

It is emphasised that (13.4.7) defines entanglement only for **pure** bipartite states.

> A pure bipartite state is entangled iff the reduced density matrix is mixed. The amount, or degree, of entanglement is defined by the von Neumann entropy of the reduced density matrix.

13.5 The Most General Maximally Entangled Two Qubit State

A "qubit" is the quantum mechanical version of the classical "bit". The latter is a two-state, or binary, variable. A qubit is any quantum system with a two-dimensional Hilbert space. Hence an observable has exactly two eigenvalues in a single qubit system, and hence can function in a similar manner to a classical bit except that there are also many superposition states for which the value of the observable is indeterminate. The obvious example is the spin state of a spin-half particle or a photon.

A two-qubit state is maximally entangled if its entanglement is 1. Examples of maximally entangled two-qubit states are the Bell states, which are conventionally defined as, $\hfill$ (13.5.1)

$$|\Phi_\pm\rangle = (|0\rangle|0\rangle \pm |1\rangle|1\rangle)/\sqrt{2} \quad \text{and} \quad |\Psi_\pm\rangle = (|0\rangle|1\rangle \pm |1\rangle|0\rangle)/\sqrt{2}$$

In passing we note that, if the states $|0\rangle$ and $|1\rangle$ are interpreted as spin-up and spin-down wrt some "z" axis, then state $|\Psi_-\rangle$ is the spin-singlet state of well-defined zero angular momentum, and $|\Psi_+\rangle$ is the spin triplet state of well-defined zero z-axis angular momentum (see §2.10). In contrast, the states $|\Phi_\pm\rangle$ have indeterminate angular momentum wrt the z-axis.

The reduced density matrix for all four of the bipartite Bell states is,

$$\hat{\rho}_{red} = \begin{pmatrix} 1/2 & 0 \\ 0 & 1/2 \end{pmatrix} \hfill (13.5.2)$$

Hence the entanglement of the Bell states is maximal, i.e. 1. Are there any other two qubit states with maximal (unity) entanglement? The answer is yes. The condition for maximal entanglement is very simple to state when expressed in terms of a basis slightly modified from the Bell basis, namely the e-basis defined by,

$$|e_1\rangle = |\Phi_+\rangle, \quad |e_2\rangle = i|\Phi_-\rangle, \quad |e_3\rangle = i|\Psi_+\rangle, \quad |e_4\rangle = |\Psi_-\rangle \hfill (13.5.3)$$

The necessary and sufficient condition for maximal entanglement of a pure two qubit state is that it has **real** coefficients when expressed in the e-basis, i.e.,

$$|\psi_{AB}\rangle = \sum_{i=1}^{4} \alpha_i |e_i\rangle \qquad (13.5.4)$$

where the four coefficients, α_i, are real, and required to obey $\sum_{i=1}^{4} \alpha_i^2 = 1$. (To be more precise, the α_i may be complex but must all have the same phase, so that they can be taken to be real as the overall phase is irrelevant). This will now be proved. The general two qubit state can be written,

$$|\psi_{AB}\rangle = a|0\rangle|0\rangle + b|0\rangle|1\rangle + c|1\rangle|0\rangle + d|1\rangle|1\rangle \qquad (13.5.5)$$

where a, b, c and d will, in general, be complex. Hence,

$$\hat{\rho}_A = (a|0\rangle + c|1\rangle)(a^*\langle 0| + c^*\langle 1|) + (b|0\rangle + d|1\rangle)(b^*\langle 0| + d^*\langle 1|)$$
$$= \begin{pmatrix} |a|^2 + |b|^2 & ac^* + bd^* \\ ca^* + db^* & |c|^2 + |d|^2 \end{pmatrix} \qquad (13.5.6)$$

To find the vN entropy of this mixed reduced state we need to find the eigenvalues of the above matrix. Elementary manipulation shows these to be,

$$\lambda = \frac{1}{2}\left\{1 \pm \sqrt{1 - 4|ad - bc|^2}\right\} \qquad (13.5.7)$$

The coefficients are subject to the normalisation $|a|^2 + |b|^2 + |c|^2 + |d|^2 = 1$. It must be that $|ad - bc| \leq 1/2$ since the eigenvalues of an Hermitian matrix must be real. When this inequality is saturated both eigenvalues are ½ and hence the vN entropy of the reduced density matrix, which is the entanglement of the bipartite state, is maximum (i.e., 1). Hence, the problem reduces to finding the most general coefficients which have $|ad - bc| = ½$.

The simplest way of solving this is to note that the state (13.5.5) can also be written in terms of the e-basis, (13.5.4). The relationship between the coefficients is then,

$$a = \frac{1}{\sqrt{2}}(\alpha_1 + i\alpha_2), \qquad d = \frac{1}{\sqrt{2}}(\alpha_1 - i\alpha_2),$$

$$b = \frac{1}{\sqrt{2}}(\alpha_4 + i\alpha_3), \qquad c = -\frac{1}{\sqrt{2}}(\alpha_4 - i\alpha_3)$$

From this it follows that,

$$|ad - bc| = \frac{|S|}{2}, \text{ where, } S = \alpha_1^2 + \alpha_2^2 + \alpha_3^2 + \alpha_4^2 \qquad (13.5.8)$$

Now the normalisation requirement for the state (13.5.4) is that $\sum_{i=1}^{4} |\alpha_i|^2 = 1$. So, if all the α_i are real, (13.5.8) gives $S = 1$ and hence (13.5.8) also gives $|ad - bc| = \frac{1}{2}$, as required.

The converse also follows: maximal entanglement requires $|ad - bc| = \frac{1}{2}$ and hence $|S| = 1$, but this is only possible if all the α_i share the same phase (and so can be taken to be real). QED.

The other important result which follows from (13.5.7) is that the entanglement is zero iff $|ad - bc| = 0$, because this is equivalent to the eigenvalues of the reduced density matrix being 0 and 1 (which gives zero vN entropy). But we have already noted that the condition $ad = bc$ is precisely that which ensures that the bipartite state can be factorised as a product state of the two sub-systems (see the reader exercise, §13.9.1). Hence, this establishes that the entanglement measure defined by (13.4.7) is sensible in that it produces zero entanglement for product states.

13.6. The Entanglement of Random States

If we choose a state at random, how likely is it to be entangled? The answer depends crucially upon whether we mean a randomly chosen pure state or a random mixture. The answer is,

- If a pure state is chosen at random from a Hilbert space describing a bipartite system, then it is essentially certain to be entangled;
- If a very large number of pure states are chosen at random and combined with random probabilities to form a mixture, then the resulting mixture will be separable (i.e., not entangled) in almost all cases.

In the first bullet, "essentially certain" means that the product states form a set of measure zero in the Hilbert space. In the second bullet, "almost all cases" tends to "all cases" in the limit that the number of states in the mixture becomes infinite. Both of these observations surprised me when I first stumbled across them. It is additionally unexpected, perhaps, that the answers are diametrically opposite. This serves to emphasise how different pure states are from mixed states.

We shall now prove the assertion for pure states. The assertion for mixed states will be proved in chapter 17 when we discuss the entanglement of mixed states in more detail. Consider a randomly chosen pure bipartite state. It suffices to consider a pair of qubits. The most general product state of two qubits can be written $(\alpha|0\rangle + \beta|1\rangle)(\gamma|0\rangle + \delta|1\rangle)$, where α, β, γ, δ are arbitrary complex numbers apart from normalisation, $|\alpha|^2 + |\beta|^2 = 1$ and $|\gamma|^2 + |\delta|^2 = 1$. Consequently there are six continuum degrees of freedom involved in the choice of a random pair of qubits: four phase angles, and two magnitudes.

In contrast, the most general state in the 2 x 2 product Hilbert space is,

$$a|0\rangle|0\rangle + b|0\rangle|1\rangle + c|1\rangle|0\rangle + d|1\rangle|1\rangle \qquad (13.6.1)$$

This is required to respect only $|a|^2 + |b|^2 + |c|^2 + |d|^2 = 1$. Hence, there are seven continuum degrees of freedom involved in choosing a random state from the 2×2 product space of two qubits. It is clear, therefore, that it would be a fluke if a randomly chosen state happened to be a product state. Specifically, the random a, b, c, d would have to be such that $a = \alpha\gamma, b = \alpha\delta, c = \beta\gamma$ and $d = \beta\delta$. It is easily seen that this can be true only if $ad = bc$. Choosing a, b, c, d at random, this relation will "almost never" be exactly obeyed. Hence, "essentially all" random $\mathcal{H}_2 \otimes \mathcal{H}_2$ states are entangled, or, more accurately, separable states are of measure zero in parameter space and hence have a vanishing probability of occurrence if $\mathcal{H}_2 \otimes \mathcal{H}_2$ is randomly sampled.

Does this generalise to a Hilbert spaces of arbitrary dimension? Yes it does. Suppose the dimensionality is D_1 x D_2. Consider the coefficients of each state in the product state. The number of phases is $D_1 + D_2$. The number of magnitudes is $D_1 + D_2 - 2$, because of the two constraint equations. Hence, the total number of degrees of freedom is $2(D_1 + D_2 - 1)$. Now consider the most general state from the product space. There are $D_1 D_2$ terms, and hence $2D_1 D_2 - 1$ degrees of freedom, because of the one constraint equation. But it is clear than $2D_1 D_2 - 1 > 2(D_1 + D_2 - 1)$ for D_1 and D_2 both ≥ 2. Hence, "essentially all" pure bipartite states, of any dimensionality, are entangled.

The morale of this story is that entangled pure quantum states in bipartite systems are not something exotic. On the contrary, they are naturally entangled. This is the norm, whereas separable states are exceptional. One might be tempted to conclude that entanglement is therefore not an especially exotic property were it not for the fact that being in a pure quantum state is itself often quite exotic.

13.7 Purification

We have seen that a pure entangled bipartite state gives rise to a mixed reduced density matrix when one part is traced out. This raises a question: can any mixed state (say of part A) be expressed as the reduced density matrix of some bipartite state by introducing an "auxiliary" part B? The answer is yes, and it is easy to construct such a bipartite state explicitly. Suppose the mixed single-part state is written in its spectral decomposition (see §2.5.4) as,

$$\hat{\rho}_A = \Sigma_i\, p_i\, |A_i\rangle\langle A_i| \qquad (13.7.1)$$

Now consider the pure state formed as a bipartite state with part B, thus,

$$|\varphi\rangle = \Sigma_k\, \sqrt{p_k}\, e^{i\delta_k}|A_k\rangle|B_k\rangle \qquad (13.7.2)$$

where we assume the basis states $|A_i\rangle$ and $|B_i\rangle$ are orthonormal. In (13.7.2) the δ_i are any real numbers, indicating that the relative phases between the terms in $|\varphi\rangle$ are arbitrary. We can now form the reduced density matrix by tracing out part B from the pure state density matrix $|\varphi\rangle\langle\varphi|$, thus,

$$\hat{\rho}^{red} = \Sigma_k\langle B_k|\varphi\rangle\langle\varphi|B_k\rangle$$

$$= \Sigma_{i,j,k}\, \sqrt{p_i p_j}\, e^{i(\delta_i - \delta_j)}\langle B_k|(|A_i\rangle|B_i\rangle)\rangle\langle A_j|\langle B_j|B_k\rangle$$

$$\Sigma_{i,j,k}\, \sqrt{p_i p_j}\, e^{i(\delta_i - \delta_j)}\delta_{ik}\delta_{jk}\, |A_i\rangle\langle A_j| = \Sigma_k\, p_k\, |A_k\rangle\langle A_k| = \hat{\rho}_A \qquad (13.7.3)$$

which establishes the equivalence. The state (13.7.2) is thus said to be a "purification" of the mixed state (13.7.1). That any mixed state may be purified in this way raises the interesting, if unanswerable, ontological question: does the purifying auxiliary part B, and its associated states, actually physically exist? Are all mixed states actually an illusion brought about by having discarded the auxiliary parts which would render them pure, but multipartite? Can Schrodinger's cat be described by a pure state after all, if only we could find the (enormous number of) other items and their states, somewhere in the world, in terms of which the cat's basis states are entangled in an ultimately pure Hilbert vector?

13.8 Entanglement of Pure States: A Summary

- The concept of entanglement applies to systems which can be considered as composed of two or more sub-systems, illustrated here for the cases of systems consisting of two parts (bipartite states);
- A pure bipartite quantum state is entangled iff it is not separable, i.e., it cannot be expressed as the product of a quantum state of part A and a quantum state of part B;
- Given a bipartite state, a density matrix for one part, A, called the "reduced density matrix", can be found by "tracing out" the other part, B, from the density matrix for the combined state;

- The reduced density matrix of a separable bipartite state is a pure state. The reduced density matrix of an entangled bipartite state is a mixed state;
- Some states are more entangled than others. The degree of entanglement of a pure bipartite state is quantified by the von Neumann entropy (S_{vN}) of the reduced density matrix. It makes no difference which part is traced-out.
- If a pure state is chosen at random from a given Hilbert space describing a bipartite system, then it will almost always be entangled (the separable states are of measure zero in parameter space).
- If a large number (N) of pure states are chosen at random and combined with random probabilities to form a mixture, then the resulting mixture will be separable (i.e., not entangled) in almost all cases. This becomes exactly all cases as $N \to \infty$. This will be demonstrated in chapter 17.

13.9 Exercises for the Reader

13.9.1 Condition for Separability

Show that $A|\varphi_1\rangle|\psi_1\rangle + B|\varphi_1\rangle|\psi_2\rangle + C|\varphi_2\rangle|\psi_1\rangle + D|\varphi_2\rangle|\psi_2\rangle$ is separable if and only if $AD = BC$ (normalisation assumed). Proof: If separable it can be written in the form,

$$\left(\alpha|\varphi_1\rangle + \sqrt{1 - |\alpha|^2}|\varphi_2\rangle\right)\left(\beta|\psi_1\rangle + \sqrt{1 - |\beta|^2}|\psi_2\rangle\right)$$

This immediately implies,

$$A = \alpha\beta \qquad\qquad D = \sqrt{(1 - |\alpha|^2)(1 - |\beta|^2)}$$

$$B = \alpha\sqrt{1 - |\beta|^2} \qquad C = \beta\sqrt{1 - |\alpha|^2}$$

and these give $AD = BC$. This establishes that $AD = BC$ is a necessary condition for separability. To establish that it is also a sufficient condition to guarantee separability, note that taking the ratio of B and D allows us to solve for $|\alpha|^2$, as does taking the ratio of A and C, and $|\beta|^2$ can be solved in a similar manner from the ratio of A and B, or from the ratio of C and D, giving,

$$|\alpha|^2 = \frac{|A|^2}{|A|^2 + |C|^2} = \frac{|B|^2}{|B|^2 + |D|^2} \quad\text{and}\quad |\beta|^2 = \frac{|A|^2}{|A|^2 + |B|^2} = \frac{|C|^2}{|D|^2 + |C|^2}$$

These expressions are consistent by virtue of $AD = BC$, and they allow us to find an α and a β, and hence confirm separability, assuming only the condition $AD = BC$. Note that α and β are not uniquely determined; only their magnitudes are determined but there is ambiguity up to a phase factor (i.e., a factor of the form $e^{i\theta}$, for real θ). **QED**

13.9.2 Prove $S_{vN}\left(Tr_A(\hat{\rho}_{AB})\right) = S_{vN}\left(Tr_B(\hat{\rho}_{AB})\right)$

Prove that the definition of pure state entanglement, (13.4.7), is independent of which state is traced-out. An arbitrary bipartite pure state can be written $\sum_{i,j} a_{ij} |\varphi_i\rangle_A |\psi_j\rangle_B$ where the single particle states are taken to be an orthonormal basis (and some of the a_{ij} coefficients may be zero). The corresponding density matrix is thus (dropping the A, B subscripts),

$$\hat{\rho}_{AB} = \sum_{i,j,m,n} a_{ij} a_{mn}^* |\varphi_i\rangle |\psi_j\rangle \langle\varphi_m| \langle\psi_n| \tag{13.9.2.1}$$

Tracing out the B-states by taking the matrix element $\langle\psi_k| \dots |\psi_k\rangle$ and summing over k gives,

$$\hat{\rho}_A = Tr_B(\hat{\rho}_{AB}) = \sum_{i,j,m,n,k} a_{ij} a_{mn}^* |\varphi_i\rangle \delta_{kj} \langle\varphi_m| \delta_{kn}$$

Hence,

$$\hat{\rho}_A = \sum_{i,m,k} a_{ik} a_{mk}^* |\varphi_i\rangle \langle\varphi_m| \tag{13.9.2.2}$$

But we can write $\sum_k a_{ik} a_{mk}^*$ in matrix notation as $[(a)(a)^+]_{im}$ so it follows that in the basis being used for the A-states, the reduced density matrix for A is just $(a)(a)^+$. In the same way the reduced density matrix for B is found to be $(a)^+(a)$. Note that both these matrices are Hermitian, as they must be, but they are different. In fact, in general, they will not even have the same dimension.

The vN entropy of the reduced A-density matrix is given by $-\sum_i p_i log_2 p_i$ where the p_i are the eigenvalues of $(a)(a)^+$ and the index i runs over the dimension of this matrix, i.e., the dimension of $\mathcal{H}_A$, which we write as N_A. The eigenvalues and eigenvectors of $(a)(a)^+$ are such that,

$$(a)(a)^+ \bar{v}_i = \lambda_i \bar{v}_i \tag{13.9.2.3}$$

The eigenvalues and eigenvectors of $(a)^+(a)$, the B-density matrix, are such that,

$$(a)^+(a) \bar{u}_j = \xi_j \bar{u}_j \tag{13.9.2.4}$$

But multiplying (13.9.2.4) on the left by (a) gives,

$$(a)(a)^+(a)\bar{u}_j = \xi_j(a)\bar{u}_j \qquad (13.9.2.5)$$

which reveals that $(a)\bar{u}_j$ must be one of the eigenvectors $\bar{v}_i$ of $(a)(a)^+$ and ξ_j must be the corresponding eigenvalue. So $(a)(a)^+$ and $(a)^+(a)$ have the same eigenvalues, and so the reduced density matrices $\hat{\rho}_A = Tr_B(\hat{\rho}_{AB})$ and $\hat{\rho}_B = Tr_A(\hat{\rho}_{AB})$ therefore have the same vN entropy, as this depends only upon their spectra, i.e., their set of eigenvalues.

The alert reader will spot a potential difficulty when the dimension of the B-subspace does not equal that of the A-subspace, say $N_B > N_A$. In this case (a) is rectangular, not square, though $(a)(a)^+$ is square and of dimension N_A and $(a)^+(a)$ is square of dimension N_B. In this case, as the reduced B density matrix is of greater dimension, and hence has a larger number of eigenvalues than the A matrix, how can the set of eigenvalues be identical? In fact the additional eigenvalues of the reduced B density matrix are all zero, and hence contribute nothing to the vN entropy, so this circumstance does not alter the conclusion. That this must be true is most easily seen by artificially increasing the dimension of the A-subspace to equal that of B by adding dummy vectors which are not actually used, and having zeros in the corresponding positions of the enhanced $(a)(a)^+$ matrix. The theorem now goes through without a problem as we have made $N_A = N_B$, and the additional eigenvalues are zero by construction. **QED**.

13.10 References

[13.1] Einstein, A; B Podolsky, N Rosen (1935). "*Can Quantum-Mechanical Description of Physical Reality be Considered Complete?*". Physical Review **47**, 777–780.

[13.2] Bohm, D. 1951. In Quantum Theory. Chap. 22, p. 611. Englewood Cliffs, NJ: Prentice Hall.

[13.3] D. Bohm, Y. Aharonov (1957), Discussion of Experimental Proof for the Paradox of Einstein, Rosen, and Podolsky, Phys. Rev. 108(4) 1070–1076

14

There Are No Hidden Variables: Part 2

In this chapter we hammer the final nail into the coffin lid of hidden variables. The hammer and nails were supplied by John Bell, but the coffin itself is, as always, experimental evidence. That Bell was able to provide a means by which the question of hidden variables could be resolved by tractable experimental means is a remarkable achievement.

14.1 Reprise: The Hidden Variable Story So Far

Let us start by reviewing the story on hidden variables so far. In chapter 13 we discussed experimental arrangements of the EPR type which produce puzzling results unless some "hidden variables" are assumed to rationalise them in classical terms. In chapter 9 we looked at von Neumann's incorrect proof that there are no hidden variables, a false start which fooled a generation of physicists. We then presented the BKS theorem which shows that it is mathematically incompatible to require, (i) that all Hermitian operators in a Hilbert space of dimension 3 or greater can be assigned deterministic values for all states, and, (ii) that these values are eigenvalues of the operators, whilst, (iii) also respecting a homomorphism relationship for commuting operators, (9.2.2). Hence, so long as we accept that observables are Hermitian operators, and that measured values of observables must be one of their eigenvalues, and that the value assigned to a sum of commuting operators always equals the sum of the values assigned separately to each one, then there is no possibility of making the theory deterministic (which is what we mean by "hidden variables").

This is a very strong case against hidden variables. There are loopholes, however. One is that the non-contextuality of observables is assumed (i.e., that observables are innate properties of the system and do not depend upon the measurement context). And BKS does not apply for 2D Hilbert spaces (i.e., qubits). Another assumption implicit in the proof is that all Hermitian operators can be regarded as observables, in principle. Actually that can be weakened to "sufficiently many". Yet we are still left with the conundrum of EPR-type experiments.

As always with results of such importance, it is desirable to establish them in diverse ways. We now turn to Bell's Theorem which establishes that any local, realistic hidden variable theory must disagree with quantum mechanical predictions in some instances. Bell derived an inequality which must be obeyed by any local, realistic hidden variable theory in an experiment of EPR type – an inequality not respected by QM. Hence, by carrying out experiments to determine if the inequality is, or is not, obeyed, it has become possible to decide empirically between QM and local, realistic, deterministic hidden variable theories.

14.2 Bell's Theorem

Bell, Ref.[14.1], considered an EPR-type experiment, as described in chapter 13. A pair of spin ½ particles emerge from the decay of a spinless precursor in the singlet spin state. Bell envisaged the spin of one particle being measured in a direction given by unit vector $\hat{a}$, and the spin of the other particle being measured in direction $\hat{b}$. These vectors can be oriented arbitrarily in 3D space. The angle between them is θ_{ab}. At issue is the correlation between the two spin measurements, for which it suffices to consider the expectation value of their product. Measuring spins in units of $\hbar/2$, the quantum mechanical expectation value of the product is,

$$\langle(\bar{\sigma}_1 \cdot \hat{a})(\bar{\sigma}_2 \cdot \hat{b})\rangle = -\hat{a} \cdot \hat{b} = -\cos\theta_{ab} \qquad (14.2.1)$$

where $\bar{\sigma}_{1,2}$ are the Pauli matrices for the two particles (see §2.10). The Pauli matrices are the same for the two particles, of course, the subscripts simply serving to indicate which state vectors they operate upon. The proof of (14.2.1) is an exercise for the reader (see §14.4 for my answer). In the particular case that $\hat{a} = \hat{b}$ this produces an expectation value for the product of -1 as it should, i.e. the spins are always opposed.

Bell argued that if the outcome of an individual spin measurement is determined by a hidden variable λ, then the first particle could be predicted with certainty to have a spin of either +1 or -1. In other words there must be some deterministic function $A(\hat{a}, \lambda)$ which takes only the values +1 or -1 and which value would be determined uniquely if λ were known. Similarly, the second particle also has determinate spin, $B(\hat{b}, \lambda) = \pm 1$. We take the same hidden variable to be carried by both particles because we are seeking to explain how the results of spin measurements on the two particles can be correlated in the absence of causal connection. The expectation value of the product of spins, averaged over many measurements, is thus, according to the hidden variable theory,

$$E(\hat{a}, \hat{b}) = \int A(\hat{a}, \lambda) \, B(\hat{b}, \lambda)\rho(\lambda)d\lambda \qquad (14.2.2)$$

where $\rho(\lambda)$ is some probability density of the hidden variable so that $\int \rho(\lambda)d\lambda = 1$. Bell, Ref.[14.1], showed that (14.2.2) is inconsistent with (14.2.1), i.e., that any such hidden variable theory must predict results at variance with QM. Specifically, Bell derived an inequality which must be respected by any such hidden variable theory, i.e.,

$$1 + E(\hat{b}, \hat{c}) \geq |E(\hat{a}, \hat{b}) - E(\hat{a}, \hat{c})| \tag{14.2.3}$$

But the quantum mechanical expectation value, (14.2.1), does **not** obey (14.2.3). For example, consider $\hat{a}$ and $\hat{c}$ to be perpendicular with $\hat{b}$ at 45° to both. Then, assuming the quantum mechanical result (14.2.1), the LHS of the inequality is $1 - \frac{1}{\sqrt{2}} = 0.292$ whereas the RHS is $\frac{1}{\sqrt{2}} = 0.707$, so the inequality is clearly false.

Note that it is not even a close miss. The quantum expectation value disrespects the Bell inequality quite radically. This is important because it means that experiments to discriminate between the two need not necessarily be of very great precision.

The proof of Bell's inequality, (14.2.3), is remarkably simple. Any theory must predict that the two particles have opposite spins when measured in the same direction since this is required by the conservation of angular momentum. Hence we have,

$$B(\hat{a}, \lambda) = -A(\hat{a}, \lambda) \tag{14.2.4}$$

So (14.2.2) becomes, $\quad E(\hat{a}, \hat{b}) = -\int A(\hat{a}, \lambda) A(\hat{b}, \lambda)\rho(\lambda)d\lambda \tag{14.2.5}$

Hence,

$$E(\hat{a}, \hat{b}) - E(\hat{a}, \hat{c}) =$$
$$- \int A(\hat{a}, \lambda) A(\hat{b}, \lambda)\rho(\lambda)d\lambda + \int A(\hat{a}, \lambda) A(\hat{c}, \lambda)\rho(\lambda)d\lambda$$
$$= \int A(\hat{a}, \lambda) A(\hat{b}, \lambda)[A(\hat{b}, \lambda)A(\hat{c}, \lambda) - 1]\rho(\lambda)d\lambda \tag{14.2.6}$$

because $A(\hat{b}, \lambda)^2 = 1$. But the value of $A(\hat{a}, \lambda)A(\hat{b}, \lambda)$ can only be ± 1 so that (14.2.6) implies,

$$|E(\hat{a}, \hat{b}) - E(\hat{a}, \hat{c})| \leq \int \left(1 - A(\hat{b}, \lambda)A(\hat{c}, \lambda)\right)\rho(\lambda)d\lambda \tag{14.2.7}$$

Using (14.2.5) this gives,

$$|E(\hat{a}, \hat{b}) - E(\hat{a}, \hat{c})| \leq 1 + E(\hat{b}, \hat{c}) \tag{14.2.8}$$

which is Bell's inequality, (14.2.3). QED.

Stronger inequalities than (14.2.3) have now been proved, and also forms of inequality which are better suited to certain experimental arrangements. However, (14.2.3) suffices to demonstrate the force of the strategy which permits experiments to refute local deterministic hidden variable theories rather than algebraic arguments alone.

14.3 Experimental Tests of Bell-Type Inequalities

The literature on Bell-type inequalities is large, as is the number of experiments to test such inequalities. Most experiments have used photons, though some now use electron spins. The purist would claim, quite rightly, that these experiments are not strictly conclusive. Experimental loopholes exist through which hidden variable theories might still escape final discreditation. However, experiments are constantly restricting the wriggle room to an ever greater extent.

Dozens of experiments have been performed over the last 40 years. The earliest were associated with the names Clauser and Aspect, e.g., Refs.[14.2, 14.3]. All the earliest experiments were plagued by low detector efficiency, which was manifest in practice by the two detectors failing to register coincident photons most of the time. Since most of the produced pairs were not detected, this left an obvious open door for doubters to claim (rightly) that violation of Bell's inequality had not been rigorously demonstrated.

Nevertheless, taken at face value, the large majority of experiments appear to indicate violation of a Bell-type inequality and are consistent with quantum mechanical expectations. Those few experiments which are exceptions have tended to come under suspicion of inaccuracy or have not proved reproducible. Although now rather out of date, Redhead, Ref.[14.4], presents a Table of experimental tests.

More recent experiments have concentrated on overcoming the limitations and defeating the residual loopholes. One of the first with high-efficiency detection was Ref.[14.5]. Perhaps the best test at the time of writing used electrons spins in diamond over a distance of 1.3 km, Ref.[14.6]. It is probably fair to say that local realistic hidden variable theories are now dead, though loopholes exist through which more exotic types of hidden variables could still be proposed. Whether there is any longer a great motivation for doing so is another matter.

14.4 Exercises for the Reader

14.4.1 Prove (14.2.1) for the singlet spin state of two spin ½ particles, i.e.,

$$\langle(\bar{\sigma}_1 \cdot \hat{a})(\bar{\sigma}_2 \cdot \hat{b})\rangle = -\hat{a} \cdot \hat{b} = -\cos\theta_{ab}$$

Using the usual z-basis, the Pauli matrices are

$$\bar{\sigma} = \begin{pmatrix} 0 & 1 \\ 1 & 0 \end{pmatrix}, \begin{pmatrix} 0 & -i \\ i & 0 \end{pmatrix}, \begin{pmatrix} 1 & 0 \\ 0 & -1 \end{pmatrix}$$

Hence,
$$\bar{\sigma}_1 \cdot \hat{a} = \begin{pmatrix} a_z & a_x - ia_y \\ a_x + ia_y & -a_z \end{pmatrix} \equiv \begin{pmatrix} a_z & a_- \\ a_+ & -a_z \end{pmatrix}$$

In the same basis, the singlet state can be written,

$$\frac{1}{\sqrt{2}}\left[\begin{pmatrix} 1 \\ 0 \end{pmatrix} \otimes \begin{pmatrix} 0 \\ 1 \end{pmatrix} - \begin{pmatrix} 0 \\ 1 \end{pmatrix} \otimes \begin{pmatrix} 1 \\ 0 \end{pmatrix}\right]$$

The expectation value of the operator $(\bar{\sigma}_1 \cdot \hat{a})(\bar{\sigma}_2 \cdot \hat{b})$ in this state is the sum of four terms, times ½, namely,

$$\begin{pmatrix} 1 & 0 \end{pmatrix} \begin{pmatrix} a_z & a_- \\ a_+ & -a_z \end{pmatrix} \begin{pmatrix} 1 \\ 0 \end{pmatrix} \times \begin{pmatrix} 0 & 1 \end{pmatrix} \begin{pmatrix} b_z & b_- \\ b_+ & -b_z \end{pmatrix} \begin{pmatrix} 0 \\ 1 \end{pmatrix}$$

$$\begin{pmatrix} 0 & 1 \end{pmatrix} \begin{pmatrix} a_z & a_- \\ a_+ & -a_z \end{pmatrix} \begin{pmatrix} 0 \\ 1 \end{pmatrix} \times \begin{pmatrix} 1 & 0 \end{pmatrix} \begin{pmatrix} b_z & b_- \\ b_+ & -b_z \end{pmatrix} \begin{pmatrix} 1 \\ 0 \end{pmatrix}$$

$$-\begin{pmatrix} 1 & 0 \end{pmatrix} \begin{pmatrix} a_z & a_- \\ a_+ & -a_z \end{pmatrix} \begin{pmatrix} 0 \\ 1 \end{pmatrix} \times \begin{pmatrix} 0 & 1 \end{pmatrix} \begin{pmatrix} b_z & b_- \\ b_+ & -b_z \end{pmatrix} \begin{pmatrix} 1 \\ 0 \end{pmatrix}$$

$$-\begin{pmatrix} 0 & 1 \end{pmatrix} \begin{pmatrix} a_z & a_- \\ a_+ & -a_z \end{pmatrix} \begin{pmatrix} 1 \\ 0 \end{pmatrix} \times \begin{pmatrix} 1 & 0 \end{pmatrix} \begin{pmatrix} b_z & b_- \\ b_+ & -b_z \end{pmatrix} \begin{pmatrix} 0 \\ 1 \end{pmatrix}$$

The first two of these both give $-a_z b_z$, the third gives $-a_- b_+$ and the last gives $-a_+ b_-$. Adding and dividing by 2 gives the desired result. <u>QED</u>.

14.5 References

[14.1] Bell, J S, (1964), "*On the Einstein-Podolsky-Rosen Paradox*", Physics **1**, 195-200.
https://journals.aps.org/ppf/pdf/10.1103/PhysicsPhysiqueFizika.1.195

[14.2] Freedman, S J and Clauser, J F (1972). *"Experimental test of local hidden-variable theories"*. Phys. Rev. Lett. **28** (938): 938–941. doi:10.1103/PhysRevLett.28.938

[14.3] Aspect, Alain; Philippe Grangier, Philippe, and Roger Gérard (1981). "*Experimental Tests of Realistic Local Theories via Bell's Theorem*". Phys. Rev. Lett. **47** (7): 460–3. doi:10.1103/PhysRevLett.47.460

[14.4] Redhead, Michael (1987), "*Incompleteness, Nonlocality, and Realism*", Clarendon Press, Oxford.

[14.5] Rowe, M A, et al (2001). *"Experimental violation of a Bell's inequality with efficient detection"*. Nature. **409** (6822): 791–94. doi:10.1038/35057215

[14.6] Hensen, B, et al. (2015). *"Loophole-free Bell inequality violation using electron spins separated by 1.3 kilometres"*. Nature. **526** (7575): 682–686. arXiv:1508.05949

15

Decoherence and Measurement

Decoherence is the process whereby a pure quantum state becomes a mixture and loses its uniquely quantum mechanical properties, such as interference effects. In contrast, a pure state, even if expressed as a superposition of states, will have its time-dependence determined by Schrodinger's equation. Quantitative decoherence theory can be used to calculate the time for decoherence to occur. Perhaps surprisingly, the concept of decoherence also sheds some light on the nature of quantum measurement, though it does not provide a complete resolution of the issues raised by that process.

15.1 What is Decoherence?

Decoherence theory is two things. It is an attempt to address some of the issues raised by the measurement problem (§2.6) and the classical limit of quantum mechanics. But it is also a procedure for the practical calculation of decoherence, i.e., whether a pure state degenerates into a mixed state, and if so how long it takes. As regards the former, it is significantly enlightening but not the whole solution to the measurement problem. However, as regards calculations of decoherence times, or determining what states may be robust against decoherence, it is an indispensable tool. This chapter introduces decoherence at an elementary level and discusses its relevance to measurement and the classical limit following the approach of Zurek, Refs.[15.1-3]. The calculation of decoherence times is illustrated later in chapter 21 and we shall have more to say about the associated issue of pointer states in chapter 22.

What do we mean by decoherence? What it can be is the disappearance of properties characteristic of quantum mechanical behaviour: the vanishing of the weirdness. But this does not always happen, even though decoherence is technically taking place. Atomic electrons retain indefinitely their discrete energy states, which is essentially quantum mechanical. Moreover, electrons in such states are not localised, which is also uniquely quantum mechanical. More specifically, decoherence can cause a pure quantum state, which is initially in a superposition of states, to become a mixture of those states, e.g.,

$$|\psi\rangle = \alpha|a\rangle + \beta|b\rangle \qquad (15.1.1)$$

decoheres to
$$\hat{\rho} = |\alpha|^2|a\rangle\langle a| + |\beta|^2|b\rangle\langle b| \qquad (15.1.2)$$

One signature of this may be that a system loses its former ability to produce interference effects. Such effects result from the system being in a superposition state, like (15.1.1). In contrast, a 'mixture', like (15.1.2) means the system is either in state a or state b, and its former ability to cause interference effects between the two states has been lost.

The similarity with quantum measurement theory is immediately apparent. Suppose that $|a\rangle$ and $|b\rangle$ are orthogonal eigenstates of some observable, $\hat{Q}$. After exposing the system in state (15.1.1) to an apparatus which measures $\hat{Q}$, the system is left in precisely the mixed state (15.1.2), i.e., it is actually in either state $|a\rangle$ or $|b\rangle$, with Born Rule probabilities, no longer in a superposition. So, is decoherence just another name for measurement? No, decoherence theory provides a framework within which measurement can be understood a little better but it is more general, applying also to situations which are not measurements. Pure quantum states often decohere quite apart from measurement situations.

The central problem of quantum mechanical measurement theory is understanding the nature of the reduction of the state vector, or the "collapse of the wavefunction", as the transformation $\alpha|a\rangle + \beta|b\rangle \rightarrow |a\rangle$ or $|b\rangle$ is called. It is indeterminate which of these two outcomes will occur (assuming α and β are both non-zero). In the usual interpretation of quantum mechanics, states evolve in time according to the Schrodinger equation, $\hat{H}|\psi\rangle = i\hbar\partial_t|\psi\rangle$, except during special events called 'measurements' when this irreversible collapse of the wavefunction occurs. Schrodinger, Ref.[15.4], had a great deal of trouble accepting that there was a completely different dynamic in operation on special occasions denoted as measurements – and so do most physicists. Decoherence theory provides a partial clarification by involving both the apparatus and the environment in the measurement process. The transformation of the system's pure quantum state into a mixture is clarified by attributing the transformation to the interaction with an environment whose precise quantum state is unknown. This provides a degree of insight into the mechanism of wavefunction collapse, but the outcome remains indeterminate.

15.2 What is a Measurement?

The transition from pure state (15.1.1) to a mixture, (15.1.2), can be written,

$$\hat{\rho} = |\psi\rangle\langle\psi| = \begin{pmatrix} |\alpha|^2 & \alpha\beta^* \\ \alpha^*\beta & |\beta|^2 \end{pmatrix} \rightarrow (\rho) = \begin{pmatrix} |\alpha|^2 & 0 \\ 0 & |\beta|^2 \end{pmatrix} \qquad (15.2.1)$$

However, this omits the measurement apparatus – without which we have no "read out" to tell us the outcome of the measurement. We saw back in chapter 3 how important it is to include the apparatus in the formalism – in that case because doing so explained why obtaining "which path" information destroys interference effects. Moreover, the apparatus must have two distinct states corresponding to the two possible outcomes of the measurement. These states will be denoted $|A(a)\rangle$ and $|A(b)\rangle$. The apparatus starts, prior to interaction with the system, in some 'resting' state $|A(0)\rangle$. To perform a measurement the apparatus must interact with the system, and the resulting evolution of the joint state occurs initially in accord with the Schrodinger equation. Hence,

$$|\psi\rangle|A(0)\rangle = (\alpha|a\rangle + \beta|b\rangle)|A(0)\rangle \rightarrow (\alpha|a\rangle|A(a)\rangle + \beta|b\rangle|A(b)\rangle) \quad (15.2.2)$$

where $\rightarrow$ denotes the unitary time evolution due to physical interaction under control of the Schrodinger equation. But (15.2.2) is still a pure quantum state, albeit a bipartite state. It has not yet decohered into a mixture. The complete measurement has not yet happened: (15.2.2) represents only the first step of the measurement. But note that the interaction of the apparatus with the system has turned the initial product state into an entangled state. The first step in a measurement is that the system becomes entangled with the apparatus.

15.3 How does Decoherence Happen?

According to decoherence theory, systems are never truly isolated in practice. There is always some interaction with their environment. Suppose the environment starts in some state $|e\rangle$. We are assuming that the environment will interact with the system and/or the apparatus (otherwise it can be of no relevance). But this means that the state of the environment will change due to this interaction. Thus, if the system were in state $|a\rangle$ the environment is taken to evolve into, say, state $|e(a)\rangle$, whereas if the system were in state $|b\rangle$ the environment evolves into $|e(b)\rangle$. Now these changes take place in accord with the Schrodinger equation, i.e., they are unitary in nature. So (15.2.2) can be written to include the environment thus,

$$|\psi\rangle|A(0)\rangle|e\rangle \rightarrow |\Psi\rangle = (\alpha|a\rangle|A(a)\rangle|e(a)\rangle + \beta|b\rangle|A(b)\rangle|e(b)\rangle) \quad (15.3.1)$$

The density matrix of the combined system-plus-environment is thus,

$$\hat{\rho} = |\Psi\rangle\langle\Psi| \quad (15.3.2)$$

How can we describe the system and apparatus alone, uncluttered by the environment? Because we are not bothered about the state of the environment, we wish to find the expectation value of $\hat{\rho}$ wrt all possible environment states. In other words we need to trace-out the environment, leaving the reduced density matrix for the combined system-apparatus, thus,

$$\hat{\rho}^{red} = \sum_e \langle e| \hat{\rho} |e \rangle \qquad (15.3.3)$$

where the sum extends over some orthonormal set of environment basis states $\{|e\rangle\}$. We shall now assume that the states $|e(a)\rangle$ and $|e(b)\rangle$ are orthogonal, so (15.3.3) gives,

$$\hat{\rho}^{red}_{S+A} = |\alpha|^2 |a\rangle\langle a| \otimes |A(a)\rangle\langle A(a)| + |\beta|^2 |b\rangle\langle b| \otimes |A(b)\rangle\langle A(b)| \qquad (15.3.4)$$

Recalling chapter 13 we understand what has happened here: because the initial state was entangled, the reduced density matrix is a mixture. Moreover, the result is subtly different from (15.2.1) because (15.3.4) includes the apparatus. (15.3.4) is actually an entangled mixture in which, if the apparatus reads "state a" then the system will indeed be in state a, and if the apparatus reads "state b" then the system will indeed be in state b. A further reduction of the density matrix of (15.3.4) by tracing out the apparatus yields the RHS of (15.2.1).

In the language of information and entropy, the evolution (15.3.1) consists of the system sharing information about its state with the environment – because its state becomes imprinted on the environment by virtue of the distinct environment states $|e(a)\rangle$ and $|e(b)\rangle$. But when we trace-out the environment we are wantonly throwing away this information. I say "wantonly", but in practice this information has become too difficult to identify once it has become diluted in a large and complex environment. So really we have no practical choice. This throwing away of information is reflected in the increased entropy of the system. Prior to its interaction with the environment the system was in a pure state and thus had zero entropy. Afterwards, the mixed state, (15.3.4), has entropy $-\{|\alpha|^2 log_2|\alpha|^2 + |\beta|^2 log_2|\beta|^2\} > 0$ (see chapter 12).

Another way of looking at this is to regard the environment as having carried out a measurement on the system+apparatus and hence "collapsed its wavefunction". The system is left in either state $|a\rangle$ or state $|b\rangle$, with the usual Born rule probabilities, in accord with (15.3.4). But in contrast to simply imposing collapse of the wavefunction by *fiat*, the decoherence viewpoint provides an explanation – of sorts – as to how (15.3.4) occurs. But it has shortcomings. Whilst "tracing out the environment" is the correct way to average over the unknown states of the environment, this "tracing out" does not obviously correspond to a physical process, and so "tracing out" does not actually provide a mechanistic explanation for state vector reduction. Also, decoherence does not tell us which measurement result, $|a\rangle$ or $|b\rangle$ actually occurs; it does not make QM measurements any more determinate.

The alert reader will have noted that we have treated the apparatus and the environment as if they were in pure quantum states. But these large, classical systems will actually be in mixed states. We committed the same over-simplification in chapters 3 and 6, though we had the excuse then that we had not yet developed the density matrix formalism for mixed states. An exercise for the reader is to reproduce the final mixed state of the system, post measurement, i.e., (15.1.2), using a mixed state formulation for the apparatus-environment.

15.4 The Pointer States

So far we have glossed over a key point. The initial quantum state, (15.1.1), could equally well be expressed in some other orthogonal basis, such as $|\psi\rangle = \alpha'|a'\rangle + \beta'|b'\rangle$. The unitary transformation which relates these two equivalent descriptions can be written,

$$\begin{pmatrix} |a'\rangle \\ |b'\rangle \end{pmatrix} = \begin{pmatrix} c & d \\ -d^* & c^* \end{pmatrix} \begin{pmatrix} |a\rangle \\ |b\rangle \end{pmatrix} \quad \text{and} \quad \begin{pmatrix} \alpha' \\ \beta' \end{pmatrix} = \begin{pmatrix} c^* & d^* \\ -d & c \end{pmatrix} \begin{pmatrix} \alpha \\ \beta \end{pmatrix} \qquad (15.4.1)$$

for any complex c, d obeying $|c|^2 + |d|^2 = 1$. But why should the apparatus-environment choose to measure the states $|a\rangle$ and $|b\rangle$ rather than $|a'\rangle$ and $|b'\rangle$? The point is that whilst $|\psi\rangle = \alpha|a\rangle + \beta|b\rangle = \alpha'|a'\rangle + \beta'|b'\rangle$ are equivalent descriptions of the pure state, the reduced density matrix derived from the transformed basis by tracing out the environment would be,

$$\rho''^{red}_{S+A} = |\alpha'|^2|a'\rangle\langle a'|\otimes|A(a')\rangle\langle A(a')| + |\beta'|^2|b'\rangle\langle b'|\otimes|A(b')\rangle\langle A(b')|$$

$$(15.4.2)$$

and this is *not* equal to (15.3.4). Further tracing out the apparatus, and assuming the states $|A(a')\rangle$ and $|A(b')\rangle$ are orthogonal, gives,

$$\hat{\rho}''^{red} = |\alpha'|^2|a'\rangle\langle a'| + |\beta'|^2|b'\rangle\langle b'| \qquad (15.4.3)$$

and this is not equal to (15.1.2). This may be established by substitution of (15.4.1) which shows that $\hat{\rho}''^{red}$ has off-diagonal elements in the original a, b basis, namely the coefficient of $|a\rangle\langle b|$ is $\{(|\alpha|^2 - |\beta|^2)(|c|^2 - |d|^2) +$

$4\Re(cd^*\alpha^*\beta)\}cd^*$. Whilst this is zero for the trivial transformations with either $c = 0$ or $d = 0$, it is clear that it will not be zero in general.

Consequently, if $\hat{\rho}^{red}$ is diagonal then in general the density matrix $\hat{\rho}'^{red}$ for other orthogonal bases will not be diagonal. The effect of the apparatus+environment on the system must be to select a particular basis out of the myriad possibilities to "decohere into". In effect the apparatus+environment break the unitary symmetry of the pure quantum state, i.e., that an arbitrary unitary transformation of basis provides as good a basis as any other, by selecting a preferred basis. This special basis is called the pointer basis. The pointer basis is determined by the combined system-apparatus-environment, not by the system alone in general.

Suppose the apparatus+environment selects (a', b') as the pointer basis, so that decoherence makes $\hat{\rho}'^{red}$ diagonal, but $\hat{\rho}^{red}$ is not diagonal. It must be that, in the (a, b) basis, either $|e(a)\rangle$ and $|e(b)\rangle$ are not orthogonal or that $|A(a)\rangle$ and $|A(b)\rangle$ are not orthogonal, because otherwise the diagonal nature of $\hat{\rho}^{red}$ would follow from the argument leading to (15.4.3). Conversely, in the pointer basis, $|e(a')\rangle$ and $|e(b')\rangle$ **are** orthogonal, as are $|A(a')\rangle$ and $|A(b')\rangle$. Note that decoherence **has** occurred despite the fact that $\hat{\rho}^{red}$ is not diagonal. Non-zero off-diagonal components in the density matrix do not immediately indicate whether a state is pure or mixed. In a general basis, as we saw in chapter 10, the signature that decoherence has occurred and that the state is no longer pure is that $\hat{\rho}^2 \neq \hat{\rho}$ or, equivalently, $Tr(\hat{\rho}^2) \neq 1$. In the eigen-basis of the density matrix, in which it is necessarily diagonal, this means more than one of its diagonal components (eigenvalues) is non-zero.

On reflection, this is just as it should be. A measuring device is specially designed to measure a particular quantity for a certain class of systems. It is the measuring device, and its immersion in the environment, which determine the preferred basis for the type of system in question and hence the particular basis in which the density matrix ends up diagonal. This preferred "pointer" basis is the eigen-basis of the observable being measured. Thus, a device to measure momentum decoheres the system into the momentum basis, and a device to measure the z-component of spin decoheres into the z-spin basis, not, say, the x-spin basis. This is the fundamental relationship between decoherence and measurement, and the reason why measurement results in the system's state vector reducing to one of the eigenstates of the observable being measured, now identified with the "pointer basis" defined by the combined system+apparatus+environment.

This selection of the pointer basis happens during the unitary evolution given by (15.3.1). It happens because the apparatus+environment selects those states of the system for which the $|e(a)\rangle$ and $|e(b)\rangle$ are orthogonal and $|A(a)\rangle$ and $|A(b)\rangle$ are orthogonal. This pointer basis has been described by Zurek, Refs.[15.1-3], as being "einselected" (environmental interaction selected) by the interaction between the system and the apparatus+environment.

Suppose the interaction Hamiltonian, $\hat{H}_{int}$, between the system and apparatus/environment is strong and dominates the dynamics. Then the eigenstates of $\hat{H}_{int}$ are approximately stationary states (see §2.4.1). Moreover, any observable which commutes with $\hat{H}_{int}$ is therefore approximately a constant of motion and shares the same eigenstates. Hence the eigenstates of $\hat{H}_{int}$ are the pointer states, see Ref.[15.1]. For example, suppose we measure the z-component of spin using a Stern-Gerlach apparatus. If the magnetic field and its gradient are in the z-direction, then the interaction Hamiltonian is,

$$\hat{H}_{int} = -mz\frac{\partial B_z}{\partial z}\sigma_z \qquad (15.4.4)$$

where m is the particle's magnetic moment and σ_z is the Pauli matrix for the z-spin (assuming a spin ½ particle). Hence the pointer states are the eigenstates of σ_z since this commutes with $\hat{H}_{int}$, whereas, say, σ_x and σ_y do not. The Stern-Gerlach apparatus does not measure the z-spin by accident. We have set it up to do so and it does so because its interaction Hamiltonian with the system is of the form (15.4.4), and that interaction is sufficiently strong to dictate the pointer basis.

15.5 What Decides the Pointer States? A Preview of Things to Come

We have seen how decoherence clarifies the measurement process, at least to some extent. But so far we have not given a complete account of how the all-important pointer states are determined. To measure an observable $\hat{Q}$ the apparatus must be contrived so that its pointer basis coincides with the basis formed by the eigenvectors of $\hat{Q}$. That much is clear, but when will that be the case? And what does decoherence theory have to say about the emergence of classical behaviour? In part this is already evident. Decoherence reduces a superposition to a classical mixture described by a diagonal density matrix – albeit in a particular preferred basis. But cannot this happen just as readily for a small system as a large system? Why do large things generally behave classically, rather than exhibiting interference effects, entanglement properties, etc? The answer lies in the readiness, and speed, with which larger objects (generally) decohere, together with the specific basis into which they decohere.

Decoherence is best illustrated by example. In chapter 21 a simple model will be used to illustrate that a key factor controlling how quickly a pure quantum state decoheres is the magnitude of the interaction energy with the apparatus-environment (V). When this is large compared with the ***spacing*** of the system's energy eigenstates, ΔE, then decoherence is rapid and the pointer states into which it decoheres approximate to eigenstates of the interaction Hamiltonian. Where this is proportional to, say, a spin operator, as in (15.4.4), the pointer states will approximate to the corresponding spin eigenstates. Commonly, the interaction Hamiltonian will depend upon some spatial separation, in which case the pointer states are likely to approximate to eigenstates of position (i.e., to be spatially localised states). The state may therefore take on the characteristics of a classical state.

In contrast, if the interaction of the system with the apparatus-environment is weak compared with the spacing of the system's energy levels $(V \ll \Delta E)$ then the pointer states will approximate to the energy eigenstates of the isolated system. This was first recognised by Paz and Zurek, Ref.[15.5], and is illustrated for a simple system in chapter 22. This has the implication that if a system is prepared in one of its energy eigenstates, and if the environmental interaction is weak in the sense that $V \ll \Delta E$, then the system will remain in an energy eigenstate indefinitely. This may mean that spatial delocalisation persists indefinitely. The state of electrons in atoms and molecules provide an example: atomic orbitals have well defined energy but are distributed spatially. Hence, in such cases, despite the conditions for decoherence being present, the state remains unchanged and specifically it retains its characteristically delocalised "quantum" nature indefinitely.

Where, then, does size come into it? Pure quantum states are associated with tiny sizes, usually involving only a small number of atoms or subatomic particles. Whilst, as we shall see in chapter 24, ever larger molecules have been coaxed to display quantum characteristics, e.g., interference, nevertheless it is true that it is far rarer, and far more difficult, to get larger systems to behave "quantumly". Why so? The relevance of the size of the object is that a large object has a very large number of degrees of freedom, N. The spacing of the system's energy eigenstates generally decreases steeply as N increases (e.g., $\Delta E \propto 1/N^2$ for some simple systems, see chapter 23). Consequently, large systems are doomed to decohere extremely quickly into classical mixed states because $V \gg \Delta E$ is almost inevitable. And, in as far as the interaction with the environment is spatially controlled, the pointer states into which the system decoheres will be spatially localised and hence will appear classical in nature.

Note the liberal use of phrases like "tends to" and "generally" in the above. The tendency of big things to be classical is not unavoidable. Examples of essentially quantum mechanical behaviour on a macroscopic scale, superconductivity and superfluidity, have been studied experimentally since the early decades of the twentieth century. The achievement of Bose-Einstein condensation is a more recent example. It is increasingly the case that quantum behaviour is being coaxed into the macroscopic range, even now at room temperature, e.g., energy eigenstates which are delocalised between two diamonds of macroscopic size (3mm), Ref.[15.6]. It is thought that coherent delocalised states play a part in the efficient collection of light energy in photosynthesis, Ref.[15.7]. Big things behaving quantumly are discussed further in chapter 24. But the exceptions prove the rule – and the rule is that large systems with many degrees of freedom generally decohere rapidly to classical behaviour.

15.6 Exercises for the Reader

Derive the post-measurement mixed state $\hat{\rho} = |\alpha|^2 |a\rangle\langle a| + |\beta|^2 |b\rangle\langle b|$ from a formulation using a mixed state for the environment-apparatus. For simplicity I shall not distinguish between the environment and the apparatus, but trace-out both together. Consider their joint state to be initially the mixed state $\sum_i p_i |A_i\rangle\langle A_i|$ so that the total state combined with the system is initially $\sum_i p_i |\psi\rangle |A_i\rangle\langle\psi|\langle A_i|$ where $|\psi\rangle = \alpha|a\rangle + \beta|b\rangle$. In general we may imagine each of the many $|A_i\rangle$-states to evolve into distinct, in fact orthogonal, states $|A_{i(a)}\rangle$ or $|A_{i(b)}\rangle$ according to which system state they coevolve alongside. The evolved state thus becomes,

$$\sum_i p_i \left(\alpha|a\rangle|A_{i(a)}\rangle + \beta|b\rangle|A_{i(b)}\rangle \right)(ditto)^+$$

Tracing out the apparatus-environment by taking $\sum_k \langle A_k| \ldots |A_k\rangle$ where the range of the subscript includes all the $|A_{i(a)}\rangle$ and $|A_{i(b)}\rangle$ states and gives,

$$\sum_{i,k} p_i \left(|\alpha|^2 |a\rangle\langle a| + |\beta|^2 |b\rangle\langle b| \right)\delta_{ik}$$

which is $\qquad \sum_i p_i \left(|\alpha|^2 |a\rangle\langle a| + |\beta|^2 |b\rangle\langle b| \right)$

which is $\qquad |\alpha|^2 |a\rangle\langle a| + |\beta|^2 |b\rangle\langle b| \qquad$ **QED**

15.7 References

[15.1] W.H.Zurek (1981) "Pointer basis of quantum apparatus: Into what mixture does the wave packet collapse?" Phys. Rev. D24, 1516. https://journals.aps.org/prd/abstract/10.1103/PhysRevD.24.1516

[15.2] W.H.Zurek (1982) "Environment-induced superselection rules", Phys. Rev. D26, 1862. http://conf.kias.re.kr/~brane/wc2005/topic/zurek1982.pdf

[15.3] W.H.Zurek (1991), "Decoherence and the Transition From Quantum to Classical. Phys. Today 44 (10): 36". See also an updated version on arXiv:quant-ph/0306072.

[15.4] E.Schrodinger, "Are There Quantum Jumps? Part 1", Br J Philos Sci (1952) **III** (10): 109-123. http://www.psiquadrat.de/downloads/schroedinger52_jumps1.pdf

[15.5] J.P.Paz and W.H.Zurek (1999), "Quantum limit of decoherence: Environment induced superselection of energy eigenstates", Phys.Rev.Lett.**82**:5181-5185,1999. https://arxiv.org/pdf/quant-ph/9811026.pdf

[15.6] K.C.Lee, et al (2011) "Entangling Macroscopic Diamonds at Room Temperature", Science **334** No.6060, 1253-1256. https://tinyurl.com/rkufsxm

[15.7] Panitchayangkoon G, et al (2010) Long-lived quantum coherence in photosynthetic complexes at physiological temperature *Proc. Natl. Acad. Sci.* **107** 12766. https://tinyurl.com/s8xkfqd

16

The Whole is Less Than the Sum of Its Parts

This chapter discusses some of the many inequalities which are obeyed by the von Neumann entropy when the system as a whole can be regarded as comprised of parts or elements. Whether the entropy of the whole is greater than, or less than, the sum of the entropies of its parts/elements depends upon whether the 'whole' is a mixture or a bipartite state.

16.1 Ensembles Versus Bipartite States

An ensemble is a mixture of states, each of which can be described by a density matrix, $\hat{\rho}_i$, and these component states may be pure or mixed themselves. Thus the ensemble has the density matrix $\hat{\rho} = \sum_i p_i \hat{\rho}_i$. I will refer to the component states, $\hat{\rho}_i$, as the elements of the ensemble. I avoid the word "parts" in this context because this is used in a different sense. For example, we may be dealing with bipartite states of two particles, A and B. Every one of the elements, $\hat{\rho}_i$, of the ensemble will consist of two such parts. Note that the decomposition of a given mixed state in an ensemble of elements is not unique. The combined $\hat{\rho}$ can be expressed as $\sum_i p_i \hat{\rho}_i$ in many (actually an infinite number) of ways. An arbitrary example for a two-dimensional Hilbert space is,

$$0.6 \begin{pmatrix} 0.7 & 0.1 \\ 0.1 & 0.3 \end{pmatrix} + 0.4 \begin{pmatrix} 0.5 & 0.3 \\ 0.3 & 0.5 \end{pmatrix} = 0.7 \begin{pmatrix} 0.6 & 0.1 \\ 0.1 & 0.4 \end{pmatrix} + 0.3 \begin{pmatrix} 0.666 & 0.366 \\ 0.366 & 0.333 \end{pmatrix}$$

$$= \begin{pmatrix} 0.62 & 0.18 \\ 0.18 & 0.38 \end{pmatrix} \tag{16.1.1}$$

In contrast, bipartite states involve superposition. That is, they involve adding state vectors not merely density matrices. Recapping what we have learnt in earlier chapters: the Hilbert space of pure bipartite states is the tensor product of the constituent spaces, $\mathcal{H} = \mathcal{H}_A \otimes \mathcal{H}_B$. The general pure bipartite state is therefore a linear superposition of product states, $\sum_{i,k} a_{ik} |i\rangle_A |k\rangle_B$. The general bipartite pure state cannot be written as a product state, $|\varphi\rangle_A |\omega\rangle_B$. Similarly, the density matrix of the most general mixed bipartite state cannot be written as a product of density matrices for the parts separately. If it can be written as $\hat{\rho}_{AB} = \hat{\rho}_A \otimes \hat{\rho}_B$ then it is said to be a separable mixed state. In general the density matrix for an arbitrary bipartite mixed state can be written,

$$\hat{\rho}_{AB} = \sum_{i,j,k,l} C_{ijkl} |i\rangle_A |j\rangle_B \langle k|_A \langle l|_B \tag{16.1.2}$$

Now we have reminded ourselves of these important distinctions we can start to discuss the relationship between the entropy of the 'whole' and the entropy of its 'parts' or 'elements', as appropriate. We shall denote by $H(p_i)$ the usual Shannon or Boltzmann expression for entropy in terms of independent probabilities, $\{p_i\}$,

$$H(p_i) = -\sum_i p_i \log_2 p_i \qquad (16.1.3)$$

where it is understood that $\sum_i p_i = 1$. Finally, recall that the von Neumann entropy of a quantum state expressed as a density matrix, $S_{vN}(\hat{\rho})$, is also given by (16.1.3) if the $\{p_i\}$ are interpreted as the eigenvalues of the density matrix. The entropy of a pure quantum state is therefore zero.

16.2 The Entropy of Ensembles

We have already met one of the inequalities involving ensembles in chapter 12, which is restated below,

$$S_{vN}\left(\sum_i p_i \hat{\rho}_i\right) \leq H(p_i) + \sum_i p_i S_{vN}(\hat{\rho}_i) \qquad (16.2.1)$$

You can find a proof in Ref.[16.1], §4.3.4). Here I am content to provide an example. When all the contributing states $\hat{\rho}_i$ are pure, the second term on the right is zero (because the von Neumann entropy of a pure state is zero). Hence the "ensemble inequality", (16.2.1), reduces in this case to $S_{vN}\left(\sum_i p_i \hat{\rho}_i\right) \leq H(p_i)$ and we gave an example of this in §12.3. But in the general case, when the $\hat{\rho}_i$ may be mixed states, the second term on the RHS of (16.2.1) is non-zero and is required to ensure that (16.2.1) is a valid upper bound to the ensemble entropy.

The two terms on the RHS of (16.2.1) have an intuitive interpretation. The second term is the weighted average of the entropies of the component states. The first term is an upper bound for the entropy associated with the mixing of those states. We might expect the second term alone to be a lower bound for the ensemble entropy. A heuristic argument for this goes as follows. Suppose we imagine physically compiling the ensemble by mixing together the individual states $\hat{\rho}_i$ in the relative proportions p_i. Knowing only the resulting $\hat{\rho}$ we cannot re-create the original set of states and their mixing ratios, $\{p_j, \hat{\rho}_j\}$, because there are many ways the ensemble $\hat{\rho}$ could have been created by mixing sub-states. Hence the mixing involves loss of information about how the mixture was composed and hence an increase of entropy. The argument is heuristic but the result is rigorous. For a proof see Ref.[16.1], §4.3.6. Combining this with (16.2.1) we get,

$$\sum_i p_i S_{vN}(\hat{\rho}_i) \le S_{vN}(\sum_i p_i \hat{\rho}_i) \le H(p_i) + \sum_i p_i S_{vN}(\hat{\rho}_i) \qquad (16.2.2)$$

The LHS (lower bound) in (16.2.2) is known as the "concavity" property of quantum entropy. It expresses the fact that the entropy of an ensemble exceeds the sum of the entropies of its component elements (when the latter are weighted by their contributing frequencies). Alternatively expressed, the ensemble can be decomposed into elemental states in more ways than the one that was used to construct it, and hence the ensemble has greater entropy: the whole is greater than the sum of its parts. Hence, where ensembles – which are mixtures – are concerned, the result conforms to the usual dictum.

The density matrix mixtures given in (16.1.1) provide examples of (16.2.2). A reader exercise is to confirm that both mixtures in (16.1.1) obey both inequalities in (16.2.2).

16.3 The Entropy of Bipartite States

If a bipartite state is separable, that is if its density matrix can be written as the tensor product of density matrices for its two parts separately, i.e., $\hat{\rho}_{AB} = \hat{\rho}_A \otimes \hat{\rho}_B$, then the combined entropy is just the sum of the entropies of its parts,

$$S_{vN}^{AB} = S_{vN}^{A} + S_{vN}^{B} \qquad (16.3.1)$$

This applies even if the separate states $\hat{\rho}_A$ and $\hat{\rho}_B$ are mixed states (i.e., their entropies are non-zero). This is the additive property of von Neumann entropy. It results from the uncorrelated nature of the A and B parts when the combined state is separable – together with the logarithm in the definition of entropy. It is the quantum mechanical generalisation of $log(N_A \times N_B) = log\,N_A + log\,N_B$, which would apply in the classical case of N_A equally likely possibilities for A and N_B possibilities for B, and the assumption that they are not correlated. The proof is simple,

$$S_{vN}^{AB} = -\sum_{i,j} p_i^A p_j^B\, log_2\, p_i^A\, p_j^B = -\sum_{i,j} p_i^A p_j^B \left(log_2\, p_i^A + log_2\, p_j^B\right)$$
$$= -\sum_{i,j} p_i^A p_j^B \left(log_2\, p_i^A\right) - \sum_{i,j} p_i^A p_j^B \left(log_2\, p_j^B\right)$$
$$= -\sum_i p_i^A \left(log_2\, p_i^A\right) - \sum_j p_j^B \left(log_2\, p_j^B\right) = S_{vN}^A + S_{vN}^B \qquad (16.3.2)$$

16.3.1 Subadditivity

In the general case the density matrix of a bipartite system is not a product state, $\hat{\rho}_{AB} \ne \hat{\rho}_A \otimes \hat{\rho}_B$, but of the general form (16.1.2). In this case, if we are given a bipartite state $\hat{\rho}_{AB}$, what do we mean by the individual states $\hat{\rho}_A$ and $\hat{\rho}_B$ of the sub-systems? We can only mean the reduced density matrix obtained by "tracing out" the other part (see Chapter 13),

$$\hat{\rho}_A = Tr_B(\hat{\rho}_{AB}) \equiv \sum_B \langle B|\hat{\rho}_{AB}|B \rangle \qquad (16.3.3\text{a})$$

$$\hat{\rho}_B = Tr_A(\hat{\rho}_{AB}) \equiv \sum_A \langle A|\hat{\rho}_{AB}|A \rangle \qquad (16.3.3\text{b})$$

A state which is not separable is entangled (and we will see in the next chapter that this holds for mixed, as well as pure, states). We have seen that entangled states imply a correlation between the parts (and a non-local correlation at that). Since additivity, (16.3.2), results from the absence of correlation between A and B, which maximises the number of joint possibilities, it is reasonable to expect that any correlation between the parts will reduce the entropy of the bipartite state. This is indeed the case and the following inequality, referred to as 'subadditivity', is the result,

$$S_{vN}^{AB} \leq S_{vN}^{A} + S_{vN}^{B} \qquad (16.3.4)$$

Equality holds iff $\hat{\rho}_{AB}$ is separable. The strict inequality applies when the state is entangled (i.e., not separable). Subadditivity, (16.3.4), is what the title of this chapter refers to: the whole is less than the sum of its parts. Note that 'the whole' refers to a bipartite state, not an ensemble (which, as we have seen, is just the opposite – the whole is greater than the sum of its parts).

Proof of the subadditivity inequality, (16.3.4), is as follows. The proof is an immediate consequence of Klein's Inequality, proof of which is a rather hard exercise for the reader (answer in §16.4.2). It states,

$$S_{vN}(\hat{\rho}) = -\sum_i p_i \, log_2 p_i \leq -Tr(\hat{\rho} \, log_2 \hat{\sigma}) \qquad (16.3.5)$$

where $\hat{\rho}$ and $\hat{\sigma}$ are any valid density matrices and the p_i are the eigenvalues of $\hat{\rho}$. Hence, if we define $\hat{\sigma}$ as the separable state $\hat{\rho}_A \otimes \hat{\rho}_B$, where the individual states are defined from the bipartite state via (16.3.3a,b), then,

$$-Tr(\hat{\rho} \, log_2 \hat{\sigma}) = -Tr(\hat{\rho}_{AB} \, log_2 \hat{\rho}_A \otimes \hat{\rho}_B)$$
$$= -Tr(\hat{\rho}_{AB}[log_2 \hat{\rho}_A + log_2 \hat{\rho}_B])$$
$$= -Tr(\hat{\rho}_A \, log_2 \hat{\rho}_A + \hat{\rho}_B \, log_2 \hat{\rho}_B) = S_{vN}^{A} + S_{vN}^{B}$$

And hence (16.3.5) yields (16.3.4). QED.

16.3.2 The Triangle Inequality

Just as we found both an upper bound and a lower bound for the ensemble entropy, (16.2.2), so also there is a lower bound to go with the upper bound of (16.3.4) for the bipartite entropy. And just as the upper bound is surprising from the classical point of view (i.e., that the whole is less than, rather than more than, the sum of its parts), so also the lower bound for bipartite states differs from its classical counterpart. For a classical system, the maximum correlation between components A and B would be if, given any state of A then the state of B was fully determined, or vice-versa. This leads to a minimum entropy in the classical case equal to the greater of that of system A and B. Hence, the classical entropy obeys $S_{AB} \geq Max(S_A, S_B)$. This corresponds to the very reasonable notion that, for a classical system, the combined system contains at least as much information as either of its components. It may be perfectly reasonable in the classical world, but in quantum mechanics it is wrong in general. In quantum theory, the von Neumann equivalent is the Araki-Lieb inequality,

$$S_{vN}^{AB} \geq |S_{vN}^A - S_{vN}^B| \tag{16.3.6}$$

This was first proved only in 1970 by Araki and Lieb, Ref.[16.2]. Together (16.3.4) and (16.3.6) constitute the triangle inequality,

$$|S_{vN}^A - S_{vN}^B| \leq S_{vN}^{AB} \leq S_{vN}^A + S_{vN}^B \tag{16.3.7}$$

The name derives from the fact that if the entropies of the individual sub-systems are regarded as the lengths of two sides of a triangle, the entropy of the combined system is restricted to the possible lengths of the third side.

There is a cunning proof of the Araki-Lieb lower bound, (16.3.6), which makes use of the concept of purification (see §13.7). We can introduce a notional third part, C, which is a purification of the bipartite state $\hat{\rho}_{AB}$. Thus the tripartite state ABC is pure. But from the subadditivity inequality we have,

$$S_{vN}^{AC} \leq S_{vN}^A + S_{vN}^C \tag{16.3.8}$$

But, since ABC is pure we have $S_{vN}^{AC} = S_{vN}^B$ and $S_{vN}^{AB} = S_{vN}^C$. (These follow from the fact that whichever part of a pure bipartite state is traced-out, the vN entropy of what remains is the same – see §13.9.2). Substituting these in (16.3.8) gives,

$$S_{vN}^B \leq S_{vN}^A + S_{vN}^{AB} \tag{16.3.9a}$$

By symmetry we have also,

$$S_{vN}^A \leq S_{vN}^B + S_{vN}^{AB} \tag{16.3.9b}$$

Together (16.3.9a,b) imply (16.3.6). QED.

The Araki-Lieb lower bound entropy is remarkable and displays essentially quantum features. The remarkable thing, of course, is that the lower bound can actually be achieved. A system with the Araki-Lieb lower bound entropy has less entropy (i.e. less information) than at least one of its components. So, not only is the whole less than the sum of its parts, but in some cases the whole can also be less than either individual part. Classically this is incomprehensible. In quantum mechanics it comes about due to entanglement.

It is not hard to produce a simple example which achieves the Araki-Lieb lower bound. The maximally entangled two-qubit state $[|\uparrow\rangle_A|\uparrow\rangle_B + |\downarrow\rangle_A|\downarrow\rangle_B]/\sqrt{2}$ achieves the lower bound entropy, namely zero as it is a pure state. Considered as separate sub-systems, each part can be in one of two states: the reduced density matrix of each part is $\frac{1}{2}\begin{pmatrix} 1 & 0 \\ 0 & 1 \end{pmatrix}$. So the entropy of each sub-system is 1. The Araki-Lieb inequality is respected because $0 \geq |1 - 1|$. Each sub-system has greater entropy (i.e., 1) than the combined system (i.e., 0). This non-classical behaviour of quantum entropy is responsible for the quantum weirdness of entangled states, such as the EPR paradox.

16.3.3 Other Multipartite Entropy Inequalities

There are many other inequalities obeyed by multipartite quantum entropies. Here I shall merely state a few, without proof or commentary. For both proofs and commentary see Refs.[16.1,2].

$$S_{vN}^{ABC} + S_{vN}^{B} \leq S_{vN}^{AB} + S_{vN}^{BC} \tag{16.3.10a}$$

This is known as 'strong subadditivity' and is far more of a challenge to prove. It may also be written in more symmetrical fashion as,

$$S_{vN}^{X \cup Y} + S_{vN}^{X \cap Y} \leq S_{vN}^{X} + S_{vN}^{Y} \tag{16.3.10b}$$

Closely associated inequalities are,

$$S_{vN}^{A} + S_{vN}^{C} \leq S_{vN}^{AB} + S_{vN}^{BC} \tag{16.3.11}$$

$$S_{vN}^{AB} \leq S_{vN}^{BC} + S_{vN}^{AC} \tag{16.3.12}$$

$$S_{vN}^{BC} \leq S_{vN}^{AB} + S_{vN}^{AC} \tag{16.3.13}$$

$$S_{vN}^{AC} \leq S_{vN}^{AB} + S_{vN}^{BC} \tag{16.3.14}$$

In every case S_{vN}^{AB} means the von Neumann entropy of density matrix $\hat{\rho}_{AB}$, and the density matrices are related by trace-out, e.g., $\hat{\rho}_{AB} = Tr_C(\hat{\rho}_{ABC})$, etc.

16.4 Reader Exercises

16.4.1: Confirm that (16.1.1) respects (16.2.2). Consider firstly the mixture,

$$0.6 \begin{pmatrix} 0.7 & 0.1 \\ 0.1 & 0.3 \end{pmatrix} + 0.4 \begin{pmatrix} 0.5 & 0.3 \\ 0.3 & 0.5 \end{pmatrix} = \begin{pmatrix} 0.62 & 0.18 \\ 0.18 & 0.38 \end{pmatrix} \qquad (16.4.1)$$

By solving the secular equations, the eigenvalues of the first matrix in (16.4.1) may be found to be 0.2764 and 0.7236, and hence its vN entropy is 0.8505. Similarly, the second matrix has eigenvalues 0.2 and 0.8, and so an entropy of 0.7219. Hence $\sum_i p_i S_{vN}(\hat{\rho}_i)$ is 0.7991. The eigenvalues of the ensemble density matrix are 0.2837 and 0.7163, giving an entropy of 0.8604 which exceeds $\sum_i p_i S_{vN}(\hat{\rho}_i)$ thus confirming the lower bound part of (16.2.2). The function $H(p_i)$ in this case, with mixing probabilities 0.6 and 0.4, is 0.9710. Hence the total $H(p_i) + \sum_i p_i S_{vN}(\hat{\rho}_i)$ is 1.7701 which exceeds the ensemble entropy, thus confirming the upper bound part of (16.2.2). QED.

Following the same process for the alternative decomposition of the same ensemble state given in (16.1.1), i.e.,

$$0.7 \begin{pmatrix} 0.6 & 0.1 \\ 0.1 & 0.4 \end{pmatrix} + 0.3 \begin{pmatrix} 0.666 & 0.366 \\ 0.366 & 0.333 \end{pmatrix} = \begin{pmatrix} 0.62 & 0.18 \\ 0.18 & 0.38 \end{pmatrix} \qquad (16.4.2)$$

By solving the secular equations, the eigenvalues of the first matrix in (16.4.1) may be found to be 0.3586 and 0.6414, and hence its vN entropy is 0.9415. Similarly, the second matrix has eigenvalues 0.0972 and 0.9028, and so an entropy of 0.4602. Hence $\sum_i p_i S_{vN}(\hat{\rho}_i)$ is 0.7971. The eigenvalues of the ensemble density matrix are 0.2837 and 0.7163, giving an entropy of 0.8604 which exceeds $\sum_i p_i S_{vN}(\hat{\rho}_i)$ thus confirming the lower bound part of (16.2.2). The function $H(p_i)$ in this case, with mixing probabilities 0.7 and 0.3, is 0.8813. Hence the total $H(p_i) + \sum_i p_i S_{vN}(\hat{\rho}_i)$ is 1.6784 which exceeds the ensemble entropy, thus confirming the upper bound part of (16.2.2). QED.

16.4.2: Prove Klein's Inequality, i.e., $S_{vN}(\hat{\rho}) \leq -Tr(\hat{\rho}log_2\hat{\sigma})$ where $\hat{\rho}$ and $\hat{\sigma}$ are density matrices. If the $\{p_i\}$ are the eigenvalues of a density matrix $\hat{\rho}$ and the $\{q_i\}$ are the eigenvalues of a density matrix $\hat{\sigma}$, then the theorem to be proved is $S_{vN}(\hat{\rho}) = H(p_i) = -\sum_i p_i \, log_2 p_i \leq -Tr(\hat{\rho}log_2\hat{\sigma})$ and we can write $\hat{\sigma} = \sum_k q_k |\varphi_k\rangle\langle\varphi_k|$ where the $|\varphi_k\rangle$ are the eigenstates of $\hat{\sigma}$. Similarly we can put $\hat{\rho} = \sum_i p_i |u_i\rangle\langle u_i|$ where the $|u_i\rangle$ are the eigenstates of $\hat{\rho}$. Recall that $log_2\hat{\sigma}$ can be identified with $\sum_k log_2 q_k |\varphi_k\rangle\langle\varphi_k|$. The on-diagonal components of $log_2\hat{\sigma}$ in the eigen-basis of $\hat{\rho}$ are thus,

$$[log_2\sigma]_{ii} = \sum_k A_{ik} log_2 q_k \qquad (16.4.3)$$

where $A_{ik} = \langle u_i |\varphi_k\rangle\langle\varphi_k|u_i\rangle$ which must be positive, real numbers. Note also that $\sum_i A_{ik} = \sum_k A_{ik} = 1$. We can therefore write,

$$-Tr(\hat{\rho}log_2\hat{\sigma}) = -\Sigma_i\, p_i\, \Sigma_k\, A_{ik}log_2q_k \qquad (16.4.4)$$

Now we appeal to the fact that log is a concave function (or -log is convex). This means that,

$$-\Sigma_k\, A_{ik}log_2q_k \geq -log_2(\Sigma_k\, A_{ik}q_k) \qquad (16.4.5)$$

So that $-Tr(\hat{\rho}log_2\hat{\sigma}) \geq -\Sigma_i\, p_i\, log_2(\Sigma_k\, A_{ik}q_k) = -\Sigma_i\, p_i\, log_2\tilde{q}_i$ where we have put $\tilde{q}_i = \Sigma_k\, A_{ik}q_k$. Thus it will suffice to show that,

$$-\Sigma_i\, p_i\, log_2\tilde{q}_i \geq -\Sigma_i\, p_i\, log_2p_i \qquad (16.4.6)$$

in order to establish the desired result. Since the original q_k and A_{ik} are arbitrary, (16.4.6) must be established for arbitrary $\tilde{q}_i$, other than the requirement that they are all positive or zero and sum to unity. This can be done by showing that the RHS of (16.4.6) is the absolute minimum of the LHS. The turning point(s) can be established by imposing the constraint $\Sigma_i\, \tilde{q}_i = 1$ via the method of Lagrange multipliers, i.e., by setting to zero the unconstrained derivatives of the function,

$$f(\tilde{q}_i) = -\Sigma_i\, p_i\, log_2\tilde{q}_i + \mu\,\Sigma_i\, \tilde{q}_i \qquad (16.4.7)$$

which gives, $\qquad \dfrac{\partial f}{\partial\tilde{q}_i} = -\dfrac{p_i}{\tilde{q}_i}log_2e + \mu = 0 \qquad (16.4.8)$

(16.4.8) means that the LHS of (16.4.6) has a turning point when the ratios $p_i/\tilde{q}_i$ are the same for all i, and since both the p_i and the $\tilde{q}_i$ sum to unity, this means that $\tilde{q}_i = p_i$ for all i at the turning point. Hence the RHS of (16.4.6) is the value of the LHS at its turning point. Moreover there is just that one turning point. The second derivative is,

$$\dfrac{\partial^2 f}{\partial\tilde{q}_i^{\,2}} = \dfrac{p_i}{\tilde{q}_i^{\,2}}log_2e \geq 0 \qquad (16.4.9)$$

Hence the turning point is a minimum. This establishes (16.4.6) which in turn therefore establishes the desired result, $-Tr(\hat{\rho}log_2\hat{\sigma}) \geq -\Sigma_i\, p_i\, log_2p_i$.

16.5 References

[16.1] M.A.Nielsen (1998), "Quantum Information Theory", Doctoral dissertation, The University of New Mexico, Albuquerque. arXiv: quant-ph/0011036.

[16.2] Huzihiro Araki and Elliott H. Lieb, *Entropy Inequalities*, Communications in Mathematical Physics, vol 18(2), 160-170 (1970). https://link.springer.com/article/10.1007/BF01646092

17

The Entanglement of Mixed States

In the case of pure states, chapter 13 defined what was meant by entanglement and also specified how the degree of entanglement could be uniquely quantified in a simple manner. In the case of mixed states, however, whilst it is simple to define what is meant by entanglement, identifying whether a given mixed state is entangled, and quantifying the degree of entanglement, is far more problematic. No effective algorithm of general applicability has yet been given to determine whether a given mixed state is entangled or to quantify its degree of entanglement. However, there are special cases in which such an algorithm is known, and some are presented in this chapter. A simple algorithm which is sufficient to positively identify mixed state entanglement is given, but a simple effective algorithm which provides a necessary criterion is lacking. In chapter 13 we showed that a randomly chosen pure bipartite state is 'essentially certain' to be entangled. Here we show the converse: if a very large number of pure bipartite states are chosen at random and combined with random probabilities in the form of a mixture, then the resulting mixed state will be separable (i.e., not entangled) in 'almost all' cases. This chapter is rather detailed and more challenging than most and may be omitted on first reading.

17.1 Defining Entanglement in Mixed States

What do we mean by an entangled mixed state? The most obvious definition is that an entangled mixture is a mixture of pure states at least one of which is entangled. However, this will not do at all. The reason is that it is possible to decompose a given density matrix which represents a mixed state into an explicit mixture of pure states in many different ways (as we saw in the last chapter). Moreover, it is possible that one decomposition may involve entangled pure states, but another decomposition involves only product states. It is possible for entanglement to effectively 'cancel out' between two entangled pure states which contribute to a mixture. A simple example is to consider a 50/50 mixture of the two maximally entangled 'Bell' states,

$$\frac{(|0\rangle|0\rangle+|1\rangle|1\rangle)}{\sqrt{2}} \text{ and } \frac{(|0\rangle|0\rangle-|1\rangle|1\rangle)}{\sqrt{2}} \tag{17.1.1}$$

Naively, one might imagine that this must constitute an entangled mixture. But the density matrix of this mixture is,

$$\hat{\rho} = \frac{1}{4}[(|0\rangle|0\rangle+|1\rangle|1\rangle)(\langle0|\langle0|+\langle1|\langle1|) + (|0\rangle|0\rangle-|1\rangle|1\rangle)(\langle0|\langle0|-\langle1|\langle1|)]$$
$$= \frac{1}{2}[|0\rangle|0\rangle\langle0|\langle0| + |1\rangle|1\rangle\langle1|\langle1|] \tag{17.1.2}$$

Hence, the same density matrix (and hence the same physical state) now appears as a 50/50 mixture of two product states, so we can hardly claim there is any entanglement. The 'entangled' parts of the Bell states, which give rise to the cross-products in the density matrix, have cancelled between the two contributing Bell states. Hence, this mixture is, in fact, separable: it has no entanglement.

We conclude that it not sufficient to identify entanglement in a mixture that the density matrix *can* be expressed as a mixture of pure states some or all of which are entangled. Rather we need to reverse the logic. If there is any decomposition of the density matrix into a mixture of pure states, all of which are product states (i.e., all of which have zero entanglement) then the mixed state is not entangled. Conversely, a mixed state is entangled if there is no decomposition into pure states, all of which are product states.

Expressing this differently for emphasis: the correct definition of entanglement in a mixed state is that it is the opposite of a separable mixed state, whilst a separable mixed state is defined as having a density matrix which *can* be decomposed into a mixture of pure product states. The most general pure product state is $|\psi\rangle_A|\varphi\rangle_B$, and hence the most general separable mixture can be written $\hat{\rho}_{AB} = \sum_{ik} p_{ik} |\psi_i\rangle_A|\varphi_k\rangle_B\langle\psi_i|_A\langle\varphi_k|_B$ where $\sum_{ik} p_{ik} = 1$.

The reader should now readily appreciate that entanglement in mixtures is more problematic than in pure states. For pure states entanglement can be recognised, and quantified, by an effective algorithm, namely the evaluation of the vN entropy of the reduced density matrix. For mixtures, in contrast, we are faced with the mathematically less tractable problem of proving the non-existence of any decomposition which could put the density matrix into the above stated separable form.

Let's express separability in mixed states in another way. The states $|\psi_i\rangle_A$ and $|\varphi_k\rangle_B$ are any normalised pure states of the A and B sub-systems, labelled by some index i, and need not be orthogonal. Also, there can be an arbitrary number of them in the mixture (not restricted by the Hilbert space dimension). Since we can write $\hat{\rho}_{Ai} = |\psi_i\rangle_A\langle\psi_i|_A$ for each contributing A sub-system state, and ditto for B, it follows that a separable mixture can equivalently be defined by the existence of a decomposition of the density matrix as a sum over direct products of density matrices of its parts, i.e., $\hat{\rho}_{AB} = \sum_{ik} p_{ik} \hat{\rho}_{Ai} \otimes \hat{\rho}_{Bk}$. This defines a separable mixture. Conversely, an entangled mixture is any mixed state which cannot be cast into this form. Recall that the range of the indices i and k are arbitrary (not restricted by the dimensionality of either Hilbert space).

Another example is a 50/50 mixture of two other Bell states,

$$|\psi_1\rangle = \frac{(|0\rangle|0\rangle+|1\rangle|1\rangle)}{\sqrt{2}} \quad \text{and} \quad |\psi_2\rangle = \frac{(|0\rangle|1\rangle+|1\rangle|0\rangle)}{\sqrt{2}} \qquad (17.1.3)$$

It is less obvious that this is separable because there are no cancellations in this case. However, writing single particle density matrices as,

$$\hat{\rho}_1 = \frac{1}{\sqrt{2}}(|0\rangle\langle0| + |1\rangle\langle1|) \quad \text{and} \quad \hat{\rho}_2 = \frac{1}{\sqrt{2}}(|0\rangle\langle1| + |1\rangle\langle0|) \qquad (17.1.4)$$

it is readily seen that a 50/50 mixture of the two Bell states in (17.1.3) can be written as,

$$\frac{1}{2}|\psi_1\rangle\langle\psi_1| + \frac{1}{2}|\psi_2\rangle\langle\psi_2| = \frac{1}{2}\hat{\rho}_{1A} \otimes \hat{\rho}_{1B} + \frac{1}{2}\hat{\rho}_{2A} \otimes \hat{\rho}_{2B} \qquad (17.1.5)$$

This may be expressed in matrix form, adopting the conventional 'computational basis' for the product states, defined by,

1	2	3	4
$|0\rangle_A|0\rangle_B$	$|0\rangle_A|1\rangle_B$	$|1\rangle_A|0\rangle_B$	$|1\rangle_A|1\rangle_B$

Hence,

$$\frac{1}{4}\begin{pmatrix} 1 & 0 & 0 & 1 \\ 0 & 1 & 1 & 0 \\ 0 & 1 & 1 & 0 \\ 1 & 0 & 0 & 1 \end{pmatrix} = \frac{1}{2}\begin{pmatrix} 1 & 0 \\ 0 & 1 \end{pmatrix} \otimes \frac{1}{2}\begin{pmatrix} 1 & 0 \\ 0 & 1 \end{pmatrix} + \frac{1}{2}\begin{pmatrix} 0 & 1 \\ 1 & 0 \end{pmatrix} \otimes \frac{1}{2}\begin{pmatrix} 0 & 1 \\ 1 & 0 \end{pmatrix} \qquad (17.1.6)$$

17.2 Detecting Entangled Mixed States

It is one thing to define what is meant by an entangled mixed state, but it is another to be able to decide, for a given bipartite mixed state, whether it is entangled. We need to know if its density matrix can be expressed in the form $\sum_{ik} p_{ik} \hat{\rho}_{Ai} \otimes \hat{\rho}_{Bk}$.

The most general pure bipartite state can be written in terms of some orthonormal basis as $\sum_{I,J} a_{I J}|I\rangle_A|J\rangle_B$, where $|I\rangle_A$ and $|J\rangle_B$ are a complete set spanning their respective Hilbert sub-spaces. It follows that the most general mixture can be written,

$$\sum_{I,J,I',J',k} p_k a_{kIJ} b^*_{kI'J'}|I\rangle_A|J\rangle_B \langle I'|_A\langle J'|_B \qquad (17.2.1)$$

where the range of k is arbitrary (simply the number of states which are being mixed). Thus, in the product basis, $|I\rangle_A|J\rangle_B$, the mixed state will generally have a filled density matrix with off-diagonal elements.

However, the mere fact that the density matrix has off-diagonal components when written in terms of some arbitrary basis does not mean that it is inseparable (entangled). A separable density matrix will have off-diagonal elements when expressed in an arbitrary basis, (17.1.6) is an example. Thus, if we write $|\psi_k\rangle = \sum a_{kl}|l\rangle_A$ and $|\varphi_k\rangle = \sum b_{kJ}|J\rangle_B$ then the most general **separable** density matrix is seen to be,

$$\sum_{I,J,I',J',k} p_k a_{kI} a_{kI'}^* b_{kJ} b_{kJ'}^* |I\rangle_A |J\rangle_B \langle I'|_A \langle J'|_B \qquad (17.2.2)$$

The reader should ensure that the difference between (17.2.1) and (17.2.2) is observed. Hence, the problem of the separability of an arbitrary density matrix, which can be specified by the coefficients $C_{IJI'J'}$, is whether these coefficients can be written in the form $C_{IJI'J'} = \sum_k p_k a_{kI} a_{kI'}^* b_{kJ} b_{kJ'}^*$. If such coefficients a_{kI}, b_{kJ}, p_k exist, then the mixture is separable. If it seems like a tricky problem to determine whether such a_{kI}, b_{kJ}, p_k exist, that's because it is. Before saying more about attempts at the general solution, firstly we remark on an obvious hypothesis which fails.

17.2.1 Entropy Does Not Identify Entanglement for Mixed States

Following chapter 16 we may be tempted to try to identify entanglement with any state whose entropy has the non-classical property of being less than that of its parts. Classically the entropy obeys $S^{AB} \geq Max(S^A, S^B)$ so one might guess that the signature of entanglement in a general mixed state might be that,

$$S_{vN}^{AB} < Max(S_{vN}^A, S_{vN}^B) \qquad (17.2.3)$$

Would it were that simple! It is not. The condition (17.2.3) is indeed a sufficient condition to ensure that the composite state is entangled. But it is not necessary. Entangled states exist which violate this inequality.

Nevertheless, (17.2.3) is important since it may well identify the majority of entangled states in a given situation, and it will tend to fail only for those states which are rather less entangled (assuming some quantification of the degree of entanglement which we have not yet defined).

Note that the condition (17.2.3) lies within the range permitted by the triangle inequality, (16.3.7), which requires for any bipartite state that

$$|S_{vN}^A - S_{vN}^B| \leq S_{vN}^{AB} \leq S_{vN}^A + S_{vN}^B \qquad (17.2.4)$$

Hence, the non-classical, definitely entangled, regime is,

$$|S_{vN}^A - S_{vN}^B| \leq S_{vN}^{AB} < MAX[S_{vN}^A, S_{vN}^B] \qquad (17.2.5)$$

and the remaining regime, which may or may not be entangled, is,

$$MAX[S_{vN}^A, S_{vN}^B] \leq S_{vN}^{AB} \leq S_{vN}^A + S_{vN}^B \tag{17.2.6}$$

The residual problem of identifying entangled mixed states therefore consists of sorting the entangled from the separable states in the regime (17.2.6).

17.2.2 Rigorous Theorems Identifying Entangled Mixtures

In (17.2.3) we have a sufficient, but not a necessary, condition for a mixture to be entangled. We are seeking a necessary and sufficient condition. Actually, we already have one. The most general density matrix for a bipartite system is,

$$\hat{\rho}_{AB} = \Sigma_{I,J,I',J'}\, C_{I\,J\,I'\,J'} |I\rangle_A |J\rangle_B \,\langle I'|_A \langle J'|_B \tag{17.2.7}$$

The necessary and sufficient condition that this is entangled is that there exists coefficients a_{kI}, b_{kJ}, p_k such that,

$$C_{I\,J\,I'\,J'} = \Sigma_k\, p_k a_{kI} a_{kI'}^* b_{kJ} b_{kJ'}^* \tag{17.2.8}$$

and also such that, $0 \leq p_k \leq 1$, $\Sigma_k p_k = 1$, and $\Sigma_I |a_{kI}|^2 = \Sigma_J |b_{kJ}|^2 = 1$, where k runs over the arbitrary number of terms in the mixture, and I and J run over the dimension of the A and B Hilbert sub-spaces. Unfortunately, whilst this is a correct necessary and sufficient condition, it does not provide an effective algorithm for deciding whether a given density matrix is separable, it merely begs the question because we have no simple procedure for determining if such a_{kI}, b_{kJ}, p_k exist.

However, Asher Peres, Ref.[17.1], devised a criterion which he showed to be another sufficient condition to ensure that a mixture is entangled. It was then shown by Horodecki et al, Ref.[17.2], that the Peres criterion is also a necessary condition for a mixture to be entangled in the special case of systems of two qubits, and also systems comprising a qubit and a three-state sub-system (a "qutrit"). The Peres criterion provides a simple effective algorithm, so a simple decision procedure exists for the 2 x 2 and 2 x 3 cases. Unfortunately, Horodecki et al, Ref.[17.2], also showed that the Peres criterion is *not* a necessary criterion for entanglement in Hilbert spaces of larger dimensions. (Warning: Refs.[17.1,2] apply the terms 'necessary' and 'sufficient' to the criterion for separability, which correspond respectively to 'sufficient' and 'necessary' as regards the criterion for entanglement).

Despite its limited range of applicability as a necessary and sufficient condition, the Peres criterion is important in practice, since two-qubit systems are of wide applicability. We have seen that the density matrix of the most general bipartite mixture can be written as (17.2.1) in terms of an orthonormal basis. Its components wrt this basis are obviously $\rho_{I\,J\,I'\,J'} = \sum_k p_k a_{kIJ} a^*_{kI'J'}$. Peres considers the matrix formed by "partial transposition", i.e. by transposing only the indices of the A sub-system but not the B subsystem. This defines $\sigma_{I'\,J\,IJ'} = \rho_{I\,J\,I'\,J'}$. If the original state is separable, the partially transposed matrix must necessarily have only non-negative eigenvalues. In other words, the existence of one or more negative eigenvalues of the partially transposed matrix is sufficient to imply that the original mixture is entangled. (This is far from obvious – see the original papers for further enlightenment).

17.3 Example: The Lowest Excited State of N Identical Bosons

In this section we give an example where the entropy criterion, (17.2.3), fails to identify an entangled state, but the Peres criterion successfully does so. A bipartite mixture can be formed by starting with a state of N identical bosons (see §24.2 for the meaning of "boson"), and then tracing-out $N - 2$ of the particles. This provides an example of a mixture which is clearly entangled and hence provides a test case for the various criteria.

Suppose the single particle states are, (i) a ground state $|0\rangle$, and, (ii) the lowest excited state $|e\rangle$. More highly excited states will not be relevant here. Consider initially the pure state of three of these identical bosons in which just one particle is in the first excited state, and the others are in the ground state. Because it is a state of identical bosons, it must be fully symmetric[3]. It is,

$$|\psi\rangle = \frac{1}{\sqrt{3}}\{|0\rangle_A|0\rangle_B|e\rangle_C + |0\rangle_A|e\rangle_B|0\rangle_C + |e\rangle_A|0\rangle_B|0\rangle_C\} \qquad (17.3.1)$$

The symmetrisation of the state leads to it being naturally entangled. Hence we see that,

$$\langle 0|_C\psi\rangle\langle\psi|0\rangle_C = (|0\rangle_A|e\rangle_B + |e\rangle_A|0\rangle_B)(\langle 0|_A\langle e|_B + \langle e|_A\langle 0|_B)/3 \qquad (17.3.2)$$

Similarly,

$$\langle e|_C\psi\rangle\langle\psi|e\rangle_C = (|0\rangle_A|0\rangle_B)(\langle 0|_A\langle 0|_B)/3 \qquad (17.3.3)$$

The reduced density matrix for the bipartite state consisting of particles A and B only is the sum of (17.3.2) and (17.3.3). It is clear that this mixture is entangled since the state (17.3.2) is clearly entangled, and the state (17.3.3)

---

[3] See any standard quantum text for the reason why, and also §24.2

involves only different states and hence cannot combine with (17.3.2) to 'cancel' the entanglement. For readers who are struggling to see why (17.3.2) cannot be written in the form $\sum_{ik} p_{ik} \hat{\rho}_{Ai} \otimes \hat{\rho}_{Bk}$ it may help to remark that a term like, say, $|0\rangle_A \langle e|_A$, is not a valid density matrix. Most obviously it is not Hermitian, but nor does it have eigenvalues which sum to unity.

We adopt the obvious matrix notation in which the indices are defined as representing the product states of A and B as follows,

1	2	3	4
$\|0\rangle_A\|0\rangle_B$	$\|0\rangle_A\|e\rangle_B$	$\|e\rangle_A\|0\rangle_B$	$\|e\rangle_A\|e\rangle_B$

The density matrix of the bipartite AB system is thus,

$$\hat{\rho}_{AB}^{(3)} = \frac{1}{3}\begin{pmatrix} 1 & 0 & 0 & 0 \\ 0 & 1 & 1 & 0 \\ 0 & 1 & 1 & 0 \\ 0 & 0 & 0 & 0 \end{pmatrix} \tag{17.3.4}$$

By considering the same development for 4, 5, etc particles it is readily seen that for N particles, tracing out all but particles A and B results in,

$$\hat{\rho}_{AB}^{(N)} = \frac{1}{N}\begin{pmatrix} N-2 & 0 & 0 & 0 \\ 0 & 1 & 1 & 0 \\ 0 & 1 & 1 & 0 \\ 0 & 0 & 0 & 0 \end{pmatrix} \tag{17.3.5}$$

Tracing out particle B to leave the density matrix for particle A alone yields,

$$\hat{\rho}_A^{(N)} = Tr_B\left(\hat{\rho}_{AB}^{(N)}\right) = \frac{N-1}{N}|0\rangle\langle 0| + \frac{1}{N}|e\rangle\langle e| \equiv \frac{1}{N}\begin{pmatrix} N-1 & 0 \\ 0 & 1 \end{pmatrix} \tag{17.3.6}$$

The vN entropy of the reduced (single particle) state is therefore,

$$S_{vN}\left(\hat{\rho}_A^{(N)}\right) = \frac{N-1}{N}\log_2 \frac{N}{N-1} + \frac{1}{N}\log_2 N \tag{17.3.7}$$

The vN entropy of the 2-particle mixture, (17.3.5), is simply found by diagonalising the matrix, which gives,

$$\hat{\rho}_{AB}^{(N)} \rightarrow \frac{1}{N}\begin{pmatrix} N-2 & 0 & 0 & 0 \\ 0 & 0 & 0 & 0 \\ 0 & 0 & 2 & 0 \\ 0 & 0 & 0 & 0 \end{pmatrix} \tag{17.3.8}$$

Hence,
$$S_{vN}\left(\hat{\rho}_{AB}^{(N)}\right) = \frac{N-2}{N} \log_2 \frac{N}{N-2} + \frac{2}{N} \log_2 \frac{N}{2} \qquad (17.3.9)$$

Thus we find that $S_{vN}\left(\hat{\rho}_{AB}^{(N)}\right) \geq S_{vN}\left(\hat{\rho}_{A}^{(N)}\right)$, as follows,

N	$S_{vN}\left(\hat{\rho}_{AB}^{(N)}\right)$	$S_{vN}\left(\hat{\rho}_{A}^{(N)}\right)$
3	0.9183	0.9183
4	1	0.8113
5	0.9710	0.7219
6	0.9183	0.6500
10	0.7219	0.4690
100	0.1414	0.0808

Hence, the entropy test fails to identify the state (17.3.5) as entangled because the inequality (17.2.3) is not met.

We now investigate the Peres criterion to see if that is successful in identifying the entanglement of the bipartite state, (17.3.5). To do so we first note that the partial transpose matrix is given in term of the matrix elements of (17.3.5) as follows,

$$\sigma = \begin{pmatrix} \rho_{11} & \rho_{12} & \rho_{31} & \rho_{32} \\ \rho_{21} & \rho_{22} & \rho_{41} & \rho_{42} \\ \rho_{13} & \rho_{14} & \rho_{33} & \rho_{34} \\ \rho_{23} & \rho_{24} & \rho_{43} & \rho_{44} \end{pmatrix} = \frac{1}{N} \begin{pmatrix} N-2 & 0 & 0 & 1 \\ 0 & 1 & 0 & 0 \\ 0 & 0 & 1 & 0 \\ 1 & 0 & 0 & 0 \end{pmatrix} \qquad (17.3.10)$$

For all $N \geq 3$ this matrix has one negative eigenvalue. The reader should be able to check that this negative eigenvalue is $\left(N - 2 - \sqrt{(N-2)^2 + 4}\right)/2N$. The existence of a negative eigenvalue is sufficient to guarantee that the 2-particle state, (17.3.5), is entangled – as indeed we already know. So the Peres criterion succeeds in identifying the entanglement in this case, whereas the entropy criterion, (17.2.3), fails to do so. In fact, the Peres criterion will always work if the dimensions are 2 x 2 or 2 x 3, but it is not guaranteed to work in higher dimensions.

17.4 Quantifying the Entanglement of Mixtures

In the case of pure states we were lucky: the means of identifying entanglement naturally also provided a means of quantifying it. Thus, for pure bipartite states, the reduced density matrix is mixed iff the initial state is entangled. Since the vN entropy is non-zero only for mixed states, it follows that a reduced density matrix with a non-zero entropy is the necessary and sufficient condition for entanglement in a bipartite pure state. The vN entropy then naturally provides a measure of the degree of entanglement.

Unfortunately, there is no single, universally agreed, measure of the degree of entanglement for mixed states. Moreover, some entanglement measures, whilst well-defined, are hard to evaluate in the general case. In this section I present one entanglement measure, the so-called "Entanglement of Formation" (EOF). One of the seminal papers in this area is Bennett et al, Ref.[17.3], which is strongly recommended.

The Entanglement of Formation of a mixed state is defined as the minimised weighted sum of the von Neumann entropies of the states contributing to the mixture. Thus, for a bipartite density matrix expressed in diagonal form as $\sum_i p_i |\psi_i\rangle\langle\psi_i|$, the EOF is defined by,

$$EoF = MIN[\sum_i p_i E(|\psi_i\rangle\langle\psi_i|)] \qquad (17.4.1)$$

where each of the states $|\psi_i\rangle$ is bipartite and the entanglement of these pure states, $E(|\psi_i\rangle\langle\psi_i|)$, is defined as the von Neumann entropy of the reduced state, $E(|\psi_i\rangle\langle\psi_i|) = S_{vN}[Tr_A(|\psi_i\rangle\langle\psi_i|)]$, noting that it does not matter which sub-system, A or B, is chosen to be traced out. For a pure state, the EoF reduces to the usual definition of the entanglement of a pure state, as it should. Bennett et al, Ref.[17.3], showed that the EoF can be interpreted as the number of pure Bell states (i.e., maximally entangled two-qubit states) which Alice and Bob would have to share in order to be able to prepare the mixed quantum state without the exchange of further quantum states. This justifies the name 'entanglement of formation'.

Unfortunately, the innocent seeming definition, (17.4.1), does not provide an effective algorithm. The difficulty lies in the meaning, and implementation, of the 'minimisation'. Given any state $|\psi_i\rangle$ there is a simple effective algorithm to find $E(|\psi_i\rangle\langle\psi_i|)$. The problem is that the set of states $\{|\psi_i\rangle\}$ such that $\hat{\rho} = \sum_i p_i |\psi_i\rangle\langle\psi_i|$ is not unique. The Entanglement of Formation is defined in (17.4.1) by minimising the result with respect to any decomposition, i.e., any $\{p_i\}$ and $\{|\psi_i\rangle\}$, for which $\hat{\rho} = \sum_i p_i |\psi_i\rangle\langle\psi_i|$ reproduces the density matrix in question. In general this is a difficult optimisation to perform. However, a simple algorithm for EoF exists for an arbitrary mixture of two qubits, which we will state explicitly in §17.4.1.

Before stating this algorithm, it is worth emphasising how crucial is the minimisation in (17.4.1). Choosing an inappropriate (i.e. non-optimal) basis, $\{|\psi_i\rangle\}$, will give a grossly misleading result. A simple example is provided by the 50/50 mixture of Bell states, $|\psi_\mp\rangle = (|0\rangle|0\rangle \mp |1\rangle|1\rangle)/\sqrt{2}$. We know that this is a separable mixture because the density matrix is $\hat{\rho} = [|0\rangle|0\rangle\langle0|\langle0| + |1\rangle|1\rangle\langle1|\langle1|]/2$. In the same basis, the reduced density matrices corresponding to each of the two terms separately are, respectively,

$$\begin{pmatrix} 1 & 0 \\ 0 & 0 \end{pmatrix} \text{ and } \begin{pmatrix} 0 & 0 \\ 0 & 1 \end{pmatrix} \tag{17.4.2}$$

Hence, in this basis we get,

$$\sum_i p_i E(|\psi_i\rangle\langle\psi_i|) = \frac{1}{2}E\begin{pmatrix} 1 & 0 \\ 0 & 0 \end{pmatrix} + \frac{1}{2}E\begin{pmatrix} 0 & 0 \\ 0 & 1 \end{pmatrix} = 0 \tag{17.4.3}$$

because both the reduced matrices have zero vN entropy. Hence $\sum_i p_i E(|\psi_i\rangle\langle\psi_i|)$ evaluated in this basis is zero. This is obviously the minimum and hence the EoF (which measures the entanglement) is zero, as it should be for a separable state. However, if we evaluated $\sum_i p_i E(|\psi_i\rangle\langle\psi_i|)$ in the Bell basis, the reduced density matrices for both $|\psi_+\rangle\langle\psi_+|$ and $|\psi_-\rangle\langle\psi_-|$ are,

$$\begin{pmatrix} 1/2 & 0 \\ 0 & 1/2 \end{pmatrix} \tag{17.4.4}$$

Hence, in this basis we get,

$$\sum_i p_i E(|\psi_i\rangle\langle\psi_i|) = \frac{1}{2}E\begin{pmatrix} 1/2 & 0 \\ 0 & 1/2 \end{pmatrix} + \frac{1}{2}E\begin{pmatrix} 1/2 & 0 \\ 0 & 1/2 \end{pmatrix} = 1 \tag{17.4.5}$$

So, evaluating in the wrong (i.e. non-optimal) basis gives a grossly misleading result. It suggests that the state is maximally entangled, when in fact it is separable. This is how crucial is the minimisation part of the definition in (17.4.1).

Incidentally, this example may give the false impression that the reduced density matrix is basis dependent. Of course that is not true. The reduced density matrix is $\begin{pmatrix} 1/2 & 0 \\ 0 & 1/2 \end{pmatrix}$ however it is computed. But (17.4.3) and (17.4.5) give different results because the density matrix is divided into two parts differently. As a reminder, the reduced density matrix is always basis independent because,

$$Tr_A(\hat{\rho}_{AB}) =$$
$$\sum_i \langle i|_A \hat{\rho}_{AB} |i\rangle_A = \sum_{i,I,J} U_{ij}^* \langle J|_A \hat{\rho}_{AB} |I\rangle_A U_{iI} = \sum_{I,J} \langle J|_A \hat{\rho}_{AB} |I\rangle_A \delta_{IJ} = \sum_I \langle I|_A \hat{\rho}_{AB} |I\rangle_A$$

The two bases, $|i\rangle_A$ and $|I\rangle_A$, used above do lead to the same reduced density matrix, but expressed as the sum of different terms, as follows,

$$\frac{1}{2}\begin{pmatrix} 1 & 0 \\ 0 & 0 \end{pmatrix} + \frac{1}{2}\begin{pmatrix} 0 & 0 \\ 0 & 1 \end{pmatrix} \text{ compared with } \frac{1}{2}\begin{pmatrix} 1/2 & 0 \\ 0 & 1/2 \end{pmatrix} + \frac{1}{2}\begin{pmatrix} 1/2 & 0 \\ 0 & 1/2 \end{pmatrix}$$

Thus, whilst both give the same total reduced density matrix, the expression for the EoF involves $\sum_i p_i E(|\psi_i\rangle\langle\psi_i|)$, which requires evaluating the S_{vN} for each term in the sum separately. This is how the difference arises.

17.4.1 Entanglement of Formation for Two-Qubit Mixed States

The Entanglement of Formation is a good entanglement measure, it suffers only from having no simple effective algorithm for its evaluation in the general case. However, for bipartite states of two qubits there *is* a simple effective algorithm for EoF, first stated by Hill and Wootters, Ref.[17.4]. The procedure is stated without proof as follows, noting that we are dealing with a 2×2 bipartite Hilbert space, so the density matrices are 4 x 4.

$$EoF = H\left(\frac{1}{2}\left[1 + \sqrt{1 - C^2}\right]\right) \tag{17.4.6}$$

Where H is the usual entropy function,

$$H(x) = -x \log_2 x - (1 - x) \log_2(1 - x) \tag{17.4.7}$$

and C (called the 'Concurrence') is,

$$C = MAX(0, \xi) \tag{17.4.8}$$

where, $\qquad \xi = \sqrt{\lambda_1} - \sqrt{\lambda_2} - \sqrt{\lambda_3} - \sqrt{\lambda_4} \tag{17.4.9}$

and the λ_i are the eigenvalues, in descending order, of the matrix,

$$\rho\Omega\rho^*\Omega \tag{17.4.10}$$

where ρ is the 4 x 4 density matrix in question and,

$$\Omega = \begin{pmatrix} 0 & 0 & 0 & -1 \\ 0 & 0 & 1 & 0 \\ 0 & 1 & 0 & 0 \\ -1 & 0 & 0 & 0 \end{pmatrix} \tag{17.4.11}$$

Hence, if ξ is negative or zero, the EOF achieves its minimum value of zero, indicating separability.

Exercises for the reader are to deploy the above algorithm for,

i. the density matrix (17.1.6) to confirm it indicates separability (EOF = 0).

ii. The density matrix (17.3.5) for various values of N to confirm it is entangled, and to evaluate the EOF.

17.5 Entanglement of Random Mixtures

Recall that in chapter 13 we showed that a randomly chosen pure bipartite state is 'essentially certain' to be entangled, but it was also stated without proof that the opposite was true for mixed states. If a very large number, N, of pure bipartite states are chosen at random and combined with random probabilities in the form of a mixture, then the resulting mixed state will be separable (not entangled) in 'almost all' cases. To be more precise that should read 'arbitrarily close to separable' in the sense that the EOF tends to zero as $N \to \infty$. This is proved algebraically in this section for an arbitrary Hilbert space, whilst the following section illustrates it using numerical simulation for the Hilbert space of two qubits.

Consider a set of arbitrary pure states of a bipartite system, $\{|\psi_i\rangle\}$, where i runs from 1 to some arbitrary number of contributing states in the mixture, N. Each state can be written as $|\psi_i\rangle = \sum_{j,k}^{D1,D2} C_{ijk} |j\rangle |k\rangle$, where $D1$ and $D2$ are the dimensions of the sub-system Hilbert spaces. We can assume the $|\psi_i\rangle$ are normalised but they need not be orthogonal, and N may exceed the dimensionality of the product space, $D1 \times D2$, which means that the $|\psi_i\rangle$ may not be linearly independent either. In fact we can take N to be as large as we like. If these states contribute to a mixture with respective probabilities p_i, the mixed state is,

$$\hat{\rho} = \sum_{i,j,k,m,n} p_i \, C_{ijk} C^*_{imn} |j\rangle|k\rangle\langle m|\langle n| \tag{17.5.1}$$

where i runs from 1 to N, and j and m run from 1 to $D1$, and k and n from 1 to $D2$. This can be simplified as,

$$\hat{\rho} = \sum_{j,k,m,n} D_{jkmn} \, |j\rangle|k\rangle\langle m|\langle n| \tag{17.5.2}$$

where, $D_{jkmn} = \sum_i p_i \, C_{ijk} C^*_{imn}$. The coefficients C_{ijk} are arbitrary except for the requirement that $\sum_{j,k} |C_{ijk}|^2 = 1$, for all i. The probabilities, p_i, are also arbitrary except for requiring that $\sum_i p_i = 1$ and $p_i \geq 0$. Consequently, for given values of i, j, k, m, n the two numbers C_{ijk} and C^*_{imn} will be uncorrelated, except when $j = m$ and $k = n$. Consequently, provided that there are enough randomly chosen terms in the sum over i, it is to be expected that,

$$D_{jkmn} = \sum_i p_i \, C_{ijk} C^*_{imn} \propto \delta_{jm} \delta_{kn} \tag{17.5.3}$$

Of course, this is only true in the sense of being an expectation value. For any finite sum, the value of D_{jkmn} for $j \neq m$ or $k \neq n$ will be non-zero. But its typical magnitude is expected to reduce as the number of states contributing to the mixture increases (proportional to $1/\sqrt{N}$). Substituting (17.5.3) into (17.5.2) shows that the mixture is expected to tend to become separable, i.e.,

$$\hat{\rho} \rightarrow \Sigma_{i,j,k}\, p_i\, |C_{ijk}|^2 |j\rangle|k\rangle\langle j|\langle k| \tag{17.5.4}$$

(recalling that any mixture of pure product states is a separable mixed state)

17.6 Entanglement of Random Mixtures (Numerical Exploration)

In this section random mixtures of N random two-qubit pure states are explored numerically. For each value of N considered, a large number, N_t, of such random mixtures were generated. The simulations reported used $N_t = 1000$. For each trial the vN entropy of the mixture and of its reduced density matrices were found, as was the Entanglement of Formation (using the algorithm of §17.4.1). It was convenient to use a variant of the usual definition of the vN entropy in which the dimension of the Hilbert space was used as the base of the logarithm, i.e.,

$$S_{vN} = -\Sigma_i\, p_i\, log_D\, p_i \tag{17.6.1}$$

The subadditivity inequality of (16.3.4) for a two-qubit Hilbert space therefore becomes $S_{vN}^{AB} \leq (S_{vN}^A + S_{vN}^B)/2$. These variants should be born in mind when interpreting the graphical results of the simulations.

The outcomes of the random simulations for $N = 1, 2, 3, 4, 5, 6, 7$ and 10 are plotted in Figures 17.1a,b as the average of the reduced state vN entropies, $S_{vN}^{av} = (S_{vN}^A + S_{vN}^B)/2$, against the combined state vN entropy, S_{vN}^{AB}. All the data necessarily lie, therefore, above the line of equality, in conformance with the subadditivity inequality. The Figures are colour coded to indicate the Entanglement of Formation. Five ranges of EoF are distinguished, from 0 to 0.2 (green) to 0.8 to 1 (mustard). Separable states are in deep blue.

For $N = 1$ the vN entropy of the combined state is zero, because these are pure states. The reduced states, however, are mixed and can have average entropies, S_{vN}^{av}, anywhere from 0 to 1. The EoF equals S_{vN}^{av} in this case, and no states are separable.

For $N = 2$ still no separable states emerge from the random simulations, all the states are entangled with EOF almost uniformly spread across the range $(0,1)$. Not to mislead the reader, we can – obviously – devise separable states with $N = 1$ and $N = 2$, but these are vanishingly unlikely to arise in random samples (they are "of zero measure").

Separable states first occur in the random simulations for $N = 3$, and become increasingly common as N increases further. As N increases, the proportion of states which have small vN entropies (combined or reduced) diminishes rapidly. At the same time, the proportion of states with large EOF also diminishes rapidly. For $N = 10$ only one state arose in the random sample with EoF > 0.2. Hence, as $N \to \infty$, we have S_{vN}^{AB}, S_{vN}^{A} and $S_{vN}^{B} \to 1$, whereas EoF$\to 0$, and the plots become increasingly dominated by the deep blue points which cluster at the top right hand corner. In other words, for a sufficiently large number of states contributing to the mixture, the resulting mixed state becomes separable. This is just as deduced algebraically in the previous section.

The number of separable states out of 1000 trials for N from 1 to 20 is plotted in Figure 17.2. For $N = 20$ about 99.7% of trials are separable.

17.7 Other Mixed State Entanglement Measures

Various alternative measures of the entanglement of mixed states have been proposed, motivated largely by the difficulty of evaluating the EOF in the general case, but also by shortcomings in its relevance in certain circumstances. One of these is 'distillable entanglement'. Entanglement of formation (EoF) measures how many pure Bell states are needed to prepare the mixed state in question. But it may be more indicative of the usefulness of a mixed state for physical purposes (e.g., quantum cryptography) to consider the number of pure Bell states which can be **made from** the given mixed state, rather than the reverse. This is 'distillable entanglement'. Other measures have been proposed which are specific to continuous variable states. See Refs.[17.3,5] for more details.

Figure 17.1a: Random mixtures of N two-qubit states ($N = 1, 2, 3, 4$). $(S_{vN}^A + S_{vN}^B)/2$ versus S_{vN}^{AB}. Colours indicate Entanglement (EOF).

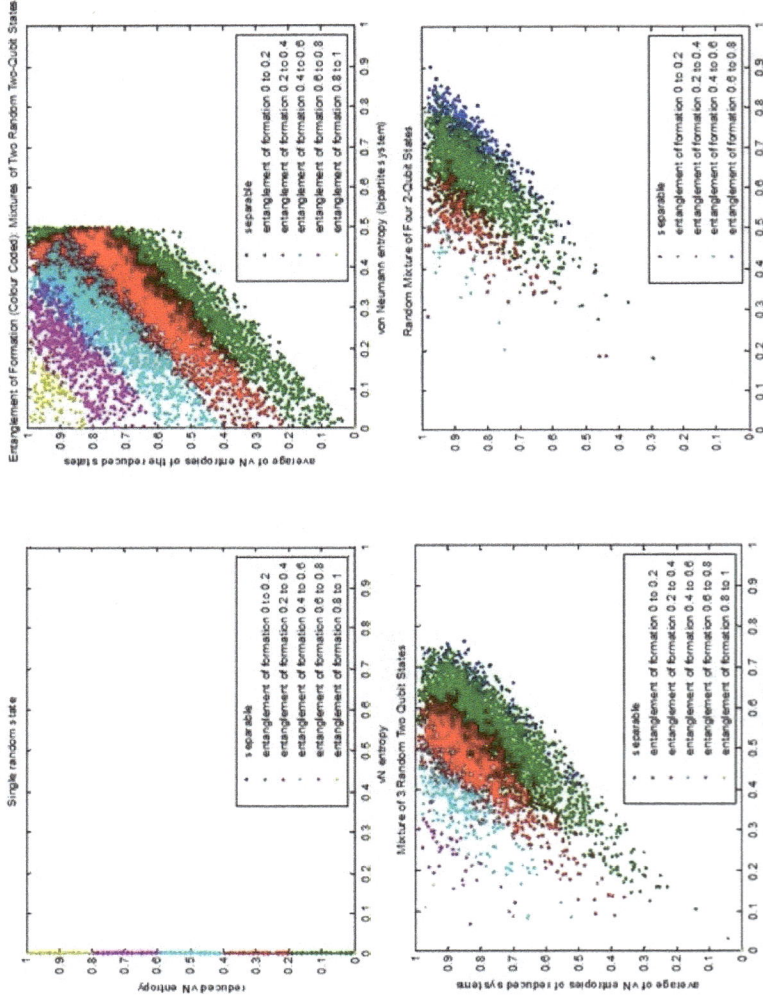

Figure 17.1b: Random mixtures of N two-qubit states ($N = 5, 6, 7, 10$), $(S_{vN}^A + S_{vN}^B)/2$ versus S_{vN}^{AB}. Colours indicate Entanglement (EOF).

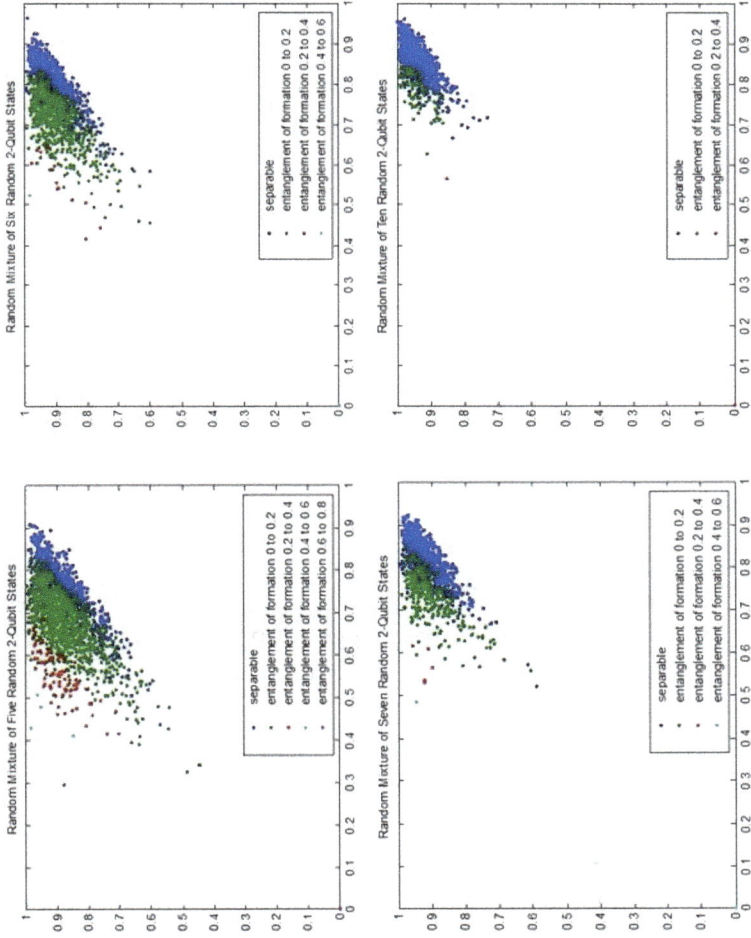

Figure 17.2: *Number of separable states out of 1000 trials versus* **N**

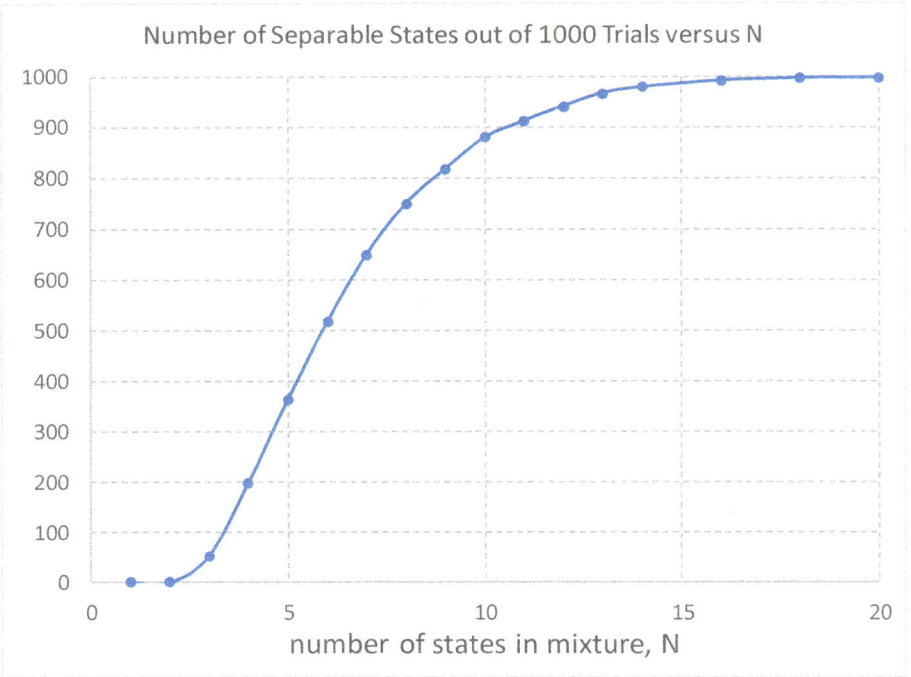

Number of Separable States out of 1000 Trials versus N

number of states in mixture, N

17.8 Exercises for the Reader

17.8.1: Use the Hill-Wootters algorithm for example 2 x 2 density matrices. Cheat – use the matrix manipulation software of your choice. For density matrix (17.1.6) the matrix $\rho\Omega\rho^*\Omega$, (17.4.10), is found to have eigenvalues 0.25, 0.25, 0, 0. Hence C is zero and the EOF equals $H(1) = 0$ and so separability is confirmed. For density matrix (17.3.5) the matrix $\rho\Omega\rho^*\Omega$, (17.4.10), is found to have three zero eigenvalues and one positive eigenvalue. Hence $C > 0$ and the EOF is evaluated as $H(z)$ where $0 < z < 1$ and so is non-zero, indicating entanglement. Specifically, for $N = 3, 4, 5, 10, 100$ the EOF evaluates to 0.550, 0.355, 0.250, 0.0815 and 0.0051 respectively, indicating reducing entanglement for larger N. Physically, this means that, despite pure states of identical bosons being maximally entangled by virtue of their necessary symmetry under interchange, the entanglement between any pair of such bosons – obtained by tracing out the states of the remainder – tends to zero for a large number of bosons.

17.9 References

[17.1] Asher Perez, "Separability Criterion for Density Matrices", June 1996, Phys.Rev.Lett.77:1413-1415,1996
https://arxiv.org/abs/quant-ph/9604005v2

[17.2] Michal Horodecki, Pawel Horodecki and Ryszard Horodecki, "Separability of mixed states: necessary and sufficient conditions", https://arxiv.org/abs/quant-ph/9605038

[17.3] Charles Bennett, David DiVincenzo, John Smolin and William Wootters "Mixed State Entanglement and Quantum Error Correction", Phys. Rev. A 54, 3824 (1996). https://arxiv.org/abs/quant-ph/9604024

[17.4] Scott Hill and William Wootters, "Entanglement of a Pair of Quantum Bits", Phys.Rev.Lett.78:5022-5025(1997)https://arxiv.org/abs/quant-ph/9703041v2

[17.5] Michal Horodecki, Pawel Horodecki and Ryszard Horodecki, "Mixed-state entanglement and quantum communication", July 2018 https://arxiv.org/pdf/quant-ph/0109124.pdf

18

Entangled Interference

Attempting interference experiments with entangled particles provides a rich source of "paradox", which is to say, confusion. Do interference experiments with entangled particles provide a means of faster-than-light communication? No.

18.1 The False Paradox

What happens when interference experiments are carried out using pairs of entangled particles? There is a greatly illuminating conundrum regarding whether a measurement on one of the entangled pair will destroy the interference observed in the other. Since the entanglement means that a measurement carried out on one of the pair is effectively also a measurement on the other, this should destroy the interference. But this appears to create the opportunity for faster-than-light (FTL) communication. This is because we could, it seems, observe the disappearance of an interference pattern as a consequence of an action at a spacelike separation. Modulating such remote actions on a stream of entangled particle-pairs would therefore allow a signal to appear instantaneously at a remote location in the form of the repeated appearance and disappearance of the interference. Of course this cannot be so if we believe transluminal communication to be impossible, but why does this scenario not work? This chapter will explain.

But let's first dispose of a cheap, and fallacious, way of escape. The lazy assertion that 'entangled particles cannot cause interference' clearly cannot always be true because, in the real world, particles will always be entangled with things you know nothing of - by virtue of their history. So, if it were true, no interference would ever be observed. More accurate guidance is that entanglement between specified degrees of freedom will prevent interference between these same degrees of freedom (as we shall see). But the fact that my particle happens to be entangled with another particle, currently perhaps in the vicinity of Alpha Centauri, will not prevent it creating interference fringes when incident upon a double slit screen here on Earth.

18.2 Entangled Photons Input to Mach-Zehnder Interferometers

Suppose we have a means of generating pairs of entangled photons, moving in opposite directions. Further suppose that these photons can be emitted only along either the x-axis or the y-axis, and these occur with equal probability from the same precursor state (and hence are a coherent superposition). This can be represented diagrammatically by,

Figure 18.1: *Entangled photons emitted as coherent pairs L+R or U+D*

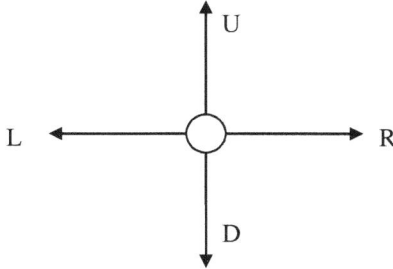

The corresponding quantum state is, in obvious notation,

$$|\psi\rangle = \tfrac{1}{\sqrt{2}}\,(|R\rangle|L\rangle + |U\rangle|D\rangle) \qquad (18.2.1)$$

This means that either the two photons emerge in the left and right directions, or in the up and down directions. I don't know whether such an arrangement would be easy to achieve experimentally, but this does not matter to the principle being illustrated. The two photons are entangled since (18.2.1) is not a product state. Now suppose we erect two Mach-Zehnder interferometers around these photons, as shown in Figure 18.2.

Figure 18.2: *Entangled photon pairs entering two interferometers*

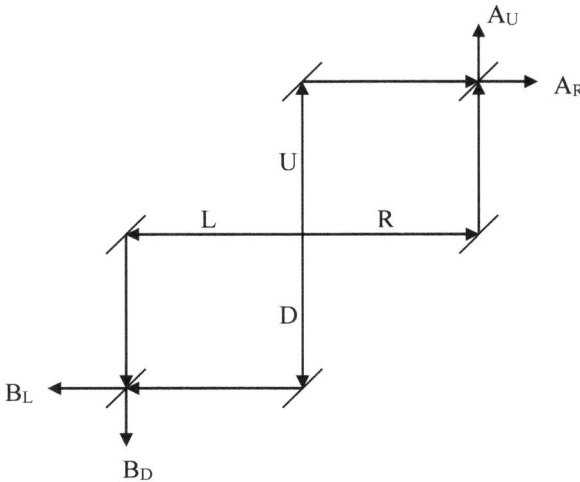

The first mirror that each photon encounters is a full mirror. The second mirror is half-silvered (i.e., a beam splitter). We refer to the top right interferometer as the A device (or the A photon), and the bottom left as B. The labels A_U, A_R, B_L, B_D are four photon detectors in an obvious notation.

The mirrors and beam splitters consist of a flat piece of glass onto one side of which has been deposited a thin layer of silver, or similar reflective metal. The terms 'silvered' and 'half-silvered' refer to different thickness of metal film, the first being sufficient to prevent any transmission of light through the mirror, whereas the latter is calibrated to allow about as much transmission as reflection. The phase shift caused by such mirrors differs according to whether the incoming beam is incident on the silvered surface (the 'front' of the mirror) or the glass surface (the 'back' of the mirror). (Note that mirrors in domestic use usually have the 'back' (the glass) facing forwards, while the 'front' is generally covered with something opaque to protect the silver film). The phase changes caused by the mirrors and beam splitters are,

- Reflection off the front (silvered) face causes a factor of -1;

- Transmission causes a factor of $e^{i\Delta}$ where Δ depends upon the wavelength and the thickness and refractive index of the glass substrate;

- Reflection from the back of the mirror causes a factor $e^{2i\Delta}$ due to the passage twice through the glass substrate (the reflection itself causing no phase change).

Derivation of these phase change rules, which differ from those for plane glass plates given in chapter 6, is an exercise for the reader, see §18.5 for my answer. In the analysis below the phase change Δ will be assumed the same for all four mirrors. This implies that the mirrors all have the same thickness to optical precision, i.e., to an accuracy much less than the wavelength of light. In practice this is improbable but this is not really important to the functioning of the interferometer. Some adjustment (calibration) of the interferometer is merely required prior to use to compensate for this practical limitation.

The four mirrors have their silvered surface (their "front") facing the beam, and hence all cause a phase change factor of -1. Since all beam paths are equally phase shifted by the mirrors, the mirrors are not significant to the interference. They are present simply to bring the beams together again. The two beam splitters both have their silvered surface (their "front") on the lower surface. Different beam paths will be phase-shifted differently by the beam splitters, and this is the cause of the interference.

Using these rules we can work out what the quantum state will be for photons entering any of the four detectors. Before we do this, though, pause to consider what we might expect based on our knowledge of Mach-Zehnder interferometers, and in the context of the conundrum of §18.1.

Consider the A device. Since the initial state is a superposition of U and R, we would expect that all the photons would emerge into just one of the detectors A_U and A_R, and none in the other (providing that we have tuned the set-up appropriately). For an explanation see chapter 6. This behaviour is the manifestation of interference in this device. Let's say all the photons would be expected in detector A_R and none in A_U (as would be deduced from the above rules if $e^{i\Delta} = -1$). Now suppose we make a "which path" measurement in device B. We can make the arms of the B interferometer very long so as to ensure that this measurement is at a spacelike separation from the A-detectors when they detect the other particle of the entangled pair. We now know which path contains the photon in the B device, and hence we also know which path contains the photon in the A device. But this must destroy the interference in the A device (as well as in the B device) and this will be apparent because the A_U detector will start registering photons. So we have achieved faster than light communication!

Of course we have not really achieved FTL communication. It turns out that where we went wrong in this analysis is in assuming that the entangled photons in this set-up behave in the same way as single, un-entangled photons in a Mach-Zehnder interferometer. They do not, as we will now show.

Consider firstly photons (potentially) entering detector A_U. We call their state $|\psi: A_U\rangle$. This state can be arrived at via either of paths R or U. Following the phase factors at the mirrors and beam splitter we find that the input photon state, (18.2.1), becomes,

$$|\psi\rangle \rightarrow |\psi: A_U\rangle = \tfrac{1}{2}(-e^{i\Delta}|A_U\rangle|L\rangle - e^{2i\Delta}|A_U\rangle|D\rangle) = -\frac{e^{i\Delta}}{2}|A_U\rangle(|L\rangle + e^{i\Delta}|D\rangle)$$

$$(18.2.2)$$

Similarly the state entering detector A_R is,

$$|\psi\rangle \rightarrow |\psi: A_R\rangle = \tfrac{1}{2}(-1 \times -1|A_R\rangle|L\rangle - e^{i\Delta}|A_R\rangle|D\rangle) = \tfrac{1}{2}|A_R\rangle(|L\rangle - e^{i\Delta}|D\rangle)$$

$$(18.2.3)$$

Assuming orthogonality, $\langle L|D\rangle = 0$, we see that $|\psi: A_U\rangle$ and $|\psi: A_R\rangle$ are also orthogonal. The square moduli of both these states is ½. Hence half the photons enter each detector. There is no interference (which would have been characterised by all photons entering just one detector). This exposes the error in the initial analysis: the entangled photons do not exhibit interference in this set-up. So there is no FTL communication.

It is worth pausing to appreciate what causes the different outcome compared to an interferometer with a single, unentangled, photon (as analysed in chapter 6). The difference is that the quantum states entering the A-detectors, (18.2.2,3), are still entangled with the other photon. And it is the presence of the state of the other photon, i.e., $|L\rangle$ or $|D\rangle$, which makes the cross-product terms in the square moduli of these states be zero, due to their orthogonality. So, this shows clearly that it is the entanglement itself which prevents interference. So far, so good.

But there is another sense in which interference ***does*** occur in this experimental arrangement – but without any possibility of FTL communication. This involves correlated behaviour between the photons in the A and B interferometers. To see this, consider how the state of photons entering detector A_U, (18.2.2), is modified by the other photon's passage through the B interferometer. The state for entry into both A_U and B_D is,

$$|\psi: A_U, B_D\rangle = -\frac{e^{i\Delta}}{2\sqrt{2}}|A_U\rangle(-e^{i\Delta}|B_D\rangle + e^{i\Delta} \times (-1)^2|B_D\rangle) = 0 \qquad (18.2.4)$$

The state for entry into both A_U and B_L is,

$$|\psi: A_U, B_L\rangle = -\frac{e^{i\Delta}}{2\sqrt{2}}|A_U\rangle(-e^{2i\Delta}|B_L\rangle + e^{i\Delta} \times (-e^{i\Delta})|B_L\rangle) = \frac{e^{3i\Delta}}{\sqrt{2}}|A_U\rangle|B_L\rangle \quad (18.2.5)$$

We conclude that if a photon registers in detector A_U then there will be no coincident photon in detector B_D, but rather there will always be a coincident photon in detector B_L. For photons entering A_R we find,

$$|\psi: A_R, B_D\rangle = \frac{1}{2\sqrt{2}}|A_R\rangle(-e^{i\Delta}|B_D\rangle - e^{i\Delta} \times (-1)^2|B_D\rangle) = -\frac{e^{i\Delta}}{\sqrt{2}}|A_R\rangle|B_D\rangle$$
$$(18.2.6)$$

$$|\psi: A_R, B_L\rangle = \frac{1}{2\sqrt{2}}|A_R\rangle(-e^{2i\Delta}|B_L\rangle - e^{i\Delta} \times (-e^{i\Delta})|B_L\rangle) = 0$$
$$(18.2.7)$$

So similarly we conclude that if a photon registers in detector A_R then there will be no coincident photon in detector B_L, but rather there will always be a coincident photon in detector B_D. Note that the four states in (18.2.4-7) maintain normalisation to unity.

Hence we see that interference ***does*** occur, but in a modified sense that depends upon coincident observations of photons in the A and B devices. Recall that the signature of interference in a Mach-Zehnder interferometer is that all the photons appear in one detector and none in the other. This is indeed found for the B device, provided that we filter out all the instances in which A_U records a photon and keep only cases when A_R detects a photon (or *vice-versa*). But this type of coincident, or correlated, interference does not provide any means of FTL communication. If the A detectors are at a space-like separation from the B detectors, we can only discover the 'interference' some time later, when a perfectly normal, sub-luminal, signal has communicated the results of the B detectors to the operator at A.

Caution is needed when reading the literature since this type of coincident, or correlated, interference is sometimes simply called 'interference' without qualification. This is horribly confusing since true 'local' interference in such situations ***would*** violate causality.

18.3 Entangled Interference with Crossed SPDC Beams

In this section we illustrate the same phenomena described in §18.2 but for what initially appears to be a very different experimental arrangement. The algebraic properties of the states, though, turn out to be essentially the same. This apparatus uses spontaneous parametric down-conversion (SPDC) which splits an input photon into two identical photons of half the energy. This can be done using, for example, a lithium iodate crystal. The arrangement is shown in Figure 18.3.

The incoming beam from the laser is split by the SPDC crystal into two beams (shown green and red) which emerge at some characteristic angle. These beams are reflected off mirrors back through the crystal and into detectors (continuous green and red lines). These beams are not significantly affected by the crystal on their second passage (due to their reduced energy). However, the crystal acts as a beam splitter also in another sense. Not all the incoming beam is initially down-converted. What is not converted is transmitted through the crystal and gets reflected back to the crystal by another mirror. On this second passage there is therefore a second down-conversion which creates the green and red *dashed* beams. These are also directed into the same detectors. (The labelling of the various beams as red or green does not reflect the true colour of the photons, of course; they actually have the same frequency).

A 'green' photon can reach the lower detector by either of two paths: by being down-converted at the first pass and following the continuous green beam, or by being down-converted at the second pass and following the dashed green beam. These two beam paths can (potentially) cause interference at the detectors. But do they? We are now alert to the possibility that they may not, by virtue of being one half of an entangled pair.

How is the state of the photon expressed algebraically? If the 'green photon' follows the continuous line, then so does the red photon. Conversely, if the green photon follows the dashed line, then so does the red photon. So the state of the entangled pair at the detectors is,

$$|\psi\rangle = \tfrac{1}{\sqrt{2}}(|G\rangle|R\rangle + |Gd\rangle|Rd\rangle) \qquad (18.3.1)$$

Here $|G\rangle$, $|R\rangle$ represent the continuous green and red beams, and $|Gd\rangle$, $|Rd\rangle$ represent the dashed beams. We can express the Hilbert states in position representation, using the variable x for the green beams and y for the red beams. So (18.3.1) becomes,

$$\psi(x,y) = \tfrac{1}{\sqrt{2}}\left(\psi_G(x)\psi_R(y) + \psi_{Gd}(x)\psi_{Rd}(y)\right) \qquad (18.3.2)$$

Figure 18.3: *Entangled interference using crossed SPDC photon beams*

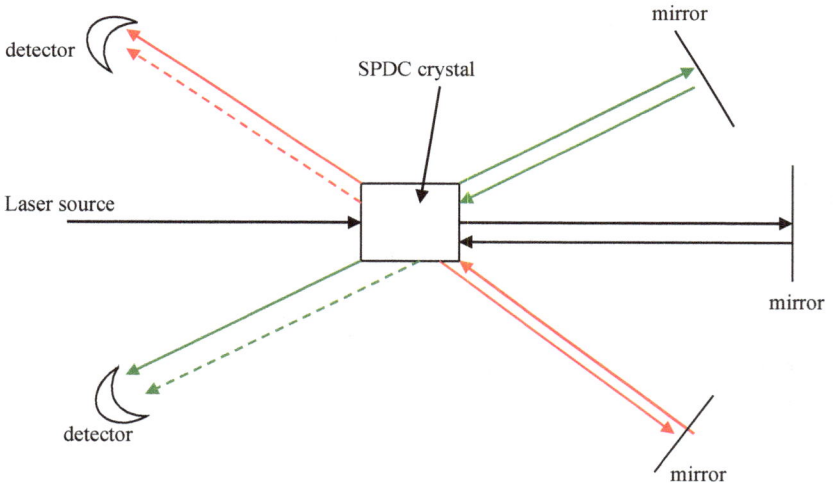

But the dashed beams differ from the continuous beams where they enter the detectors only by their phase (because they are coherent). So we can write,

$$\psi_{Gd}(x) = e^{i\theta_G}\psi_G(x) \quad \text{and} \quad \psi_{Rd}(y) = e^{i\theta_R}\psi_R(y) \qquad (18.3.3)$$

Here the phase angles will be functions of position, and so properly written $\theta_G(x)$ and $\theta_R(y)$. Hence (18.3.2) becomes,

$$\psi(x,y) = \frac{1}{\sqrt{2}}\psi_G(x)\psi_R(y)\left(1 + e^{i(\theta_G + \theta_R)}\right) \qquad (18.3.4)$$

Now for plane waves we can take $|\psi_G| = |\psi_R| = 1$ so that the square modulus of (18.3.4) is,

$$|\psi(x,y)|^2 = 1 + cos(\theta_G + \theta_R) \qquad (18.3.5)$$

If we ignore whereabouts on the y-screen the red photon is detected, do we see an interference pattern on the x-screen which is detecting the green photon? The answer is 'no' because the intensity on the x-screen is then the average of (18.3.5) over all y values, i.e.,

$$\langle|\psi(x,y)|\rangle_y = \frac{1}{2\pi}\int_0^{2\pi} d\theta_R(1 + cos(\theta_G + \theta_R)) = 1 \qquad (18.3.6)$$

This retains no θ_G dependence (i.e., no x dependence), and hence is a uniform illumination without an interference pattern.

However, just as with the Mach-Zehnder interferometer example, we can rediscover an interference pattern by considering correlated measurements. Thus, we choose any position on the y-screen (i.e., any fixed θ_R) and scan our detector slowly over the x-screen recording counts only if there is a coincident count at the fixed y-position detector. The pattern which emerges from the x-screen is just (18.3.5), for a constant value of θ_R. This is an interference pattern, varying from 0 to 2 as the cosine varies from -1 to +1.

The absence of a "local" interference pattern, i.e., one which does not require knowledge of the results from the other detector, saves us from a causality disaster. However, interference of sorts does occur but can only be revealed when the results of both detectors are brought together, or one detector is used as a veto over the other. So, no FTL signalling is possible because the only interference which can be observed requires ordinary communication of the results from the other screen.

This SPDC crossed-beam arrangement is essentially the same as the Mach-Zehnder example, despite the very different experimental arrangment. The reason is that the algebraic structure of the quantum states, given by (18.3.1) and (18.2.1) respectively, are the same.

18.4 Double-Slit Interference with Spin-Entangled Particles

The experimental arrangements considered in §18.2 and §18.3 have emphasised the difference between strictly local interference and interference which is only revealed by coincidence counting, and hence requires communication of results between detectors. This may have given the impression that local interference can never be observed using one of an entangled pair of particles. However this would be too loose a statement. Careful examination of these examples shows that the reason why local interference does not occur is due to the entangled degrees of freedom being the same as the degrees of freedom involved in the potential interference. However, it may be possible to obtain interference from one particle of an entangled pair provided that the interference involves different degrees of freedom from those which are entangled. Consider the arrangement shown in Figure 18.4.

Figure 18.4: *Double-slit interference using one beam of a spin-entangled pair*

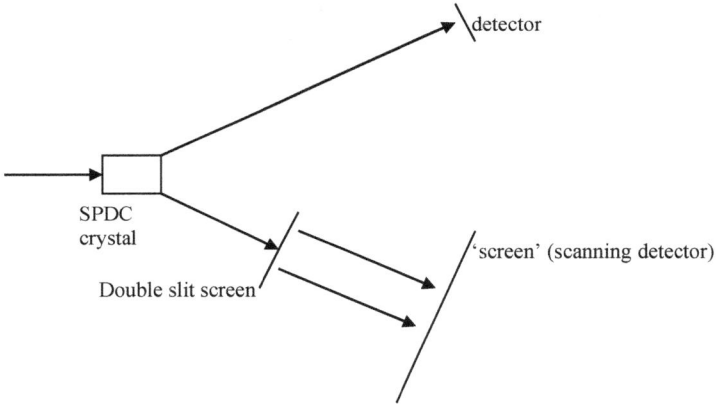

Here just one of the beams output from an SPDC crystal (e.g., beta-barium borate) is incident on a double slit screen. The other photon path plays no essential part initially. Do we see purely local interference from the lower beam? The answer is "yes", despite these photons being entangled with the photons in the other beam. This can be understood algebraically as follows.

Suppose that either photon can emerge from the SPDC crystal in a vertical or a horizontal polarisation state, but that the two must be polarised differently. We shall denote these states v or h. Also suppose we refer to the upper and lower beams emerging from the SPDC crystal as U and L. The initial state can therefore be written,

$$|\psi\rangle = \frac{1}{\sqrt{2}}(|Uv\rangle|Lh\rangle + |Uh\rangle|Lv\rangle) \tag{18.4.1}$$

The two photons are thus entangled via their spin (polarisation) states. At the double slit screen the lower beam is split into two beams which we label as L1 and L2. Hence, after the double slit the total state becomes,

$$|\psi\rangle = \frac{1}{2}(|Uv\rangle[|L1h\rangle + |L2h\rangle] + |Uh\rangle[|L1v\rangle + |L2v\rangle]) \tag{18.4.2}$$

The two horizontally polarised states emerging from the slits will differ by some phase factor $e^{i\theta}$ where θ is a function of the position on the screen, x. The same is true for the vertically polarised states. In the x-basis the overall state, (18.4.2), is therefore,

$$\begin{aligned}|\psi\rangle &= \frac{1}{2}\Big(|Uv\rangle\psi_{L1}(x)\big[|h\rangle + e^{i\theta}|h\rangle\big] + |Uh\rangle\psi_{L1}(x)\big[|v\rangle + e^{i\theta}|v\rangle\big]\Big)\\ &= \frac{1}{2}(|Uv\rangle|h\rangle + |Uh\rangle|v\rangle)\psi_{L1}(x)(1 + e^{i\theta}) \end{aligned} \tag{18.4.3}$$

where $|\psi_{L1}(x)| = 1$ for a normalised plane wave state. Note that I have here explicitly separated the spatial wavefunction from the polarisation state. Taking the absolute square of (18.4.3) we get the intensity on the x-screen to be,

$$|\psi(x)|^2 = 1 + \cos\theta\,(x) \tag{18.4.4}$$

so there *is* an interference pattern in this case. How has this happened?

The two ket terms on the second line of (18.4.3) are, of course, orthogonal – so there is no cross-term to cause interference arising from them. This corresponds to our previous observations because this term involves the entangled degrees of freedom, i.e., the polarisation states. However, the interference does not arise from this term. The interference arises from the last term, the factor of $(1 + e^{i\theta})$. And this arises from the purely lower beam based cross-terms in (18.4.2), i.e., the cross-term arising from $[|L1h\rangle + |L2h\rangle]$ and equally that with the opposite polarisation. So the interference arises from the spatial degrees of freedom – the spatial separation of the two beams emerging from the double slits, beams L1 and L2. The entanglement of the polarisation degrees of freedom does not prejudice local interference arising from the spatial degrees of freedom. Nor does it provide an opportunity for FTL communication since any measurement of the polarisation in the upper beam has no effect upon the interference on the x-screen. This can be seen by collapsing the wavefunction of the first term in

(18.4.3), to leave just one of the ket terms. It does not matter which is left since either will retain the factor of $\left(1 + e^{i\theta}\right)$ which causes the interference.

So it would be quite wrong to claim that 'entangled particles cannot produce local interference'. Whilst the examples in §18.2 and §18.3 attempted to produce local interference using the entangled degrees of freedom – and therefore failed - in the present example the spin entanglement with the upper beam is simply irrelevant to the local interference associated with the spatial degrees of freedom of the lower beam.

18.5 Exercises for the Reader

18.5.1: Derivation of Phase Change Rules for Mirrors

All the phase changes derived below are with respect to what the phase would have been at the same place if propagation had involved passage through vacuum (effectively air) alone, i.e., with respect to the phase $e^{i(\vec{k}\cdot\vec{r}-\omega t)}$ of a propagating wave. The phase change rules are the same for silvered and half-silvered mirrors.

Propagation through the glass of a mirror causes a phase change of $e^{ik'x}$ for a distance x of travel, compared with the phase change of e^{ikx} through air. The wave-numbers are related by $k' = nk$, where n is the refractive index of the glass, so the phase factor due to a thickness a of glass is $e^{i(n-1)ka} = e^{i\Delta}$, as given in §2. For reflection from the rear of the mirror (glass side facing the beam), the beam passes through the glass twice so the total phase factor is $e^{2i\Delta}$.

If a wave in air, with phase e^{ikx} ignoring the time-dependent part, meets a silvered surface so that any wave entering the glass will be $Ce^{ik'x}$, the boundary conditions at the surface are that the wavefunction and its x derivative must be continuous. We must also account for a reflected wave, Be^{-ikx}. Hence, assuming the surface is at $x = 0$ and recalling that the incident wave has datum zero phase, we require,

$$1 + B = C \quad \text{and} \quad k - Bk = k'C \tag{18.5.1}$$

These equations are readily solved to give,

$$B = -\left(\frac{n-1}{n+1}\right) \quad \text{and} \quad C = \frac{2}{n+1} \tag{18.5.2}$$

Since $n > 1$ it is clear that B is real and negative, corresponding to a phase change with respect to the incident wave of 180°, or a phase factor of -1.

This confirms the phase change rule relating to reflection from the 'front' (silvered) face of a mirror, i.e., a phase factor of -1.

Now consider a beam incident on the "back" face and hence travelling through the glass plate first before reflecting off the silvered layer on the opposite surface. If a wave in glass meets the air boundary, the above analysis still applies except that k and k' are interchanged. This means replacing n with $1/n$. So the reflection and transmission coefficients are now respectively,

$$B = + \left(\frac{n-1}{n+1}\right) \quad \text{and} \quad C = \frac{2n}{n+1} \tag{18.5.3}$$

The reflection coefficient is now real and positive, and so there is no phase change between the reflected and incident waves, consistent with the rule that the phase factor for reflection from the "back" face is just the factor $e^{2i\Delta}$ associated with two passages through the glass. QED.

19

Quantum Erasure

What is quantum erasure? When a measurement is not a measurement. Popular accounts are guilty of talking-up the weirdness of quantum erasure when, in truth, it really isn't weird at all.

19.1 Measurement, Erasure, Causality

Quantum erasure is the phenomenon whereby a "measurement", which would destroy an interference pattern, can be erased and the interference regained. How can a measurement possibly be erased? Such erasures become even more intriguing when coupled with experiments on entangled pairs of particles because this provides the opportunity to carry out the erasure of the "measurement" *after* the interference data has already been obtained, so-called "delayed erasure". One particle, it is claimed in popular accounts, appears to "know" that a measurement will be carried out in the future on its entangled partner. Hence an interference pattern is re-established because of an erasure which will be carried out after the interference data has already been collected. This is often presented in popular accounts as posing a challenge to the correct temporal order of causality. It does not, of course, and this chapter and the next explain why.

Unfortunately, popular accounts of these issues are often desperately inaccurate. Indeed I am tempted to say that, "when I hear of delayed quantum erasure I reach for my gun". The key to a sound understanding, as always with these matters, is the Hilbert state algebra. The algebra is virtually trivial and certainly far easier to follow than accounts of detailed experimental arrangements.

This chapter discusses some experimental arrangements exhibiting erasure and how they can be analysed. It will be seen that the key to understanding quantum erasure is an ambiguity in how the word "measurement" is used. Once this is appreciated, quantum erasure ceases to have any great significance.

Chapter 20 will focus on delayed erasure. The key to understanding why delayed erasure on entangled pairs does not conflict with causality will be shown to lie in the distinction between local interference and correlated interference, a distinction with which the reader should, following chapter 18, now be familiar. Popular accounts are guilty of failing to distinguish clearly between the two and this is the main source of confusion.

Experiments of the sort discussed in this chapter are of considerable importance in verifying, yet again, that the predictions of quantum mechanics are borne out. Quantum mechanics does **seem** weird. But the apparent weirdness lies in the fact that alternative outcomes are not deterministic (there are no hidden variables) and yet, despite that, causally unconnected measurements contrive to be correlated. In other words, the weirdness lies in the non-locality of entangled quantum states, the same phenomenon which exercised EPR, and everyone since. That remains a challenge to our classical intuition, right enough, but it's not as weird as popular accounts of delayed erasure generally suggest.

19.2 Quantum Erasure: A Double-Slit Example

Recall that if we measure the path by which a photon travels in an interferometer or double-slit experiment then we destroy the interference pattern. Remarkably such a "measurement" can be undone – or erased – and the interference pattern regained. There is a caveat, though, which most sources are guilty of failing to emphasise. The caveat hinges upon exactly what is meant by "measurement". We shall return to this at the end of this section. Firstly a brief reminder of how "which path" information destroys interference. Consider a double slit interference experiment as illustrated by Figure 19.1.

Figure 19.1: *Double-slit with Polaroid: example of quantum erasure*

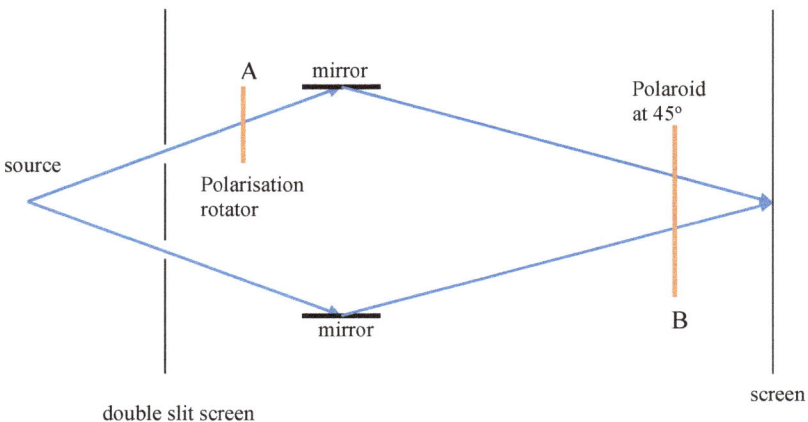

The source is assumed to supply vertically polarised light. Initially the polarisation rotator (A) and the 45º Polaroid (B) are not present. There is then an interference pattern on the screen. Algebraically this occurs as follows. The state arriving at the screen can be written,

$$|\psi\rangle = \frac{1}{\sqrt{2}}(|U\rangle + |L\rangle) \qquad (19.2.1)$$

Here $|U\rangle, |L\rangle$ represent the upper and lower beam paths. At the screen they will differ by a phase $e^{i\theta}$, so that, in the x basis, where x is the position on the screen, we have,

$$\psi(x) = \frac{1}{\sqrt{2}}(1 + e^{i\theta})\psi_U(x) \qquad (19.2.2)$$

Hence, $$|\psi(x)|^2 = 1 + cos\ \theta\ (x) \qquad (19.2.3)$$

(assuming $\psi_U(x)$ is a plane wave with $|\psi_U| = 1$). Hence, (19.2.3) displays the usual interference pattern.

We now insert the polarisation rotator (A), which we arrange to rotate the polarisation of the upper beam to the horizontal. (The 45º Polaroid is still not present). Since the upper and lower beams are now distinguishable by their distinct polarisations we expect the interference pattern to disappear. This is indeed the case, and the reason may be seen algebraically as follows. The polarisation part of the state will be written $|v\rangle$ or $|h\rangle$, for vertical and horizontal polarisation respectively. These are perfectly distinguishable states, so $\langle v|h\rangle = 0$. The state before encountering the polarisation rotator is $(|U\rangle|v\rangle + |L\rangle|v\rangle)/\sqrt{2}$ but afterwards it is,

$$|\psi\rangle = \frac{1}{\sqrt{2}}(|U\rangle|h\rangle + |L\rangle|v\rangle) \qquad (19.2.4)$$

In the x-basis at the screen this becomes,

$$\psi(x) = \frac{1}{\sqrt{2}}(1 \cdot |h\rangle + e^{i\theta} \cdot |v\rangle)\psi_U(x) \qquad (19.2.5)$$

Hence, $$|\psi(x)|^2 = 1 \qquad (19.2.6)$$

as a consequence of $\langle v|h\rangle = 0$. So the interference pattern has disappeared. From this it is algebraically transparent how the distinguishability of the two paths eradicates the interference: it makes the two terms in (19.2.4) orthogonal and hence the cross-product, which causes interference, is zero.

Now, what about erasure? Can we undo the effects of the polarisation rotator later? We insert the 45º Polaroid shown in Figure 19.1. Let us write the state emerging from this Polaroid as $|45\rangle$, and the perpendicular polarisation as $|-45\rangle$. We can express the vertical and horizontal polarisations in terms of these as,

$$|v\rangle = \frac{1}{\sqrt{2}}(|45\rangle + |-45\rangle) \tag{19.2.7}$$

$$|h\rangle = \frac{1}{\sqrt{2}}(|45\rangle - |-45\rangle) \tag{19.2.8}$$

Substituting (19.2.7,8) into (19.2.4) gives,

$$|\psi\rangle = \frac{1}{2}\left(|U\rangle(|45\rangle - |-45\rangle) + |L\rangle(|45\rangle + |-45\rangle)\right) \tag{19.2.9}$$

This is the state after the first polarisation rotator, but before encountering the 45° Polaroid. The 45° Polaroid traps the component $|-45\rangle$ and lets through the component $|45\rangle$. So the state reaching the screen is,

$$|\psi\rangle = \frac{1}{2}(|U\rangle|45\rangle + |L\rangle|45\rangle) = \frac{1}{2}(|U\rangle + |L\rangle)|45\rangle \tag{19.2.10}$$

which, in the x basis, is,

$$\psi(x) = \frac{1}{2}\left(1 + e^{i\theta}\right)\psi_U(x)|45\rangle \tag{19.2.11}$$

And so we now have interference once again,

$$|\psi(x)|^2 = \frac{1}{2}\left(1 + \cos\theta(x)\right) \tag{19.2.12}$$

The "which path" information transiently provided by the polarisation rotator has been successfully erased by the Polaroid. (Note that the reason why (19.2.12) has an average of only ½ is because the 45° Polaroid has absorbed half the photons).

Are we entitled, though, to refer to what the polarisation rotator does as being "a measurement"? I suggest not. A measurement actually consists of two things (as discussed in §2.8 and chapter 15),

(i)   the establishment of entanglement between the system being measured and the apparatus, and,

(ii) the "reduction of the state vector", i.e., the actual, irreversible, selection of just one of the possible outcomes.

The first of these, (i), is embodied by (19.2.4). Note that the transformation of (19.2.1) into (19.2.4) is purely unitary – and this is why it can be reversed.

The second of the measurement steps, (ii), never happens in this setup. If it had occurred in respect of our so-called measurement by the polarisation rotator, then, before reaching the 45° Polaroid, the state would have been either $|\psi\rangle = |U\rangle|h\rangle$ or $|\psi\rangle = |L\rangle|v\rangle$ but not $|\psi\rangle = (|U\rangle|h\rangle + |L\rangle|v\rangle)/\sqrt{2}$. It would then have been quite impossible to reverse such a *true* measurement so as to subsequently yield an interference pattern. Instead, the polarisation rotator carries out only the first step of a measurement. The state remains coherent, with no increase in entropy, and hence is reversible.

In fact, we should have been suspicious about claims of reversing a measurement from the start as we know, from chapter 12, that measurement increases entropy (for any state which is not already an eigenstate of the observable being measured). Consequently, in the absence of changes in the rest of the world, reversing a measurement would imply reducing entropy, which violates the second law of thermodynamics.

But a partial "measurement" which stops at step (i) *is* sufficient to eliminate interference, because (19.2.6) follows from (19.2.4). Merely bringing the system into entanglement with a set of orthogonal pointer states is enough to eliminate interference by virtue of the above algebra. But this is reversible, because it is a unitary evolution, unlike a completed measurement which is obviously irreversible once the wave-function has collapsed.

The moral is that the term "measurement" is used ambiguously in the literature. A true measurement could not be erased. But there is a lesson to be learnt here. Previously we have loosely stated that "which-path measurement destroys interference". But we now see that it would have been more accurate to say, "the first step of a which-path measurement (entanglement) destroys interference". It is this correction to our previously loose statement which renders the destruction of interference reversible if the previous interference was destroyed by only a partial measurement involving step (i) only. Only the effects of step (i) can be erased (i.e., reversed), not the effects of a completed measurement including step (ii).

19.3 Quantum Erasure in the Crossed SPDC Beam Setup

Recall the example of interference in chapter 18 based on spontaneous parametric down-conversion (SPDC). What happens in the crossed SPDC beam setup if we introduce a polarisation rotator, and perhaps subsequently erase its effect using a 45° Polaroid as in §19.2 (see Figure 19.2)? We assume that the two beams emerging from the lithium iodate SPDC crystal are both vertically polarised. (This is a different SPDC behaviour from that of beta-barium borate as used in §18.4 for which the polarisations of the emerging photons are orthogonal). Referring to the notation used in §18.3, but now also including the polarisation, the state entering the detectors in the absence of any polarisers/Polaroids would be,

$$|\psi\rangle = \tfrac{1}{\sqrt{2}}\left(|G\rangle|v\rangle|R\rangle|v\rangle + |Gd\rangle|v\rangle|Rd\rangle|v\rangle\right) \qquad (19.3.1)$$

where $|Gd\rangle$ and $|Rd\rangle$ refer to the green and red 'dashed' beams in Figure 19.2, which are down-converted only on the main beam's second pass through the SPDC crystal. In (19.3.1) the polarisation ket written first/second is understood to refer to the 'green' and 'red' photon respectively. (Recall that these colours are merely labels used to identify the two photon beams, not literally the physical colours of the photons which actually have the same wavelength). We have already seen that this experiment would not produce a *local* interference pattern, but would produce a correlated or coincidence interference pattern. The latter is seen provided that green and red detector correlations are taken into account. For example, if detections by the green detector are counted only if there is a simultaneous detection in the red detector, then an interference pattern in the green signal will be seen. But if all detections by the green detector are counted, irrespective of the red detector, then no interference pattern will be seen. Recall that this distinction between local interference and correlated interference is crucial. The latter presents no challenge to causality, whereas the former would – but does not occur. These conclusions follow simply by virtue of the algebraic structure of (19.3.1).

Figure 19.2: *Quantum erasure in the SPDC interferometer*

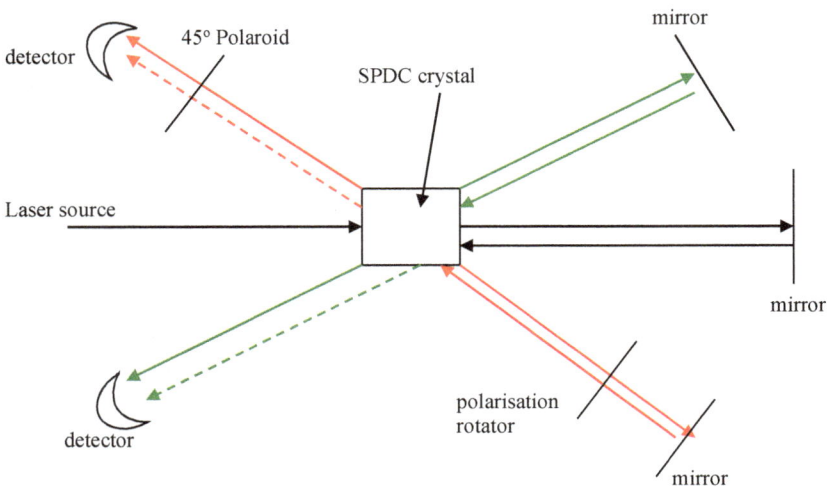

Suppose we now introduce a polarisation rotator into the red beam at the position shown in Figure 19.2 so that the affected beam emerges from it with horizontal polarisation. (The 45° Polaroid shown in the Figure is not yet present). The state entering the detectors is now,

$$|\psi\rangle = \frac{1}{\sqrt{2}}(|G\rangle|v\rangle|R\rangle|h\rangle + |Gd\rangle|v\rangle|Rd\rangle|v\rangle) \qquad (19.3.2)$$

Note that it is only the state $|R\rangle$ which is affected by the polarisation rotator. The dashed red beam does not pass through the rotator. This change is sufficient to make the two terms in (19.3.2) orthogonal and hence to eliminate any interference – even the non-local interference which might result from green/red correlations. To spell this out, if the 'green' photon hits the screen at position x, then, in the x basis, we have $\langle x|G\rangle = \psi_G(x)$. The dashed-green path is coherent with this and has a relative phase $\theta_G(x)$ so that $\langle x|Gd\rangle = \psi_G(x)e^{i\theta_G}$. Similar remarks hold for the red and dashed-red photon paths in terms of a screen position y, giving,

$$|\psi(x,y)\rangle = \frac{1}{\sqrt{2}}\psi_G(x)\psi_R(y)(|v\rangle|h\rangle + e^{i(\theta_G+\theta_R)}|v\rangle|v\rangle) \qquad (19.3.3)$$

So that, $$\langle\psi(x,y)|\psi(x,y)\rangle = 1 \qquad (19.3.4)$$

because $\langle v|h\rangle = 0$ and the planes waves have $|\psi_G(x)| = |\psi_R(y)| = 1$. Consequently even the correlated interference which previously resulted from (19.3.1) does not occur.

We know, of course, that local interference cannot occur in this entangled setup because this would provide the opportunity for FTL communication (see chapter 18). But can the correlated interference be resurrected by erasure of the effects of the polarisation rotator? Consider placing a 45° Polaroid into the red beam path as shown in Figure 19.2. Expressing the state (19.3.3) **before** encountering this Polaroid in terms of the 45° polarisation states (for the red and dashed-red beams only) gives, using (19.2.7,8),

$$|\psi(x,y)\rangle = \frac{1}{2}\psi_G(x)\psi_R(y)\left(|v\rangle(|45\rangle - |-45\rangle) + e^{i(\theta_G+\theta_R)}|v\rangle(|45\rangle + |-45\rangle)\right)$$

$$(19.3.5)$$

Because the effect of the 45° Polaroid is to filter the perpendicular state, $|-45\rangle$, the state entering the detectors is,

$$|\psi(x,y)\rangle = \frac{1}{2}\psi_G(x)\psi_R(y)(|v\rangle|45\rangle + e^{i(\theta_G+\theta_R)}|v\rangle|45\rangle)$$

$$= \frac{1}{2}\psi_G(x)\psi_R(y)\left(1 + e^{i(\theta_G + \theta_R)}\right)|v\rangle|45\rangle \qquad (19.3.6)$$

Hence, $\quad |\psi(x,y)|^2 = \frac{1}{2}(1 + cos(\theta_G(x) + \theta_R(y))) \qquad (19.3.7)$

The average of (19.3.7) over the screen is only ½ because half the photons have been absorbed by the Polaroid. (19.3.7) indicates that non-local (correlated) interference has been re-introduced by the Polaroid, in the sense that if we record x screen data only when photons are registered at a given, fixed y position on the other screen (or *vice-versa*) then we discover an interference pattern. But if we simply look at the x-screen without regard to where photons arrive on the y-screen, then there is no interference pattern. The correlated interference is reintroduced because the effect of the 45⁰ Polaroid is to make the two terms in the state vector parallel, (19.3.6), rather than orthogonal as in (19.3.3).

19.4 Summary

The example of §19.2 shows that true local interference is destroyed by "which path measurement" even if the "measurement" involves only the first, entanglement, phase of measurement. In this case, the interference can be restored by erasure of this "measurement", precisely because step (i) of a measurement is unitary and hence is itself reversible. Similarly, the example of §19.3 shows that interference of the non-local, correlated type can also be destroyed by "which path measurement" carried out on one of the entangled particles alone, even if only the first, entanglement, phase of a measurement is carried out. If this is the case, and step (ii) of the measurement (state vector reduction) has not occurred, then this effect can also be erased and correlated interference regained by appropriate interaction. But this phenomenon of "quantum erasure" does not apply to completed, irreversible (true) measurements in which the state vector has reduced. The effects of a completed measurement in destroying interference cannot be erased, whether one considers local or correlated interference.

20

Delayed Quantum Erasure

Chapter 19 explained what is meant by "quantum erasure". Delayed quantum erasure consists of erasing a "measurement" only after interference data has been obtained. Popular accounts give the impression that it is especially weird that interference is nevertheless observed, as the data was collected while the "measurement" was still in place and hence the interference would be expected to be destroyed. Somehow, the claim goes, the particle "knew" that the erasure would occur some time later. This is a false understanding, of course, and arises from a confusion between local interference and non-local, correlated interference between entangled particles (as the reader might be anticipating by now).

20.1 The Experimental Arrangement of Walborn *et al*

Recall the arrangement discussed in §18.4 where one of a pair of spin-entangled photons was incident on a double slit screen, thus producing an interference pattern. This illustrated that one of an entangled pair of particles can produce interference provided that the interfering degrees of freedom (spatial, in this case) are distinct from the entangled degrees of freedom (spin in this case). The setup is shown in Figure 20.1.

Figure 20.1: *The experimental arrangement of Walborn et al, Ref.[20.1]*

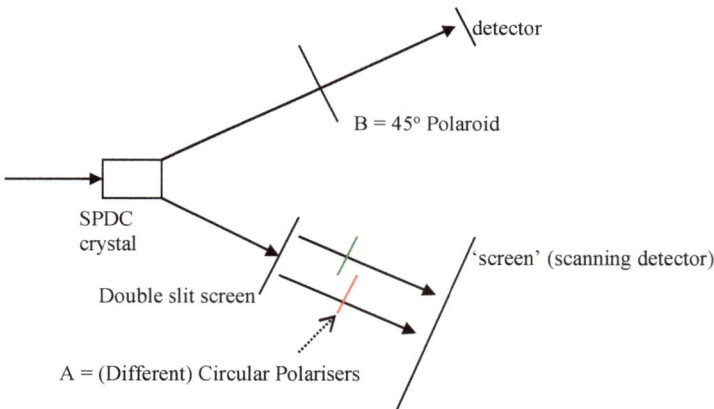

Supposing the circular polarisers and the 45° Polaroid are not present, the bipartite state after the double slits was shown in §18.4 to be,

$$|\psi\rangle = \frac{1}{2}(|Uv\rangle[|L1h\rangle + |L2h\rangle] + |Uh\rangle[|L1v\rangle + |L2v\rangle]) \quad (20.1.1)$$

where U and L refer to the upper and lower beams emerging from the SPDC crystal, v and h refer to the polarisation states, and 1 and 2 refer to the two beams emerging from the double slit. The state (20.1.1) arises because one photon emerges from the beta-barium borate SPDC crystal in a vertical polarisation state and the other in a horizontal polarisation state, but either photon can be the one vertically polarised. Chapter 18 showed that state (20.1.1) leads to (true local) interference at the screen between the $L1$ and $L2$ beams.

Now let's see what happens when we insert the two Circular Polarisers after the two slits in the lower beam (but the 45º Polaroid is not yet present in the upper beam path). These Circular Polarisers are such that an input polarisation state $|v\rangle$ becomes a clockwise polarised beam, $|C\rangle$, after passing through the 'green' polarisation rotator (aligned with the higher slit), whereas the input state $|v\rangle$ would become an anti-clockwise polarised beam, $|AC\rangle$, after passing through the 'red' rotator (aligned with the other slit). If the input polarisation state were $|h\rangle$, on the other hand, then the green rotator would produce $|AC\rangle$ and the red rotator would produce $|C\rangle$. Consequently, from (20.1.1), the state after passing through the Circular Polarisers is,

$$|\psi\rangle = \frac{1}{2}(|Uv\rangle[|L1\rangle|AC\rangle + |L2\rangle|C\rangle] + |Uh\rangle[|L1\rangle|C\rangle + |L2\rangle|AC\rangle]) \quad (20.1.2)$$

It is clear that there will be no interference now because all the terms in (20.1.2) are orthogonal, because $\langle Uv|Uh\rangle = 0$ and $\langle AC|C\rangle = 0$. So all cross-product terms are zero. Hence, the insertion of the Circular Polarisers at A has destroyed the previous interference.

Why is this? Well we could carry out a measurement of the vertical/horizontal polarisation of the upper photon, and this would then tell us which slit the lower photon had travelled through via its circular polarisation. This "which path" information provided by the two polarisation rotators destroys the interference, as we would expect. Note that it is not necessary to actually obtain this information (i.e., to reduce the state vector to determine which path was taken). To destroy the interference it is sufficient that the first phase of a true measurement, the entanglement phase, has been done. By "entanglement" in this context I mean the entanglement between the polarisation states and the 1 and 2 states in (20.1.2), which is what destroys the interference that would otherwise arise from the lower beam state in (20.1.1).

The interesting question which now arises is, "can we erase this 'measurement' and resurrect the interference by interacting only with the upper beam?" In popular accounts the answer is often given as an unqualified "yes". However, following chapter 19, the reader will be alert to the true phenomenon. The correct answer is that local interference cannot be resurrected. What can be regained is only non-local, correlated interference of the type discussed previously. Regaining local interference would provide an opportunity for FTL communication, so we are wise to anticipate that it cannot occur. Thus it is not quite accurate to refer to "erasure" in this case. It is only partial erasure. Whereas initially we had local interference, what we regain is the weaker phenomenon of correlated interference.

In the example of §19.3 the measurement on one of the entangled beams was erased by a subsequent Polaroid placed in the path of the same beam. In contrast, in this example, we will see how the "measurement" carried out on the lower beam can effectively be erased by introducing a 45° Polaroid in the upper beam path, as shown in Figure 20.1. This simply filters out the polarisation component $|-45\rangle$ and lets $|45\rangle$ through unimpeded. So we must first re-write the state emerging from the SPDC crystal, $|\psi\rangle = (|Uv\rangle|Lh\rangle + |Uh\rangle|L1v\rangle)/\sqrt{2}$, in terms of the states $|\pm45\rangle$ by substituting from (19.2.7,8). This gives,

$$|\psi\rangle = \frac{1}{2\sqrt{2}}\left\{\begin{array}{l}(|U,45\rangle + |U,-45\rangle)(|L,45\rangle - |L,-45\rangle) + \\ (|U,45\rangle - |U,-45\rangle)(|L,45\rangle + |L,-45\rangle)\end{array}\right\}$$

So, $\quad |\psi\rangle = \frac{1}{\sqrt{2}}(|U,45\rangle|L,45\rangle - |U,-45\rangle|L,-45\rangle)$ \hfill (20.1.3)

Consequently, passing the upper beam through the 45° Polaroid causes the state to become,

$$|\psi\rangle = \frac{1}{\sqrt{2}}|U,45\rangle|L,45\rangle \hfill (20.1.4)$$

Using (19.2.7,8) this can also be written,

$$|\psi\rangle = \frac{1}{2}|U,45\rangle(|L\rangle|v\rangle + |L\rangle|h\rangle) \hfill (20.1.5)$$

When the lower beam passes through the double slits and then through the Circular Polarisers the state therefore becomes,

$$|\psi\rangle = \frac{1}{2\sqrt{2}}|U,45\rangle(|L1\rangle|C\rangle + |L2\rangle|AC\rangle + |L1\rangle|AC\rangle + |L2\rangle|C\rangle)$$
$$= \frac{1}{2\sqrt{2}}|U,45\rangle(|L1\rangle + |L2\rangle)(|C\rangle + |AC\rangle) \hfill (20.1.6)$$

We thus (apparently!) get local interference on the screen onto which the lower beams are focussed. To spell this out, recall that after the double slits the two resulting beams in the spatial basis at the screen differ only by a phase factor so we can write $|L2\rangle = e^{i\theta(x)}|L1\rangle$. Thus, taking the inner product of (20.1.6) with the spatial-basis bra $\langle x|$ and then taking the absolute square, the probability of seeing a photon (of either polarisation) at screen position x is,

$$|\psi(x)|^2 = \frac{1}{2}\left(1 + \cos\theta(x)\right) \tag{20.1.7}$$

The factor of ½ is because the 45° Polaroid stops half the photons getting through (or does it?). So – apparently - we have successfully erased the effect of the polarisation rotators, which destroyed the interference, and restored the interference in the lower beam (on the x-screen) by placing a Polaroid in the path of the upper beam. From (20.1.7) this would appear to be local interference observable using the x-detector data alone – but is it? This would provide a route to FTL communication so we must suspect an error.

How does the erasure happen? It is because the "measurement" of the polarisation by the 45° Polaroid forces the lower beam into a 45° state, and this is a superposition of states with vertical and horizontal polarisation. It is only these vertical and horizontal polarisation states that the polarisation rotator turns deterministically into a definite circularly polarised state. A 45° polarisation state produces a superposition of clockwise and anti-clockwise polarisation states after passing through either rotator. Consequently knowing whether the photon at the x-screen has clockwise or anti-clockwise polarisation no longer tells us which slit it passed through. Either slit could result in either clockwise or anti-clockwise polarised photons, since the slits are illuminated by a superposition of vertical and horizontal polarised photons. This is apparent from (20.1.6), as the polarisation state is not entangled with the spatial state, i.e., whether it is beam L1 or beam L2 (the spatial and polarisation states appear as separable, i.e., as a product).

Where is the snag this time? The snag is that we have failed to notice that half the time the Polaroid will not produce a $|\psi\rangle = |U, 45\rangle|L, 45\rangle/\sqrt{2}$ state but will absorb the upper beam photon. But in these cases the lower beam photon still gets detected at the x-screen. In this case the lower beam photon is left in the state $|L, -45\rangle/\sqrt{2} = (|L\rangle|v\rangle - |L\rangle|h\rangle)/2$. After passage through the double slits and the two Circular Polarisers this becomes,

$$|\psi\rangle = \frac{1}{2\sqrt{2}}(|L1\rangle|C\rangle + |L2\rangle|AC\rangle - |L1\rangle|AC\rangle - |L2\rangle|C\rangle)$$
$$= \frac{1}{2\sqrt{2}}(|L1\rangle - |L2\rangle)(|C\rangle - |AC\rangle) \qquad (20.1.8)$$

Taking the inner product with the spatial-basis bra $\langle x|$ then forming the absolute square gives the intensity at the x-screen due to those cases when the upper beam photon is absorbed by the Polaroid,

$$|\psi(x)|^2 = \frac{1}{2}(1 - cos\,\theta\,(x)) \qquad (20.1.9)$$

The only difference between (20.1.9) and (20.1.7) is the sign of the cosine term. What is **actually** observed locally at the x-detector is not (20.1.7) – did I successfully con you? – but the **sum** of (20.1.7) and (20.1.9). So the cosine term cancels and we get a total signal of $|\psi(x)|^2 = 1$ at the x-screen, i.e., no interference pattern! Why the sum of (20.1.7) and (20.1.9)? Because, whilst the 45° Polaroid stops half the photons in the upper beam, it does not stop any in the lower beam – so both lower beam states $|L, 45\rangle/\sqrt{2}$ and $|L, -45\rangle/\sqrt{2}$ contribute.

The claim that was made above regarding observing a **local** interference pattern was incorrect. Actually the interference corresponding to Equ.(20.1.7) will be observed only if coincidence counts are used. That is, only photons arriving at the x-screen coincident with photons successfully reaching the upper detector will result in an interference pattern. (Or, equivalently, assuming perfect detectors, interference would be observed at the lower screen if only photons corresponding to **no** photon in the upper detector are counted, i.e., (20.1.9)). The interference is not local but correlated, non-local interference. There is, of course, no FTL communication.

The experiment described here has actually been carried out, by Walborn, Terra Cunha, Padua, and Monken, with results just as anticipated by the theory above, see Ref.[20.1].

20.2 Delayed Erasure

The term "delayed erasure" is given to experiments which involve erasure that is carried out *after* detection of the particles which produce the interference pattern. This is easily accomplished in the experimental setup of Walborn et al, Figure 20.1, simply by making the upper beam path long enough prior to the 45° Polaroid. Note that in §20.1 the analysis implicitly assumed that the Polaroid was encountered first and the double slits with their polarisation rotators encountered second. This order is easily reversed in the algebra. After the double slits and polarisation rotators, but before the 45º Polaroid, the state is given by (20.1.1), i.e.,

$$|\psi\rangle = \frac{1}{2}(|Uv\rangle[|L1\rangle|AC\rangle + |L2\rangle|C\rangle] + |Uh\rangle[|L1\rangle|C\rangle + |L2\rangle|AC\rangle]) \quad (20.2.1)$$

Using (19.2.7,8) to re-express the upper beam polarisation in terms of the $|\pm 45\rangle$ states this becomes,

$$|\psi\rangle = \frac{1}{2\sqrt{2}}\left(\begin{array}{l}|U\rangle(|45\rangle + |-45\rangle)[|L1\rangle|AC\rangle + |L2\rangle|C\rangle]\\+|U\rangle(|45\rangle - |-45\rangle)[|L1\rangle|C\rangle + |L2\rangle|AC\rangle]\end{array}\right) \quad (20.2.2)$$

The effect of the 45º Polaroid - as regards cases where the upper photon successfully passes through the Polaroid - is to reduce state (20.2.2) to,

$$|\psi\rangle = \frac{1}{2\sqrt{2}}|U, 45\rangle([|L1\rangle|AC\rangle + |L2\rangle|C\rangle] + [|L1\rangle|C\rangle + |L2\rangle|AC\rangle])$$
$$= \frac{1}{2\sqrt{2}}|U, 45\rangle(|L1\rangle + |L2\rangle)(|AC\rangle + |C\rangle) \quad (20.2.3)$$

This is exactly as (20.1.6), and hence also produces the intensity at the x-screen given by (20.1.7), i.e., an apparent interference pattern (but which can actually be seen only by coincidence counting with arrivals at the upper detector, as before). Consequently, the delay of the erasure makes no difference to the results, which are the same as in §20.1.

Great play is made in popular accounts of the weirdness of this outcome. The lower beam photon appears to "know" that a 45º Polaroid will be inserted into the upper beam *some time after* the lower photon has already been detected. The lower photon has to "know" this in order to know whether or not to produce an interference pattern on the x-screen.

But these protestations of weirdness forget the crucial issue: that an interference pattern is found only when the x-detector counts are vetoed according to whether there is a corresponding count in the upper detector. The interference is not local, but only a correlated, non-local interference. The interference pattern is found only when those x-screen (lower beam) photons are counted which correspond to a photon successfully emerging from the 45º Polaroid into the upper detector.

The apparent weirdness is largely the result of a conceptual error. One confuses the situation with an instantaneous effect at the x-screen due to the insertion of the Polaroid in the (possibly remote) upper beam. This would be FTL communication and would indeed be very weird – in fact, impossible. But there is no such effect. Ask yourself, when the Polaroid is inserted what exactly is the instantaneous effect on the signal in the x detector? Say that you have placed your x detector carefully at a minimum of the potential interference pattern – and that this pattern is perfect so that no photons at all would reach this point if interference did occur. Will you detect a photon or not? You cannot tell in advance because it depends upon whether the Polaroid in the upper beam happens to pass a 45° state photon or to absorb a -45° state photon – and you cannot control which will happen. So the insertion of the remote Polaroid cannot be used to propagate a FTL signal. A prediction of whether x-screen photons will be detected can be made only if one is first given the result from the upper detector.

Suppose the 45° Polaroid is so far away that we can count large numbers of photons and develop the whole x-screen whilst remaining causally disconnected from the Polaroid. What do we see? We see no interference. Whether or not the Polaroid has been inserted we see a uniformly illuminated x-screen. Look – nothing weird at all. Only when the signal received at the upper detector is used as a mask to retain or veto individual x-screen counts does the interference pattern emerge. What does this mean? It means that there is a correlation between two sets of measurements which were carried out in a causally disconnected manner. How weird is this? In itself it is not especially weird.

If two sub-systems have a common source, their properties will often be correlated. This is an objective fact and does not require subsequent causal connection between the sub-systems to manifest it. For example, if a mass breaks up into two smaller masses it will surprise no one to find that the mass of one part is (inversely) correlated with the mass of the other – even if the measurements of these masses is carried out when the parts are light years apart.

However this goes rather too far in dismissing quantum weirdness. There *is* a degree of weirdness, namely non-locality in spite of the absence of hidden variables. In the above classical example relating to conserved total mass the matter is clear because each part can be considered as possessing a definite mass before any measurement is made. This is not the case for the polarisation of the photons – they are in a superposed polarisation state initially. So the weirdness lies in the fact that the upper and lower detectors produce correlated results at space-like separations despite being in an indeterminate state before measurement. This is exactly the same weirdness first discussed by EPR, Ref.[20.2], and underlying all Bell-type situations (chapter 14). Delayed erasure experiments with entangled particles do not expose any new or deeper weirdness, just the same tension between genuine indeterminacy and non-local correlation. Quantum states do indeed exhibit non-locality (Bell, chapter 14), and quantum mechanics does not respect non-contextual realism (Kochen-Specker, chapter 9).

20.3 References

[20.1] Walborn, Terra Cunha, Padua, and Monken (2002), Physical Review A, (**65**, 033818, 2002). https://tinyurl.com/rmoscyc

[20.2] Einstein, A; B Podolsky, N Rosen (1935). "*Can Quantum-Mechanical Description of Physical Reality be Considered Complete?*". Physical Review **47,** 777–780

21

How the Magic is Lost: A Simple Example of Decoherence

The idea of decoherence was introduced in chapter 15. It is the process whereby a system loses its uniquely quantum mechanical properties, specifically the ability to form new states by superposition. In this chapter the process of decoherence is illustrated using simple examples. The first example involves an initial state with the typically "quantum" property of spatial delocalisation which decoheres into a localised state. Just as importantly, the second example illustrates how apparently "quantum" properties, such as lack of spatial localisation, can persist indefinitely. The two phenomena are not truly distinct; both are examples of the influence of the environment stabilising the system in a "pointer state". In both cases the process of decoherence is active, but they differ according to the specific pointer state which applies. These examples therefore help illustrate what determines the pointer basis. Finally, the example also illustrates how the timescale for decoherence to occur may be estimated.

21.1 Decoherence

The principles of decoherence were described in chapter 15. In this section a simple example of a decoherence calculation is presented. The simplest possible case is a two-state system. The system is considered to be in an excited state, but the excitation may reside in either of two sites within the system. The corresponding quantum states of the system are denoted $|1s\rangle$ and $|2s\rangle$, where the 's' means that these are states in which the excitation energy is spatially localised (at sites 1 and 2 respectively). However the system has some internal interaction between the two sites. As a result the states $|1s\rangle$ and $|2s\rangle$ are not energy eigenstates of the system. Under suitable circumstances the energy eigenstates may be approximated by,

$$|1\rangle = \frac{1}{\sqrt{2}}(|1s\rangle - |2s\rangle) \text{ and } \quad |2\rangle = \frac{1}{\sqrt{2}}(|1s\rangle + |2s\rangle) \qquad (21.1.1)$$

The reader may like to figure out why the states (21.1.1) arise naturally (see §21.3 for the answer). The corresponding eigen-energies are E_1 and E_2. Writing the Hamiltonian of the system as $\hat{H}$ we therefore have,

$$\hat{H}|i\rangle = E_i|i\rangle \qquad (21.1.2)$$

In this energy basis, $\{|i\rangle\}$, the free Hamiltonian in matrix notation is just,

$$(H) = \begin{pmatrix} E_1 & 0 \\ 0 & E_2 \end{pmatrix} \qquad (21.1.3)$$

The density matrix for the system may be expressed in either basis,

$$\hat{\rho} = \sum_{i,j} \rho_{ij} |i\rangle\langle j| = \sum_{i,j} \rho_{ij}^{s} |is\rangle\langle js| \qquad (21.1.4)$$

so that
$$(\rho^{s}) = \frac{1}{2}\begin{pmatrix} 1 & 1 \\ -1 & 1 \end{pmatrix}(\rho)\begin{pmatrix} 1 & -1 \\ 1 & 1 \end{pmatrix} \qquad (21.1.5)$$

We shall suppose that the system is prepared in state $|1\rangle$, the energy eigenstate with energy E_1. Assuming $E_2 > E_1$, the state $|2\rangle$ may be regarded as the excited state and $|1\rangle$ as the ground state. The quantum 'magic' in this example is the delocalisation of the excitation energy between two spatially separated sites as expressed by (21.1.1). State $|2\rangle$ is sometimes called an "exciton" state. Thus the "loss of the magic", or decoherence, will consist of the state becoming localised onto one or other of the two sites (in other words the state being reduced to either $|1s\rangle$ or $|2s\rangle$, or more generally a classical mixture of these two possibilities). Note that the initial density matrices in the two bases are,

$$(\rho)_0 = \begin{pmatrix} 1 & 0 \\ 0 & 0 \end{pmatrix} \quad \text{and} \quad (\rho^{s})_0 = \frac{1}{2}\begin{pmatrix} 1 & -1 \\ -1 & 1 \end{pmatrix} \qquad (21.1.6)$$

consistent with (21.1.5). The non-zero off-diagonal components in the density matrix expressed in the spatial basis, $(\rho^{s})_0$, are necessary in order that $|1s\rangle$ and $|2s\rangle$ combine coherently to produce the energy eigenstate $|1\rangle$. For this reason, the off-diagonal components of a density matrix are often (rather loosely) referred to as the "coherences"[4]. Consequently the signature of decoherence in this example will be the vanishing of the off-diagonal terms in (ρ^{s}). However, it is convenient initially to work in the energy basis.

To evaluate decoherence, a specific environmental interaction must be assumed. Suppose the environment couples to the system via an interaction Hamiltonian which in the unperturbed energy basis is,

$$(H_I) = \begin{pmatrix} 0 & V \\ V & 0 \end{pmatrix} \qquad (21.1.7)$$

This may be interpreted physically as follows. The system's two energy eigen-states might possess distinct magnetic dipole moments, perhaps due to

---

[4] "rather loosely" because, in general, even a diagonal density matrix (a classical mixture) can be given off-diagonal terms by a change of basis. So non-zero off-diagonal terms do not automatically ensure that the state has any quantum coherent properties. In this example we are specifically concerned with coherence between spatially localised states, so that (ρ^{s}) is in the preferred basis in which to examine this question.

opposing alignments of a spin ½ component in the conventional z-direction. Suppose the environment causes a magnetic field in the x-direction, $\bar{B} = B\hat{x}$. The magnetic field couples to the system's magnetic moment via an interaction energy $m\bar{B} \cdot \hat{\sigma}$ where $\hat{\sigma}$ are the Pauli matrices and m is the magnetic dipole moment of the system. For an x-direction field this is,

$$mB\sigma_x = \begin{pmatrix} 0 & mB \\ mB & 0 \end{pmatrix} \tag{21.1.8}$$

which is of the form proposed for (H_I) with $V = mB$.

The eigen-energies, λ, of the system are perturbed by interaction with the environment and the perturbed eigen-energies given by,

$$\left\| \begin{matrix} E_1 - \lambda & V \\ V & E_2 - \lambda \end{matrix} \right\| = 0 \tag{21.1.9}$$

This gives, $\qquad \lambda_\pm = \left\{ E_1 + E_2 \pm \sqrt{(E_2 - E_1)^2 + 4V^2} \right\}/2 \tag{21.1.10}$

In the unperturbed energy basis the new (perturbed) eigenstates are,

$$\begin{pmatrix} \alpha \\ \beta \end{pmatrix} \quad \text{and} \quad \begin{pmatrix} \beta \\ -\alpha \end{pmatrix} \quad \text{where,} \tag{21.1.11}$$

$$\alpha = \left[1 + \left(\tfrac{\lambda_+ - E_1}{V}\right)^2\right]^{-1/2} \text{and} \quad \beta = \left[1 + \left(\tfrac{V}{\lambda_+ - E_1}\right)^2\right]^{-1/2} \tag{21.1.12}$$

and where the first of the states in (21.1.11) has energy λ_+ and the second has energy λ_-. Because, at time $t = 0$, the system is prepared in free state $|1\rangle$, the initial state can be written,

$$|1\rangle \equiv \begin{pmatrix} 1 \\ 0 \end{pmatrix} = \alpha \begin{pmatrix} \alpha \\ \beta \end{pmatrix} + \beta \begin{pmatrix} \beta \\ -\alpha \end{pmatrix} \equiv \alpha|\tilde{1}\rangle + \beta|\tilde{2}\rangle \tag{21.1.13}$$

where the tilde denotes the perturbed eigenstates of the system, (21.1.11), when the environmental interaction is taken into account. But assuming that the interaction energy, V, is constant, i.e., that the environment state does not change, then the time evolution of the perturbed eigenstates is given by,

$$|\tilde{1}, t\rangle = |\tilde{1}\rangle e^{-it\lambda_+/\hbar} \quad \text{and} \quad |\tilde{2}, t\rangle = |\tilde{2}\rangle e^{-it\lambda_-/\hbar} \tag{21.1.14}$$

in accord with the Schrodinger equation. Note that in general the interaction with the system will cause the state of the environment to change, so that V would not necessarily be constant. However it suffices to assume constant V for this simple illustration. This means that the time evolution of the initial state, (21.1.13), can be written down immediately as,

$$|1, t\rangle = \alpha |\tilde{1}\rangle e^{-\frac{it\lambda_+}{\hbar}} + \beta |\tilde{2}\rangle e^{-\frac{it\lambda_-}{\hbar}} \qquad (21.1.15)$$

Hence, $\qquad |1, t\rangle = \begin{pmatrix} \alpha^2 e^{-it\lambda_+/\hbar} + \beta^2 e^{-i\lambda_- t/\hbar} \\ \alpha\beta \left(e^{-i\lambda_+ t/\hbar} - e^{-i\lambda_- t/\hbar} \right) \end{pmatrix} \qquad (21.1.16)$

where the RHS of (21.1.16) is understood to be in the unperturbed energy basis. The corresponding density matrix, again in the unperturbed energy basis, is thus,

$$(\rho) = \begin{pmatrix} \alpha^2 e^{-i\lambda_+ t/\hbar} + \beta^2 e^{-i\lambda_- t/\hbar} \\ \alpha\beta \left(e^{-i\lambda_+ t/\hbar} - e^{-i\lambda_- t/\hbar} \right) \end{pmatrix} \begin{pmatrix} \alpha^2 e^{-i\lambda_+ t/\hbar} + \beta^2 e^{-i\lambda_- t/\hbar} \\ \alpha\beta \left(e^{-i\lambda_+ t/\hbar} - e^{-i\lambda_- t/\hbar} \right) \end{pmatrix}^+ \qquad (21.1.17)$$

The off-diagonal element being,

$$(\rho)_{12} = \left(\alpha^2 e^{-i\lambda_+ t/\hbar} + \beta^2 e^{-i\lambda_- t/\hbar} \right) \alpha\beta \left(e^{+i\lambda_+ t/\hbar} - e^{+i\lambda_- t/\hbar} \right)$$
$$= \alpha\beta(\alpha^2 - \beta^2) \left(1 - \cos\frac{\Delta\lambda t}{\hbar} \right) - i\alpha\beta \sin\frac{\Delta\lambda t}{\hbar} \qquad (21.1.18)$$

noting that α, β are real and where

$$\Delta\lambda = \sqrt{(E_2 - E_1)^2 + 4V^2} \qquad (21.1.19)$$

Thus, the off-diagonal component is oscillatory. It passes through zero every time $\Delta\lambda t/\hbar$ is an integer multiple of 2π. But its averaged magnitude does not tend to reduce.

So far we have assumed a single environment state, as effectively labelled by the interaction energy, V. But the nature of the environment is that it is a complicated assemblage of many parts and hence many states. Decoherence will arise as a consequence of the involvement of more than one environment state. But this can be evaluated only if sufficient is known about the distribution of environment states. In this case we require the probability of a given interaction energy, $p(V)$, so that the reduced density matrix can be found as the average,

$$(\rho^{red}) = \Sigma_V p(V)(\rho) \qquad (21.1.20)$$

where (ρ) is given by (21.1.17). Note that the V-dependence of (ρ) comes from all the three parameters $\alpha, \beta, \Delta\lambda$ due to (21.1.12) and (21.1.19).

In general many different behaviours might result from the sum (21.1.20), depending upon the spectral density, $p(V)$, of the environmental interaction. However our purpose here is only to show that complete spatial decoherence can occur for a reasonable choice of $p(V)$. It suffices to assume that the strength of the interaction with the environment, as measured by $2V$, is far greater than the separation of the energy levels of the unperturbed system. That is,

$$2V >> |E_2 - E_1| \tag{21.1.21}$$

We then have, from (21.1.19), that $\Delta\lambda \approx 2V$, and (21.1.10) implies that,

$$\lambda_+ - E_1 \approx V \tag{21.1.22}$$

so that (21.1.12) gives $\alpha \approx \beta \approx 1/\sqrt{2}$. Hence the off-diagonal component of the reduced density matrix becomes,

$$(\rho^{red})_{12} = -\frac{i}{2}\Sigma_V\, p(V)\, \sin\frac{2Vt}{\hbar} \tag{21.1.23}$$

For instance, if we assume $p(V)$ has a flat distribution up to some maximum V_{max} then we can replace $\Sigma_V\, p(V) \ldots$ with $\int_0^{V_{max}} \ldots \frac{dV}{V_{max}}$ which leads to,

$$(\rho^{red})_{12} = \frac{i\hbar}{4tV_{max}}\left[1 - \cos\left(\frac{2tV_{max}}{\hbar}\right)\right] \tag{21.1.24}$$

Thus, the amplitude of the off-diagonal component of the density matrix, expressed in the unperturbed energy basis, reduces continuously with time, in proportion to $1/t$. It will be small for times larger than a decoherence time scale given roughly by $t_{dec} = \hbar/V_{max}$. Perhaps more realistically we could assume an environment with a Gaussian spectral density, so that (21.1.23) becomes,

$$(\rho^{red})_{12} = \frac{2}{\sqrt{\pi}V_0}\int_0^\infty (\rho)_{12}\, exp\left\{-\frac{V^2}{V_0^2}\right\}\cdot dV \tag{21.1.25}$$

This results in,

$$(\rho^{red})_{12} \propto exp\left\{-\left(\frac{V_0 t}{\hbar}\right)^2\right\} \tag{21.1.26}$$

implying an even more rapid decay of the coherence once $t > \hbar/V_0$. In both cases, though, the decoherence timescale is of the form $t_{dec} \approx \hbar/V_0$ for some characteristic strength of the environmental interaction, V_0.

Examination of the diagonal terms in (21.1.17) when $\alpha \approx \beta \approx 1/\sqrt{2}$ shows that they are both ½ plus an oscillatory cross-term. When summed over V as in (21.1.25) these cross-terms will tend to zero for times longer than the decoherence time, for the same reason that we found $(\rho^{red})_{12} \to 0$. Consequently the reduced density matrix of the system in the unperturbed energy basis becomes simply,

For $t \gg t_{dec}$ $\qquad\qquad (\rho^{red}) \rightarrow \frac{1}{2}\begin{pmatrix} 1 & 0 \\ 0 & 1 \end{pmatrix}$ $\qquad\qquad\qquad$ (21.1.27)

Recall that (21.1.27) is in the unperturbed energy basis. The final step is to note that, by virtue of (21.1.5), the reduced density matrix of the system in the spatial basis is identical to (21.1.27). Hence the spatial coherence – the off diagonal terms which were evident in the initial state, (21.1.6) - have disappeared. The system is now in a classical mixture of states for which the energy resides at site 1 or site 2, no longer a superposition of the two. The system will be found in a localised state. The quantum delocalisation has vanished - decohered.

Warning: do not imagine that decoherence times are always well approximated by the result $t_{dec} = \hbar/V$ derived for this simple example. Things are not so straightforward in every case, though similar results are typical. Also, do not imagine that it is usual for the (reduced) density matrix to be diagonal in both the energy basis and the spatial basis. This happens in this very simple example only because of the simplifying approximation (21.1.21) which leads to $\alpha \approx \beta \approx 1/\sqrt{2}$ and hence to reduced density matrices which are multiples of the unit matrix.

Note that the decoherence has been achieved in this simple model whilst assuming only unitary time evolution in accord with the Schrodinger equation. All that is required is a sufficient number of environment states with a range of interaction energies with the system. Hence, in as far as decoherence explains why measurement involves a reduction of the state vector ("collapse of the wavefunction") this can now be attributed to the environment without invoking any different type of time evolution. This provides a route for eliminating the blemish of the Copenhagen interpretation that appealed to a completely different type of temporal evolution specific to measurements. We can now interpret this as a convenient phenomenological short-hand for the highly complex, and highly variable, interaction with the environment, the formulation of which requires the formalism of QM to be set up before it can be elucidated.

21.2 Persistent Delocalisation

In the preceding example, spatial decoherence to a classical mixture of localised states occurred on a timescale of about $t_{dec} = \hbar/V$. The key assumption leading to this conclusion was that the interaction with the environment was strong compared with the system's energy level spacings, i.e., $2V \gg |E_2 - E_1|$.

Providing that the interaction, V, with the environment is non-zero, then decoherence will typically occur, perhaps on a timescale of around $t_{dec} \approx \hbar/V$ (though we emphasise that this is not invariably the case). So a weaker environmental interaction will lead to longer decoherence times. However, decoherence rates tend to be very rapid by human standards, even for very weak interactions, because of the small magnitude of $\hbar$. Hence, irrespective of whether $2V \gg |E_2 - E_1|$ or $2V \ll |E_2 - E_1|$, an interaction strength of ~1 meV will typically (but not invariably!) cause decoherence in the order of a picosecond.

So, does it matter whether we are in the regime $2V \gg |E_2 - E_1|$ or $2V \ll |E_2 - E_1|$? Yes, it does, but the distinction between these two cases lies more in the corresponding pointer states than in the timescale for decoherence. Recall that the pointer states are the preferred basis into which the system, due to the environmental interaction, decoheres. It turns out that the latter case, with a weak environmental interaction, results in pointer states which approximate to the system's energy eigenstates. Thus, if the initial state is close to an energy eigenstate then decoherence will alter it little, and spatial delocalisation will persist even after decoherence. Atomic electron orbitals are the obvious example. These remain in energy eigenstates and are spatially delocalised permanently as a result.

Our simple example can be used to illustrate how coherence can persist if the interaction with the environment is weak compared with the spacing of the system's energy levels. Thus if,

$$2V \ll |E_2 - E_1| \tag{21.2.1}$$

then $\Delta\lambda \approx |E_2 - E_1|$ and $\lambda_- \approx E_1$, $\lambda_+ \approx E_2$ and hence $\alpha \approx 0$, $\beta \approx 1$. So (21.1.13) shows that the initial state, of energy E_1, approximates to the state $|\tilde{2}\rangle$ expressed in the perturbed basis (we are assuming that $E_2 > E_1$). And, with this approximation, (21.1.17) indicates that the density matrix in the unperturbed energy basis becomes time-independent, being about $\begin{pmatrix} 1 & 0 \\ 0 & 0 \end{pmatrix}$ at all times. Hence, in the spatial basis (21.1.5) gives $(\rho^s) \approx \frac{1}{2}\begin{pmatrix} 1 & -1 \\ -1 & 1 \end{pmatrix}$. The former means that the initial energy eigenstate persists indefinitely, and the latter is merely a reminder that this implies that coherence in the spatial basis (non-zero off-diagonal terms) persists indefinitely.

Note that it is not merely that decoherence takes a long time. The spatial coherence remains indefinitely (as for atomic electron orbitals) because the pointer states are the energy eigenstates which are spatially delocalised. This can be seen from (21.1.17) if we consider α to take some small but non-zero value $|\alpha| \ll |\beta|$. Then the density matrix element ρ_{11} is much larger than the other components of (ρ) at all times. The exact magnitudes of ρ_{ik} will oscillate, but ρ_{11} is always much larger than the other components. Consequently the 'pointer states' resulting from the environmental interaction, and given by (ρ), approximate to the energy eigenstates at all times however long. (And, in this approximation, there is little difference between the perturbed and unperturbed energies and energy eigenstates).

In fact it can be shown quite generally that if the energy available from environmental interaction is small compared with the **spacings** of the free system's energy levels, then the pointer states approximate to the free system's eigenstates. In chapter 22 we illustrate this with a further model.

21.3 Reader Exercises

How do equs.(21.1.1) arise naturally? Work in the spatial basis in which,

$$|1s\rangle \rightarrow \begin{pmatrix} 1 \\ 0 \end{pmatrix} \quad \text{and} \quad |2s\rangle \rightarrow \begin{pmatrix} 0 \\ 1 \end{pmatrix} \tag{21.3.1}$$

Suppose initially there was no interaction between these two spatial positions. Each would therefore have the same energy, say E_0. The interaction-free Hamiltonian in the spatial basis is therefore simply,

$$\hat{H}_0 \rightarrow \begin{pmatrix} E_0 & 0 \\ 0 & E_0 \end{pmatrix} \tag{21.3.2}$$

and the spatially localised states, (21.3.1), are also the energy eigenstates, which are degenerate. The degeneracy of the energy eigenstates can be broken by introducing an interaction between the two spatial sites. This arises naturally if the particles in question have charge or magnetic, or electric, dipole moments. Suppose the interaction energy associated with such a particle being present on both spatial sites is V_{int}. This is the same as saying that there is an interaction Hamiltonian as follows,

$$\hat{H}_{int} \rightarrow \begin{pmatrix} 0 & V_{int} \\ V_{int} & 0 \end{pmatrix} \tag{21.3.3}$$

That this represents an interaction between the two spatial sites can be seen from the fact that the matrix elements of $\hat{H}_{int}$ are non-zero only between different spatial states (i.e., the diagonal elements are zero). The total Hamiltonian is thus,

$$\hat{H} \rightarrow \begin{pmatrix} E_0 & V_{int} \\ V_{int} & E_0 \end{pmatrix} \tag{21.3.4}$$

It is readily seen that the eigen-energies of this Hamiltonian are,

$$E_\pm = E_0 \pm V_{int} \qquad (21.3.5)$$

and that the corresponding eigenstates are,

$$\bar{v}_\pm = \frac{1}{\sqrt{2}}\begin{pmatrix}1\\ \pm 1\end{pmatrix} \qquad (21.3.6)$$

Hence we see that states (21.1.1) are just the same as the states (21.3.6), specifically $|1\rangle = \bar{v}_-$ and $|2\rangle = \bar{v}_+$, so that (21.1.1) are indeed the energy eigenstates which arise in this rather natural situation, with $|2\rangle$ having the higher energy if V_{int} is positive (repulsive force), or the lower energy if V_{int} is negative (attractive force). In the main text we can thus identify $E_1 = E_-$ and $E_2 = E_+$.

(Aside: Note that the system's internal interaction Hamiltonian (21.3.3) is not the same as the interaction Hamiltonian with the environment examined in the main text, equ.(21.1.7), despite appearing the same, because the latter is expressed in the energy-basis, whereas the former is expressed in the spatial basis).

22

How the Magic is Retained: Another Example of Decoherence

When the environment couples to a system with typical energies which exceed the spacing between the system's eigen-energies, decoherence will give rise to pointer states which differ from the free system's energy eigenstates and which are often spatially localised states. This was illustrated using a simple model in chapter 21. But if the interaction energies with the environment are small compared with the spacing of the system's eigen-energies then the pointer states towards which the system decoheres are the system's energy eigenstates themselves. An initial energy eigenstate, and its attendant spatial delocalisation, will then persist despite decoherence. This is illustrated here using a simple model.

22.1 Introduction: The Theorem of Paz-Zurek-Gogolin

In general, interaction with the environment will cause an initially pure quantum state to decohere into a mixed state. In chapter 15 we saw that each system-environment combination defines a particular set of pointer states, these forming the preferred basis in which the decohered mixed state density matrix is diagonal. If the system-environment interaction energy is large compared with the system's energy level spacings, the pointer states approximate to the eigenstates of the interaction Hamiltonian. This was the case illustrated in chapter 21.

At the opposite extreme, if the system-environment interaction energy is small compared with the spacings between the system's free eigen-energies, then the pointer states will approximately coincide with the system's energy eigenstates. This was first recognised and demonstrated in the context of a particular model by Paz and Zurek, Ref.[22.1]. The theorem has been proved in all generality recently by Gogolin, Ref.[22.2]. Here we merely make the theorem plausible to the reader by illustrating how it arises in a very simple model. The implication is that, if a system is prepared in an energy eigenstate, often spatially delocalised, then this spatial delocalisation will persist indefinitely if the environment is only weakly coupled to the system compared to the internal coupling within the system (i.e., if $V \ll \Delta E$). On the other hand, with such weak environmental coupling, if the system starts in some other pure quantum state, then it will decohere into a mixture of eigenstates of the system's free Hamiltonian.

The general situation was summarised by Paz and Zurek, Ref.[22.1], thus,

'There are three basically distinct regimes in which one can analyze the properties of the pointer states selected through decoherence. They differ through the relative strength of the self-Hamiltonian and of the Hamiltonian of interaction. The first one is the measurement situation where the self-Hamiltonian of the system is negligible and the evolution is completely dominated by the interaction with the environment. In such cases, pointer states are eigenstates of the interaction Hamiltonian. The second, most common and complex situation, occurs when neither the self-Hamiltonian nor the interaction with the environment are clearly dominant. Then the pointer states arise from a compromise between self-evolution and interaction. The third situation corresponds to the case where the dynamics is dominated by the system's self-Hamiltonian. In this case einselection produces pointer states which coincide with the energy eigenstates of the self-Hamiltonian'.

We now illustrate this last possibility.

22.2 Illustrating the Theorem of Paz-Zurek-Gogolin

Consider a two-level system with free Hamiltonian $\hat{H}_S$ such that,

$$\hat{H}_S|I\rangle = E_I|I\rangle \qquad (22.2.1)$$

for $I = 1,2$ with $E_1 < E_2$. Suppose the system interacts with an environment such that the interaction Hamiltonian is $\hat{V}$. The environment states will be denoted $|e\rangle$. The environment is assumed to be in some mixed state given by the density matrix,

$$\hat{\rho}_{env} = \Sigma_e\, p(e)|e\rangle\langle e| \qquad (22.2.2)$$

for real p with $0 \leq p(e) \leq 1$ and $\Sigma_e\, p(e) = 1$. If the environment were in the state $|e\rangle$ then the interaction Hamiltonian acting on the system's sub-space alone is written $\hat{V}(e)$. In this case the system's energy eigenstates are perturbed from their free states by the interaction and are given by,

$$\left(\hat{H}_S + \hat{V}(e)\right)|I,e\rangle = E_I^e|I,e\rangle \qquad (22.2.3)$$

The dependence of the perturbed eigenstates/eigenvalues on the environment state is denoted by the label "e". Suppose the system starts in an arbitrary pure state, which can be expressed in either basis thus,

$$|\psi_0\rangle = \alpha|1\rangle + \beta|2\rangle = \tilde{\alpha}|1,e\rangle + \tilde{\beta}|2,e\rangle \qquad (22.2.4)$$

where $\tilde{\alpha}, \tilde{\beta}$ will depend upon the environment state, e. We now make the simplifying assumption that the environment states do not change due to the interaction with the system, so that the time evolution of a combined pure state is,

$$|\psi_0\rangle|e\rangle \rightarrow |\psi_t(e)\rangle|e\rangle \qquad (22.2.5)$$

We have already seen in chapter 21 that the restriction to a constant environment state, whilst not realistic, does not prevent decoherence. The evolution of the system state, allowing for the interaction with the environment, follows from Schrodinger's equation, which is just the time-dependent version of (22.2.3),

$$\left(\hat{H}_S + \hat{V}(e)\right)|\psi_t\rangle = i\hbar \frac{\partial}{\partial t}|\psi_t\rangle \qquad (22.2.6)$$

from which we get (see §2.4),

$$|\psi_t(e)\rangle = \tilde{\alpha} \, exp\left\{-i\frac{E_1^e t}{\hbar}\right\}|1,e\rangle + \tilde{\beta} \, exp\left\{-i\frac{E_2^e t}{\hbar}\right\}|2,e\rangle \qquad (22.2.7)$$

The combined system-plus-environment density matrix evolves as, $\qquad$ (22.2.8)

$$\hat{\rho} = \sum_e p(e)|e\rangle\langle e| \otimes |\psi_0\rangle\langle\psi_0| \rightarrow \sum_e p(e)|e\rangle\langle e| \otimes |\psi_t(e)\rangle\langle\psi_t(e)|$$

The reduced density matrix describing the system alone is obtained by tracing-out the environment,

$$\hat{\rho}_{red} = \sum_e \langle e'|\hat{\rho}|e'\rangle = \sum_e p(e) |\psi_t(e)\rangle\langle\psi_t(e)| \qquad (22.2.9)$$

The pair of states $\{|1\rangle, |2\rangle\}$ are orthogonal, as are the pair of states $\{|1,e\rangle, |2,e\rangle\}$, because both are eigenstates of an Hermitian Hamiltonian with distinct eigenvalues. So there is a unitary transformation between them, which can be written in terms of some coefficients a, b such that $|a|^2 + |b|^2 = 1$ thus,

$$\begin{pmatrix}|1,e\rangle \\ |2,e\rangle\end{pmatrix} = \begin{pmatrix} a & b \\ -b^* & a^* \end{pmatrix}\begin{pmatrix}|1\rangle \\ |2\rangle\end{pmatrix} \quad \text{and} \quad \begin{pmatrix}\tilde{\alpha} \\ \tilde{\beta}\end{pmatrix} = \begin{pmatrix} a^* & b^* \\ -b & a \end{pmatrix}\begin{pmatrix}\alpha \\ \beta\end{pmatrix} \qquad (22.2.10)$$

Note that the coefficients a, b depend upon the environment state, e. Substituting (22.2.10) into (22.2.7) and re-arranging gives,

$$|\psi_t(e)\rangle = A|1\rangle + B|2\rangle \qquad (22.2.11)$$

where,

$$A = \left[(|a|^2\alpha + ab^*\beta) \, exp\left\{-i\frac{E_1^e t}{\hbar}\right\} + (|b|^2\alpha - ab^*\beta) \, exp\left\{-i\frac{E_2^e t}{\hbar}\right\}\right]$$
$$B = \left[(a^*b\alpha + |b|^2\beta) \, exp\left\{-i\frac{E_1^e t}{\hbar}\right\} + (-a^*b\alpha + |a|^2\beta) \, exp\left\{-i\frac{E_2^e t}{\hbar}\right\}\right] \qquad (22.2.12)$$

Substitution of (22.2.11) into (22.2.9) yields an expression for the reduced density matrix. However, before doing this it is sensible to simplify (22.2.12) for the case that the environmental interaction is weak. By this we mean that the matrix elements of the interaction potential, $\hat{V}(e)$, are small compared with $E_2 - E_1$. This is exactly the condition for which we may employ first order perturbation theory. So we have,

$$b \approx -\frac{V_{21}}{E_2 - E_1} \qquad \text{where, } V_{ik} = \langle i|\hat{V}(e)|k \rangle \qquad (22.2.13)$$

and $a = \sqrt{1 - |b|^2}$. The derivation of (22.2.13) using first order perturbation theory is an exercise for the reader (for the answer, see §22.4). The weak environment condition is implicit in the derivation of (22.2.13) and is,

$$|V_{21}| << |E_2 - E_1| \Rightarrow |b| << |a| \approx 1 \qquad (22.2.14)$$

Hence, retaining only leading terms in (22.2.11) we get,

$$|\psi_t(e)\rangle \approx \left[\alpha \, exp\left\{-i\frac{E_1^e t}{\hbar}\right\}\right]|1\rangle + \left[\beta \, exp\left\{-i\frac{E_2^e t}{\hbar}\right\}\right]|2\rangle \qquad (22.2.15)$$

Substitution in (22.2.9) provides the components of the reduced density matrix in the basis of the unperturbed (free) system's energy eigenstates, to first order,

$$(\rho_{red})_{11} = |\alpha|^2 \qquad (22.2.16a)$$

$$(\rho_{red})_{22} = |\beta|^2 \qquad (22.2.16b)$$

$$(\rho_{red})_{12} = \alpha\beta^* \sum_e p(e) \, exp\left\{i\frac{(E_2^e - E_1^e)\cdot t}{\hbar}\right\} \qquad (22.2.16c)$$

Recall that α, β are the coefficients in terms of which the arbitrary initial state is expressed in the energy basis, (22.2.4). Consequently they are independent of the environment state and can be taken outside the sum over the e states. The energy eigenvalues which appear in (22.2.16c) are, however, those of the perturbed states, and hence depend upon the environment state, e. First order perturbation theory (see §22.4) tells us what these perturbed eigenvalues are, approximately, i.e., $E_1^e \approx E_1 + V_{11}$ and $E_2^e \approx E_2 + V_{22}$. Using these estimates and putting $\Delta V(e) = V_{22} - V_{11}$, (22.2.16c) becomes,

$$(\rho_{red})_{12} = \alpha\beta^* \, exp\left\{i\frac{(E_2 - E_1)\cdot t}{\hbar}\right\}\sum_e p(e) \, exp\left\{i\frac{\Delta V(e)\cdot t}{\hbar}\right\} \qquad (22.2.17)$$

For times which are long compared with $\hbar/\Delta V(e)$ the relative phases of the terms in this sum will differ by amounts which are large compared with unity. Hence, for such times, the sum in (16) is over terms with essentially random relative phases. Consequently there will be strong cancellation and it is fairly clear that the sum will reduce to zero for $t >> \hbar/\Delta V(e)$. Explicit evaluation of the sum requires some model for the environment to provide the spectrum of states, $p(e)$. In chapter 21 we considered a flat spectrum up to some energy cut-off, ΔV_{max}, and showed that a similar integral reduced to zero $\propto 1/t$ and became small compared with unity when $t \gg \hbar/\Delta V_{max}$.

Similarly, as in chapter 21, a Gaussian spectrum of width ΔV_σ leads to an exponential decay: $(\rho_{red})_{12} \propto exp\left\{-\left(\frac{\Delta V_\sigma t}{\hbar}\right)^2\right\}$.

Consequently, on a timescale $t \gg \hbar/\Delta V$, where ΔV is some characteristic environmental interaction energy, the initial pure state decoheres into a mixed state whose density matrix in the unperturbed energy basis is,

$$(\rho_{red}) = \begin{pmatrix} |\alpha|^2 & 0 \\ 0 & |\beta|^2 \end{pmatrix} \qquad (22.2.18)$$

This shows that, for this illustrative example, a weak environmental interaction, such that $|V_{21}| \ll |E_2 - E_1|$, leads to the pointer states being the energy eigenstates, because the density matrix becomes diagonal in this basis at times long compared with $\hbar/(V_{22} - V_{11})$. Note that it is the strength of the interaction with the environment which determines the timescale of decoherence both in this example, where the pointer states are the energy eigenstates, and in the example of chapter 21 where the pointer states were the spatially localised states determined by the interaction Hamiltonian.

One may reasonably ask what is special about the energy basis that this defines the pointer basis when the environmental interaction is sufficiently weak. What is special about the energy basis is that time evolution and energy are intimately related. Time evolution is governed by the Hamiltonian, which is also the energy operator. Time and energy are conjugate variables, and energy is that quantity which is conserved due to temporal homogeneity by Noether's Theorem and its quantum equivalent (see chapter 5).

Despite the weak environment in the sense that, $|V_{21}| \ll |E_2 - E_1|$, decoherence still occurs in general, and the corresponding timescale (typically, but not invariably, $\sim \hbar/\Delta V$) may be very short. Interaction strengths in the order of milli-eV produce decoherence in the order of picoseconds if $\hbar/\Delta V$ is indicative. The initial pure state, (22.2.4), becomes the mixed state, (22.2.18). However, an obvious exception is when the initial state happens to be an energy eigenstate, say $|1\rangle$ with $\alpha = 1, \beta = 0$. Then (22.2.4) and (22.2.18) are identical. Although the mechanism of decoherence is still present, i.e., the interaction with the environment is still in operation, the initial state does not change because it happens to already be a pointer state. Moreover, since an energy eigenstate will generally be spatially delocalised, this delocalisation will persist indefinitely.

22.3 Examples in Photosynthesis

A possible example of this phenomenon occurs in a most unexpected place: a biological system, namely photosynthesis. Quoting Engel et al, Ref.[22.3],

'Photosynthetic complexes are exquisitely tuned to capture solar light efficiently, and then transmit the excitation energy to reaction centres, where long term energy storage is initiated. The energy transfer mechanism….. invokes 'hopping' of excited-state populations along discrete energy levels…..spectroscopic data clearly document the dependence of the dominant energy transport pathways on the spatial properties of the excited-state wavefunctions of the whole bacteriochlorophyll complex….. electronic quantum beats arising from quantum coherence in photosynthetic complexes have been predicted and indirectly observed. Here we extend previous two-dimensional electronic spectroscopy investigations of the FMO bacteriochlorophyll complex, and obtain direct evidence for remarkably long-lived electronic quantum coherence playing an important part in energy transfer processes within this system. The quantum coherence manifests itself in characteristic, directly observable quantum beating signals among the excitons within the Chlorobium tepidum FMO complex at 77 K. This wavelike characteristic of the energy transfer within the photosynthetic complex can explain its extreme efficiency, in that it allows the complexes to sample vast areas of phase space to find the most efficient path.'

The longevity of the observed quantum beats exceeded expectation prior to 2007. The point here is that the "warm, wet" environment in biological systems is normally considered to create a strong interaction with the organelle, and hence to imply very rapid decoherence. It is feasible that the explanation for the observed sustained spatial coherence is simply that the interaction between neighbouring chromophores is strong, resulting in exciton energy level spacings which exceed the interaction energies with the environment. In this case spatial coherence would be maintained despite the potential for rapid decoherence because the pointer states are the energy eigenstates, and so an initial energy eigenstate (exciton) would be unaffected by decoherence. The case for this has been argued by Bradford, Ref.[22.4].

22.4 Exercises for the Reader

22.4.1 Derive (22.2.13). In §8.4 we derived the equations of first order time dependent perturbation theory. Here we address the simpler case of the time-independent stationary states (i.e., energy eigenvalues). I will not derive the equations of first order perturbation theory in full generality here, instead just concentrate upon (22.2.13). The eigen-equation for the perturbed state $|1, e\rangle$, essentially (22.2.3) is,

$$\left(\hat{H}_S + \hat{V}(e)\right)|1, e\rangle = E_1^e|1, e\rangle \tag{22.4.1}$$

whereas the unperturbed energy eigenstates obey,

$$\hat{H}_S|i\rangle = E_i|i\rangle \tag{22.4.2}$$

So writing $|1, e\rangle = a|1\rangle + b|2\rangle$, consistent with (22.2.10), we have,

$$\left(\hat{H}_S + \hat{V}(e)\right)(a|1\rangle + b|2\rangle) = E_1^e(a|1\rangle + b|2\rangle) \tag{22.4.3}$$

Hence,

$$\hat{V}(e)(a|1\rangle + b|2\rangle) = a(E_1^e - E_1)|1\rangle + b(E_1^e - E_2)|2\rangle \tag{22.4.4}$$

Writing the matrix elements of the perturbation as $V_{ik} = \langle i|\hat{V}(e)|k\rangle$ and taking matrix elements of (22.4.4) gives the eigen-equation,

$$\begin{pmatrix} V_{11} + E_1 & V_{12} \\ V_{21} & V_{22} + E_2 \end{pmatrix}\begin{pmatrix} a \\ b \end{pmatrix} = E_1^e\begin{pmatrix} a \\ b \end{pmatrix} \tag{22.4.5}$$

Solving the corresponding quadratic secular equation, i.e.,

$$\left\|\begin{matrix} V_{11} + E_1 - E_1^e & V_{12} \\ V_{21} & V_{22} + E_2 - E_1^e \end{matrix}\right\| = 0 \tag{22.4.6}$$

And discarding terms in the discriminant of higher order than V/E, i.e., appealing to a weak perturbation, we find to first order that,

$$E_1^e \approx E_1 + V_{11} \quad \text{and} \quad E_2^e \approx E_2 + V_{22} \tag{22.4.7}$$

The first equation in (22.4.5) cannot be used to find b because, to first order it merely gives $b \approx 0$. However, the second equation in (22.4.5) is more helpful, giving, for small V_{ik} compared with $E_{1,2}$,

$$\frac{b}{a} = \frac{V_{21}}{E_1^e - V_{22} - E_2} \approx \frac{V_{21}}{E_1 + V_{11} - V_{22} - E_2} \approx \frac{V_{21}}{E_1 - E_2} \tag{22.4.8}$$

Since this is small compared with 1, we can further approximate b/a simply as b, so that (22.4.8) is just (22.2.13). <u>QED</u>.

22.5 References

[22.1] J.P.Paz and W.H.Zurek (1999), "Quantum limit of decoherence: Environment induced superselection of energy eigenstates", Phys.Rev.Lett.**82**:5181-5185,1999.

[22.2] C.Gogolin (2010) "Einselection without pointer states", Phys. Rev. E **81**, 051127, arXiv: 0908.2921v4 [quant-ph]

[22.3] Engel G S, Calhoun T R, Read E L, Ahn T K, Mancal T, Cheng Y C, Blankenship R E and Fleming G R (2007), "Evidence for wavelike energy transfer through quantum coherence in photosynthetic systems", *Nature* **446** 782.

[22.4] R A W Bradford (2011) "The persistence of quantum coherence in dipole dimer states in an environment of dipoles: implications for photosynthetic systems?", J. Phys. A: Math. Theor. **44**, 245303.

Delineating the border between the quantum realm ruled by the Schrödinger equation and the classical realm ruled by Newton's laws is one of the unresolved problems of physics. **Figure 1**

Zurek, Physics Today (1991)

23

Why Are Big Things Classical?

Since everything is composed of particles, atoms and molecules, it follows that large things are composed of very many parts. In other words, big things have a very large number of degrees of freedom. It is this, rather than physical size, which promotes classicality in their behaviour.

23.1 What is the Problem?

We have seen in chapters 15, 21 and 22 that decoherence is due to interaction of the system with the environment. This causes the density matrix of the system to become diagonal in a particular basis called the pointer basis. The simple example in chapter 22 has illustrated that the pointer states of a system approximate to its energy eigenstates if its energy level spacings are large compared with the energies characterising the interaction with the environment. Decoherence naturally 'einselects' the energy eigenstates in this situation. The energy eigenstates are usually spatially delocalised and hence delocalisation is the natural condition in such a case. An example is the state of an atomic, or molecular, electron.

On the other hand, macroscopic objects tend not to be in delocalised states. They tend to be in classical states, characterised by being located at a definite position – not simultaneously "here and there". Why is it that being "big" causes this? To be consistent with our understanding of decoherence, "big" things must tend to have very closely spaced energy levels – so that even an extremely weak interaction with the environment will involve interaction energies greater than the natural energy spacing. Under these conditions the environment will decohere the system away from its pure energy eigenstates. Accordingly, this chapter provides some simple illustrations of how very close spacing of energy levels typically arises when the system's number of degrees of freedom is large.

Although this explains why classical behaviour is typical for "big" things, there are exceptions to this general tendency. In fact some of the most interesting physical phenomena arise when big things continue to behave quantumly, and chapter 24 will provide some examples of just that. Moreover, we will see in that chapter that the retained "quantum-like" behaviour is again associated with large system energy spacings (band gaps).

23.2 The Continuum Field Limit

Consider a lattice consisting of a large number of bonded atoms – sufficiently many that we may treat it approximately as an elastic continuum. Phonons with wave-vector $\bar{k}$ are associated with an energy $\hbar\omega$ and in the long wavelength limit we can consider the almost-continuum to have a constant velocity of sound, v_s, such that $\omega = v_s |\bar{k}|$. Hence the possible values of $\hbar\omega$ are the eigen-energies with wavevectors given by,

$$|\bar{k}| = \frac{\pi}{L}\sqrt{n_x^2 + n_y^2 + n_z^2} \tag{23.1.1}$$

where we have assumed a cube of material of side L and N_1 atoms per side, hence with lattice spacing $a = L/N_1$. In (23.1.1), the quantum numbers (n_x, n_y, n_z) take the values 1, 2, 3... to a maximum of N_1. In a true continuum the quantum numbers could be arbitrarily large, but for a discrete lattice the number of classical natural frequencies is limited to the number of degrees of freedom.

Considering a 1D line of atoms the energy spacing is thus,

$$\Delta E \approx \frac{\pi \hbar v_s}{a N_1} \tag{23.1.2}$$

The energy level spacing decreases asymptotically to zero for large numbers of atoms in inverse proportion to the number of atoms in the line, $\propto 1/N_1$. So we see immediately with this simplest example why systems with many degrees of freedom have small eigen-energy spacing.

If we consider a 2D square array of atoms, the minimum energy spacing is obtained as the difference between (23.1.1) evaluated at $n_x = 1$ and at $n_x = 2$, setting $n_y = N_1$ and $n_z = 0$ in both cases. Noting that, for large N_1,

$$\sqrt{2^2 + N_1^2} - \sqrt{1^2 + N_1^2} \approx \frac{3}{2N_1} \tag{23.1.3}$$

We find that the minimum eigen-energy spacing is,

$$\Delta E_{min} \approx \frac{3\pi \hbar v_s}{2L N_1} = \frac{3\pi \hbar v_s}{2a N_1^2} = \frac{3\pi \hbar v_s}{2a N_2} \tag{23.1.4}$$

where $N_2 = N_1^2$ is the number of atoms in the square array. Hence the result is the same as for the 1D case: the minimum energy level spacing reduces asymptotically to zero in inverse proportion to the total number of atoms. For the 3D case we get, using $n_x = 1$, $n_y = N_1$ and $n_z = N_1$, and $\Delta n_x = 1$,

$$\Delta E_{min} \approx \frac{3\pi \hbar v_s}{2\sqrt{2}L N_1} = \frac{3\pi \hbar v_s}{2\sqrt{2}a N_1^2} = \frac{3\pi \hbar v_s}{2\sqrt{2}a (N_3)^{2/3}} \tag{23.1.5}$$

The minimum energy spacing still tends asymptotically to zero as the total number of atoms, $N_3 = N_1^3$, becomes large, but now only as the 2/3 power of the total number of atoms. Because the minimum energy spacing is vanishingly small for large N, even the weakest environmental interaction will decohere the system away from its energy eigenstates, generally resulting in localisation for large objects (but not invariably, as we shall see in chapter 24).

23.3 An Exciton Example

The example of §23.2 considered only the vibrational (phonon) modes. There may be other degrees of freedom which result in smaller energy level spacings. This might happen, for example, if the atoms/molecules have an accessible excited state, and these excited states can mutually interact. In such cases the interaction can break the initial N-fold degeneracy resulting from the possibility that an excitation could be located on any of the N atoms. The weaker the interaction, the closer are the resulting split levels. Consequently, very close energy levels must be regarded as a generically common. Moreover, as we shall now show, even if the interaction is strong, the minimum eigen-energy spacing tends to zero rapidly as N becomes large.

Consider a system comprising N particles in a line. Each particle is assumed to have just two energy states, energy 0 or E. The Hilbert space of this system is of dimension 2^N. However, we suppose there is insufficient energy available to excite two or more particles to their higher energy state, but sufficient to excite a single particle. In this N dimensional sub-space the Hamiltonian in the energy basis is simply the $N \times N$ diagonal matrix with diagonal elements all equal to E. Any N dimensional vector is an eigenvector because this is an N-fold degenerate energy level. However, we will work in the basis defined by which particle is excited. We now introduce an interaction between nearest neighbours only consisting of a potential energy Δ. The Hamiltonian therefore becomes,

$$(H) = \begin{pmatrix} E & \Delta & 0 & 0 & 0 & 0 & 0 \\ \Delta & E & \Delta & 0 & 0 & 0 & 0 \\ 0 & \Delta & E & \Delta & 0 & 0 & 0 \\ 0 & 0 & \Delta & E & \Delta & 0 & 0 \\ 0 & 0 & 0 & \Delta & E & \Delta & 0 \\ 0 & 0 & 0 & 0 & \Delta & E & \Delta \\ 0 & 0 & 0 & 0 & 0 & \Delta & E \end{pmatrix} \qquad (23.3.1)$$

(illustrated here for $N = 7$). The interaction between the particles breaks the degeneracy, the perturbed energy levels being given by the solutions to the secular equation,

$$\begin{Vmatrix} E - \lambda & \Delta & 0 & 0 & 0 & 0 & 0 \\ \Delta & E - \lambda & \Delta & 0 & 0 & 0 & 0 \\ 0 & \Delta & E - \lambda & \Delta & 0 & 0 & 0 \\ 0 & 0 & \Delta & E - \lambda & \Delta & 0 & 0 \\ 0 & 0 & 0 & \Delta & E - \lambda & \Delta & 0 \\ 0 & 0 & 0 & 0 & \Delta & E - \lambda & \Delta \\ 0 & 0 & 0 & 0 & 0 & \Delta & E - \lambda \end{Vmatrix} = 0 \quad (23.3.2)$$

The easiest means of solving (23.3.2) is numerically using the proprietary matrix manipulation software of your choice. However, it is educational to do these things analytically as far as is reasonable. Using the substitution,

$$x = (E - \lambda)^2 \quad (23.3.3)$$

and denoting the $N \times N$ secular determinant as $\|N\|$, it is readily seen that,

$$\|N\| = \sqrt{x}\|N - 1\| - \Delta^2\|N - 2\| \quad (23.3.4)$$

Hence the secular equations are found to reduce to the expressions given in Table 23.1

Table 23.1: The secular equations, (23.3.2)

N	Secular Equation for the Energy Levels
2	$x - \Delta^2 = 0$
3	$x = 0$ or $x - 2\Delta^2 = 0$
4	$x^2 - 3\Delta^2 x + \Delta^4 = 0$
5	$x = 0$ or $x^2 - 4\Delta^2 x + 3\Delta^4 = 0$
6	$x^3 - 5\Delta^2 x^2 + 6\Delta^4 x - \Delta^6 = 0$
7	$x = 0$ or $x^3 - 6\Delta^2 x^2 + 10\Delta^4 x - 4\Delta^6 = 0$
8	$x^4 - 7\Delta^2 x^3 + 15\Delta^4 x^2 - 10\Delta^6 x + \Delta^8 = 0$
9	$x = 0$ or $x^4 - 8\Delta^2 x^3 + 21\Delta^4 x^2 - 20\Delta^6 x + 5\Delta^8 = 0$
10	$x^5 - 9\Delta^2 x^4 + 28\Delta^4 x^3 - 35\Delta^6 x^2 + 15\Delta^8 x - \Delta^{10} = 0$

The $x = 0$ solutions give $\lambda = E$ whereas the nth order polynomials in x have n positive roots x_r which provide $2n$ energy levels $\lambda = E \pm \sqrt{x_r}$. Relative to the mean energy E, and in units of the interaction strength, Δ, the energy levels are given in Table 23.2a. The last column gives the smallest spacing

Table 23.2a: *Eigen-energies of Hamiltonian (23.3.1)*

N	$(\lambda - E)/\Delta$	Minimum Spacing / Δ
2	± 1	2
3	$-\sqrt{2}, 0, +\sqrt{2}$	$\sqrt{2}$
4	$-\sqrt{\dfrac{(3+\sqrt{5})}{2}}, -\sqrt{\dfrac{(3-\sqrt{5})}{2}}, +\sqrt{\dfrac{(3-\sqrt{5})}{2}}, +\sqrt{\dfrac{(3+\sqrt{5})}{2}}$	1
5	$-\sqrt{3}, -1, 0, +1, +\sqrt{3}$	$\sqrt{3}-1$
6	-1.802, -1.247, -0.445, 0.445, 1.247, 1.802	0.5549
7	-1.8478, -1.4142, -0.7654, 0, 0.7654, 1.4142, 1.8478	0.4336
8	-1.8794, -1.5321, -1, 0, 1, 1.5321, 1.8794	0.3473
9	$\pm 1.90211, \pm 1.61803, \pm 1.17557, \pm 0.61804, 0$	0.28408
10	$\pm 1.91899, \pm 1.68250, \pm 1.30972, \pm 0.83083, \pm 0.28464$	0.23649

Table 23.2b: *Eigen-energies of Hamiltonian (23.3.1), cont.*

N	Minimum Spacing / Δ
11	0.19980
12	0.17097
13	0.14792
14	0.12920
15	0.11381
16	0.10100
17	0.09023
18	0.08109
19	0.073263
20	0.066516
25	0.043534
30	0.030679
35	0.022774
40	0.017571
50	0.011366
75	0.0051225
100	0.0029014

between adjacent energy levels, which in every case is between the lowest excited state and the second lowest: $E_2 - E_1 = \lambda_2 - \lambda_1$. Numerical evaluation of the eigenvalues is the simplest means of continuing to larger N, and the results for the minimum energy spacing are given in Table 23.2b.

The minimum energy spacing is plotted against N in Figure 23.1 on a log-log scale. The energy spacing is clearly asymptotic to zero as N increases. The slope of Figure 23.1 suggests that the asymptotic behaviour is that the minimum spacing decreases $\propto 1/N^{1.9}$.

The presence of this sort of "exciton" behaviour therefore promotes very small minimum energy spacing, being smaller for weaker internal coupling within the system (small Δ). But, regardless of the internal coupling strength, the minimum energy spacing becomes vanishingly small for large N, so that even the weakest environmental interaction will decohere the system away from its energy eigenstates. The common, but not invariable, result is the localisation of systems with many degrees of freedom.

Figure 23.1: *The energy level spacings of Table 23.2 plotted against N*

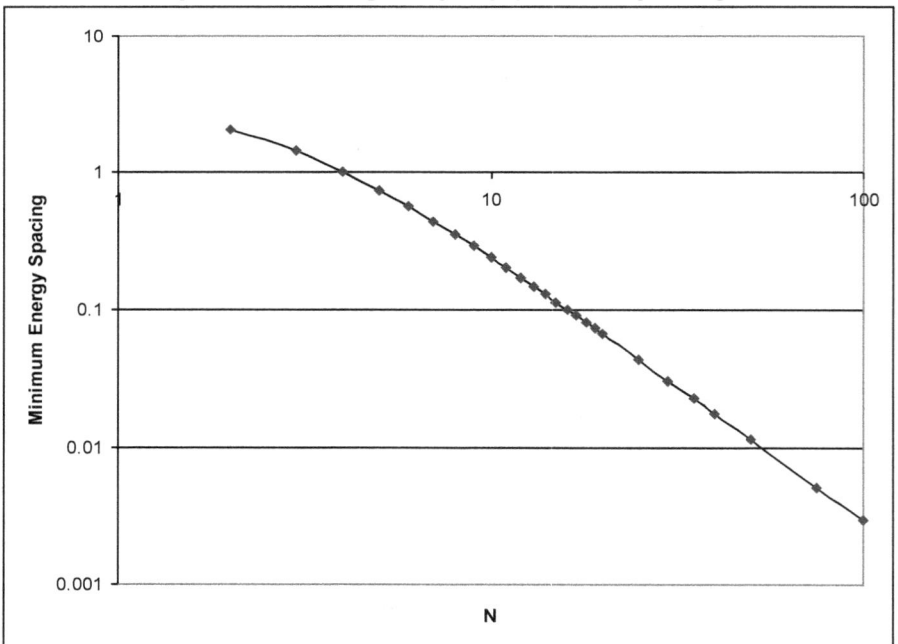

23.4 But the Exceptions are of Greater Interest

Obviously one cannot conclude anything rigorous or universal from a few very simple illustrations. The purpose of these examples has been only to make plausible the claim that "big" objects will most often tend to have some very closely spaced eigen-energies, and hence will most often decohere away from their energy eigenstates and towards localised states. In this context "big" means that the system has a large number of degrees of freedom (as opposed from actual size). Moreover, this is an absolute criterion in that "a large number of degrees of freedom" means $N \gg 1$. A few tens of degrees of freedom might be big enough, so even quite modest sized molecules may be large enough. The giant molecules of biochemistry are therefore easily big enough. This is why molecules can often be regarded as having a definite shape. Much of chemistry, especially biochemistry, is concerned with the geometrical interlocking of pairs of molecules. A successful chemical reaction may depend upon correct geometrical compatibility. This "Lego brick" view of molecular interactions would hardly be possible if atoms were so delocalised that they caused molecules to have indeterminate shape.

Nevertheless, even large molecules can be coaxed into quantum behaviour, such as interference, as we shall see in chapter 24. And it would be a mistake to think that "quantum-like" phenomena were impossible in macroscopic systems. Superfluidity, superconductivity and Bose-Einstein condensation are all examples of essentially quantum phenomena, involving sustained coherence, which are observable on the macroscopic scale. One of the key features which permit these phenomena to arise is the existence of energy gaps in their spectra. Such cases are fascinating precisely because they are unusual and unfamiliar in their behaviour, of which more in chapter 24.

24

Big Things Behaving Quantumly

In chapter 23 we saw why it is that "big things" generally behave in a classical rather than a quantum manner. We also saw that "big" in this context is an absolute because it means a system with many degrees of freedom. However, I also warned that there are exceptions. Actually, you are surrounded by them. This chapter gives some examples of such exceptions and elucidates how quantum mechanical behaviour can be retained despite being "big". Unfamiliar behaviours resulting from uniquely quantum effects typically require specially contrived conditions to occur. That is why they are unfamiliar. However, one does not have to be a pedant to note that our everyday world is full of phenomena which are essentially quantum mechanical. The physical properties of matter are a case in point. Thus, whilst superconductivity is an unfamiliar phenomenon in the everyday world, and requires quantum mechanics to explain it, the fact that metals tend to be excellent conductors under ordinary room-temperature conditions, whilst most other materials are excellent insulators, is also a phenomenon explicable only within quantum mechanics. The apparently mundane nature of the familiar properties of matter under normal conditions may cause us to overlook their non-classical origins. I am tempted to say that "everything is quantum, dammit", but I doubt that engineers will start using Hilbert space to design bridges any time soon.

My objective in this chapter is modest: it is only to correct any impression left by chapter 23 that "big" things – however defined – are necessarily always devoid of any uniquely quantum mechanical behaviour. I have been at pains to emphasise that the identification of "big" things with classical behaviour must be hedged by words like "generally". In this chapter I drive home the message by the simple expedient of giving counterexamples. It is important to understand how these examples of persistent "quantum-ness" manage to buck the general trend. It is also important to emphasise that, whilst some of these examples involve exotic phenomena which occur only under special conditions (usually very low temperatures), essentially quantum mechanical effects are also key to understanding many of the properties of matter under ordinary, room temperature, conditions. The latter observation only serves to make more pressing further clarification of when, and why, quantum mechanical behaviours are retained and when not.

There are broadly two ways of achieving persistent quantum-ness. One is to reduce environmental interactions as much as possible and contrive conditions such that the desired quantum effects can be observed before there is time for decoherence to occur. In this category are the interference experiments with large molecules discussed in §24.1. The other possibility is that decoherence itself leads to pointer states which retain non-classical features. Atomic electron orbitals are a case in point – and as the whole of chemistry rests upon their properties, one can reasonably argue that our everyday world is built upon essentially quantum mechanical behaviours. But this is not the only example of essentially quantum properties under normal conditions. Most material properties are explicable only on the basis of quantum mechanics, generally as a result of being many-body systems. However, what we also have in mind in this chapter are behaviours which are unusual – and whose unusual behaviour is also essentially quantum mechanical. Examples are superconductivity and superfluidity.

Chapter 23 has already provided a clue to how some systems have pointer states which retain quantum behaviours. Recall that if the system has a large spacing between its energy levels compared with the strength of its interaction with the environment, the pointer states will tend to be defined by the system alone, not by its environmental interaction. It turns out that, for crystalline solids, gaps in their energy spectra is a natural state of affairs and this is key to many of their properties. In particular, it is the existence of an energy gap which gives rise to the phenomena of superconductivity and superfluidity.

Much of this chapter will rely on description, using mathematics only as rather oversimplified illustrations. This is an unfortunate necessity as a big fat text dedicated to solid state physics would be required to do true justice to the topics – in fact several books. Such books exist, of course, and I hope the reader is inspired to turn to them next. Here, my purpose is only to provide an introduction, including some indications of how energy gaps arise in solids with regular crystalline lattices to wet the reader's appetite for more complete expositions.

24.1 Interference of Large Molecules

In 1930, Estermann and Stern first confirmed that atoms of helium, and molecules of hydrogen, behaved as de Broglie waves by demonstrating they were subject to diffraction, Ref.[24.1]. In 1999, Arndt *et al* showed that molecules of the fullerene C_{60} similarly displayed interference, Ref.[24.2]. With 12 nucleons per carbon atom, C_{60} is a molecule of mass 720 amu (actually slightly more if the isotopic contribution of carbon-13 is taken into account). Including also the electrons as independent particle constituents, C_{60} comprises some 1080 particles (equal numbers of protons, neutrons and electrons). But even if we disregard that and treat each carbon atom as a point particle in its own right, C_{60} still has 3 x 60 - 6 = 174 vibrational degrees of freedom. Ample scope, you might think, to result in an energy spectrum with closely spaced energy levels, and hence highly sensitive to decoherence.

Arndt *et al* sublimated C_{60} molecules in a furnace at between 900K and 1000K. The resulting molecular beam was collimated and then incident on a diffraction grating containing 50nm wide slits with a 100nm period. Downstream of this was a detector. Figure 24.1 is reproduced from Ref.[24.2]. The lower picture shows the molecular beam profile at the detector without the diffraction grating present. The upper picture shows the beam profile found with the diffraction grating in place. The interference effect is clear in the minima and maxima at around ±12 and ±24 microns.

Arndt *et al* discussed the reasons why decoherence was avoided in their experiment. Two types of interaction with the environment were identified. The first was collisions with molecules of the remnant gases within their equipment. Their interferometer operated at a moderately high vacuum of 5 x 10^{-7} mbar. The mean free path through such a near-vacuum was estimated to be ~100m, thus making molecular collisions unlikely for the great majority of fullerene molecules. The second mechanism for interaction with the environment was photon emission or absorption. Arndt *et al* argued that, for the interference to be destroyed, such photon processes would need to be capable, in principle, of determining which path the C_{60} molecule took through their interferometer. Because their diffraction grating has a spacing of 100nm, this would require photons whose wavelength was smaller than 100nm in order to resolve the "which path" information. C_{60} molecules were judged likely to emit 2 or 3 photons between the diffraction grating and the detector. However, by considering the thermal energies available, Arndt *et al* estimated that the photons in question would have wavelengths in the region of 10 to 19 microns, and hence would be incapable of resolving which path the molecules took through the grating.

A few years later, the same team under Anton Zeilinger in Vienna demonstrated interference in molecular beams of tetraphenylporphyrin and the fluorofullerene $C_{60}F_{48}$, Ref.[24.3]. The former is a molecule with biochemical significance. The latter set a new record in 2003 for the heaviest object for which interference had then been demonstrated, 1632 amu.

By 2011, that record had been beaten. Gerlich *et al*, Ref.[24.4], reported interference produced by a range of large organic molecules. The largest of these was a derivative of tetraphenylporphyrin, $C_{168}H_{94}F_{152}O_8N_4S_4$, with 430 atoms, designated TPPF152, and having mass 5,310 amu. However the most massive molecule successfully interfered by Gerlich *et al* was a perfluoroalkylated nanosphere $C_{60}[C_{12}F_{25}]_{10}$, also with 430 atoms but with mass 6,910 amu (designated PFNS10). Many experimentalists have declared their ambition to demonstrate the diffraction of a virus. However, the smallest viruses have around 60,000 atoms so this will be a leap of more than two orders of magnitude on mass if diffraction is achieved.

Figure 24.1: Diffraction of C60 from Ref.[24.2]

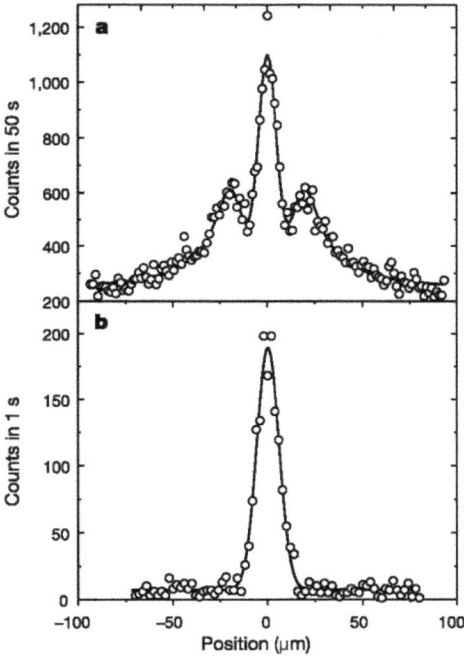

Figure 2 Interference pattern produced by C_{60} molecules. **a**, Experimental recording (open circles) and fit using Kirchhoff diffraction theory (continuous line). The expected zeroth and first-order maxima can be clearly seen. Details of the theory are discussed in the text. **b**, The molecular beam profile without the grating in the path of the molecules.

24.2 Bosons and Fermions

Particles with integral spin (i.e., with spin angular momenta of $\hbar, 2\hbar, 3\hbar$ ...) are called "bosons". Particles with half-integral spin (i.e., with spin angular momenta of $\frac{\hbar}{2}, \frac{3\hbar}{2}, \frac{5\hbar}{2}, ...$) are called "fermions". There is a crucial difference between the two types of particle. Any number of identical bosons can occupy the same quantum state. In the case of identical fermions, however, any quantum state can be occupied by at most one particle. That no two identical fermions can occupy the same quantum state is known as Pauli's Exclusion Principle. It is the reason why, as one progresses through the elements of the periodic table, each additional electron occupies a different quantum state from the electrons already present. This leads to the shell structure of atomic orbitals. If it were not for the Exclusion Principle, all the electrons in the atoms of every element would drop down into the lowest energy state and their familiar chemical properties would not exist. The Exclusion Principle is also essential to explain the stability of matter, i.e., that atoms do not all occupy the same volume leading to the collapse of bulk matter.

The different behaviour of bosons and fermions results from the behaviour of their quantum states when subject to a rotation through a full revolution. The quantum state of a boson is left unchanged by a full rotation, whereas that of a fermion changes sign. Because of the different possible quantum state occupation numbers for bosons and fermions (namely many versus just one or none) the thermodynamic distributions for the two particle types at finite temperatures are also distinct (namely Bose-Einstein versus Fermi-Dirac statistics). That integral or half-integral spin determines which thermodynamic statistics a particle obeys is known as the spin-statistics theorem. Pauli first enunciated the Exclusion Principle in 1925, but it was not until 1940 that he proved the spin-statistics theorem as a rigorous consequence of quantum field theory, Ref[24.5]. It is actually a mathematical consequence of relativistic (Lorentz) invariance.

However, all the reader need appreciate is that an arbitrarily large number of identical bosons can occupy the same quantum state, whilst for identical fermions there can be at most one particle in each quantum state.

24.3   Superfluidity, Superconductivity and Bose-Einstein Condensates

The phenomena of superfluidity, superconductivity and Bose-Einstein condensation are closely related, but different. Bose-Einstein condensation is the easiest to define though it was the last to be achieved experimentally, having been first obtained in the lab only in 1995, Ref.[24.6]. Einstein assisted in the publication of Bose's paper which introduced what we now call Bose-Einstein statistics. Einstein realised that an implication of the work was that near absolute zero all bosons would "condense" into the same quantum state, namely that of lowest energy, Ref.[24.7]. This is a Bose-Einstein condensate: a state of matter in which all the atoms are in the same quantum state, something which is possible only if the atoms in question are bosons. This state of matter is very delicate. Only tiny amounts of thermal energy will disrupt it. Temperatures of a few billionths Kelvin are used. Even then the state persists only for a few minutes, or less, at least in the earliest experiments. Moreover, it is achieved in practice only in gases of extremely low density. In the first successful experiments, only a few thousand atoms were placed in the condensed state, though later experiments have increased this substantially. Nevertheless, it is remarkable that such a large number of atoms can be completely specified by a single Hilbert state.

As a student, well before Bose-Einstein condensates had been achieved in the lab, I was rather confused about all the fuss being made at the time trying to produce them. I thought it had been done already. Specifically I thought that superfluidity and superconductivity were examples of Bose-Einstein condensation. This is not entirely wrong, but certainly isn't right either. What I now know is that I was, in my confusion, in better company than I realised.

Superconductivity was first observed by Kamerlingh Onnes as early as 1911, Ref. [24.8]. He noted that the electrical resistance of mercury dropped discontinuously to virtually zero when the temperature was decreased below ~4.2K. Since then experiments appear to indicate that the superconducting state really does have zero resistance. Currents have been observed to flow in superconducting rings for years. The decay time has been estimated to be longer than hundreds of thousands of years. However, it was not until 1933 that Meissner and Ochsenfeld, Ref.[24.9], discovered the Meissner effect, namely that when the temperature is decreased below the superconducting transition the specimen rejects all magnetic flux. Another way of expressing this is that the superconducting state acts with perfect diamagnetism, the induced magnetisation exactly cancelling the applied magnetic field. This magnetic effect is not a consequence of the vanishing electrical resistance, it is an independent effect. The superconducting state has other properties characteristic of a phase change, such as a specific heat which changes discontinuously at the transition, and decreases towards zero as the temperature is dropped further. This is an example of a "lambda transition", so called because the specific heat plotted against temperature resembles the shape of the Greek letter lambda, λ.

Superfluidity was observed by Kapitsa, and independently by Allen and Misener, in 1938 in helium-4, Ref.[24.10]. The critical temperature for the onset of superfluid behaviour is also marked by a lambda transition. The superfluid phase has zero viscosity and hence will flow freely, without resistance, through the tiniest of capillaries or crevices. This can be a pain for experimentalists as a vessel previously "helium tight" will suddenly start leaking when the superfluid state is achieved. Zero viscosity is demonstrated by flows in ring shaped containers which persist indefinitely. In this respect superconductivity is analogous to superfluidity, the "frictionless fluid" in the former case being the conduction electrons. The parallels between the two phenomena do not stop there. The Meissner effect has an analogue in superfluidity, though you would be forgiven for finding it less than obvious. Just as the Meissner effect is the rejection of magnetic flux, superfluids have a tendency, as the temperature is reduced, to reject angular momentum. Thus, the superfluid phase in a rotating container will decline to share in the rotation if the temperature is low enough.

The nucleus of helium-4, also called an alpha particle, comprises two protons and two neutrons. The two protons have anti-aligned spins, as do the two neutrons. The He-4 atom's two electrons also have anti-aligned spins (thus permitting both to be in the lowest energy state). None of these particles has orbital angular momentum. Thus, helium-4 has zero angular momentum and is thus a boson. This led Fritz London, soon after superfluidity had first been observed, to postulate that it may be due to Bose-Einstein condensation. So, you see, I was indeed in good company in my student misapprehension. Nor was I completely wrong. London was, of course, more knowing. He was well aware that the superconducting samples obtained experimentally could not be entirely Bose-Einstein condensate (BEC). Instead he proposed a two-fluid model, in which only one was BEC phase. The two fluids would be distinguished by their velocities. This is broadly the correct picture, though the details are more intricate and beyond this elementary account.

The early history of superconductivity was confused by a theory put forward by Lev Landau, who was very influential. It would appear there was personal animosity between Landau and Fritz London, as the former steadfastly refused to mention the latter's ideas on superfluidity at all. In particular, Landau never accepted that BEC played a part in superfluidity. The history of those early days has been engagingly recorded by Balibar in Ref.[24.10]. He concludes, "*in summary, the superfluidity is certainly linked to the Bose statistics, contrary to Landau's statement*". In practice, in a sample exhibiting superfluid behaviour at finite temperature, only a proportion of the sample will be in the superfluid phase, although this proportion will increase as 0K is approached more closely. Similarly, in a superconducting specimen, only a fraction of the electrons will be in the superconducting phase. It is tempting, therefore, to imagine that the superfluid phase can be identified as BEC. But even this is too simplistic. To quote Ref.[24.11], "*even at the lowest temperatures where the superfluid fraction is almost 100%, the strong correlations between the atoms forming the liquid deplete the population of the Bose-Einstein condensate state to only 10%*".

Nevertheless, it is certainly the case that the collective quantum behaviour of many atoms – in a manner analogous to BEC if not actually identified with it – underlies superfluidity, and hence that the phenomenon is related to bosonic behaviour. In the early years this position was bolstered by the failure to observe superfluidity in helium-3 down to about 1K. As helium-3 has only one neutron it is necessarily spin-half and hence a fermion. The absence of superfluidity when the bosonic helium-4 is replaced by the fermionic helium-3 appeared to support the relevance of Bose-Einstein statistics. And yet this neat picture was not quite right. In the 1970s, superfluidity in helium-3 was discovered at temperatures below the order of milliK, Ref.[24.12],.

If bosonic behaviour is so important to superfluidity, how come superfluid behaviour occurs in a system of fermions? This problem was, of course, already apparent in that analogue of superfluidity: superconductivity. The system in question in the case of superconducting metals consists of electrons, i.e., spin-half fermions. So the same problem is apparent there: seemingly bosonic behaviour being displayed by fermions. But by the time helium-3 was discovered to be superconducting, this was no longer a puzzle. The BCS theory of superconductivity had already addressed the matter, Ref.[24.13]. In §24.5 we shall return to the BCS theory of superconductivity, but first...

24.4 The Band Gap Under Normal Conditions

Before considering exotic phenomena like superconductivity we shall look at how gaps in the energy spectra of crystalline solids are related to more familiar properties of matter. Specifically we shall consider the electrical conductivity of metals versus that of insulators – and semiconductors. Having made the point that gaps in the energy spectra of solids, so called "band gaps", are crucial to familiar material properties, we shall then address how these band gaps arise.

Why is it that a good electrical conductor, such as copper, has a conductivity as high as 6 x 10^7 S/m whereas a good insulator, such as the polymer Teflon, might have a conductivity as low as 10^{-23} S/m? (The unit S is a siemen, which is the reciprocal of the unit of resistance, the ohm). Why is there 30 orders of magnitude, or more, between the best and worst of electrical conductors? And why do semiconductors have an intermediate conductivity (typically between 10^{-6} and 10^3 S/m), making them neither good conductors nor good insulators?

The reader will be familiar with the idea that the conduction of electricity in solids consists of a flow of valence electrons, i.e., the electrons from the outermost shell of the atoms in question. To cause an electron to start moving with some drift velocity requires its energy to be increased by the associated kinetic energy. Hence, in order for an applied potential difference (emf) to initiate a flow of electrons there must be allowed quantum states with energies immediately above the energy of the valence electrons with no current flowing. At this point the fact that electrons are fermions obeying the Exclusion Principle is key.

Imagine all the quantum states of well-defined energy (i.e., the eigenstates of the Hamiltonian) arranged in order of increasing energy. Ignoring for now the disruption caused by random thermal energies, the electrons will fill up the available states from the lowest upwards. This is displayed in schematic form by Figure 24.2. In this Figure I have assumed for now that the energy eigenstates have a spectrum which has bands separated by band gaps. Why this might be so we will address shortly. Within a band there is (to a very good approximation) a continuum of energy levels possible, between some minimum and some maximum. But between bands there are no valid quantum states, i.e., the Schrodinger equation has no energy eigenvalues in the range indicated by the band gap.

In Figure 24.2a the valence electrons fill up completely the highest occupied band. There are no available energy states immediately above the highest states normally occupied and hence the electrons cannot be induced to flow (unless abnormally large energies are supplied). Figure 24.2a indicates an insulator. In contrast, Figure 24.2b illustrates the occupancy of the bands for a metal, in which the valence electrons only half-fill the highest energy band. There is thus an available continuum of energy states for electrons to move into from their normal state, facilitating electron flows even for very modest emfs and hence only tiny increases in energy. This is how the extreme difference between good conductors and good insulators arises.

For a given material, the maximum energy of occupied states is referred to as the Fermi surface, the term "surface" relating to energy-space, or, more properly, momentum-space. In reality it is a three-dimensional surface, not just one specific energy. However, such complications aside, an insulator is a substance in which the Fermi surface coincides with the top of an energy band, whereas in a metal the Fermi surface is contained within a band and has plenty of "head room" for increasing electron energies within the band.

Figure 24.2: Schematic illustration of band gaps and the Fermi surface, contrasting (a) insulators with (b) conductors. Blue shading indicates occupied states.

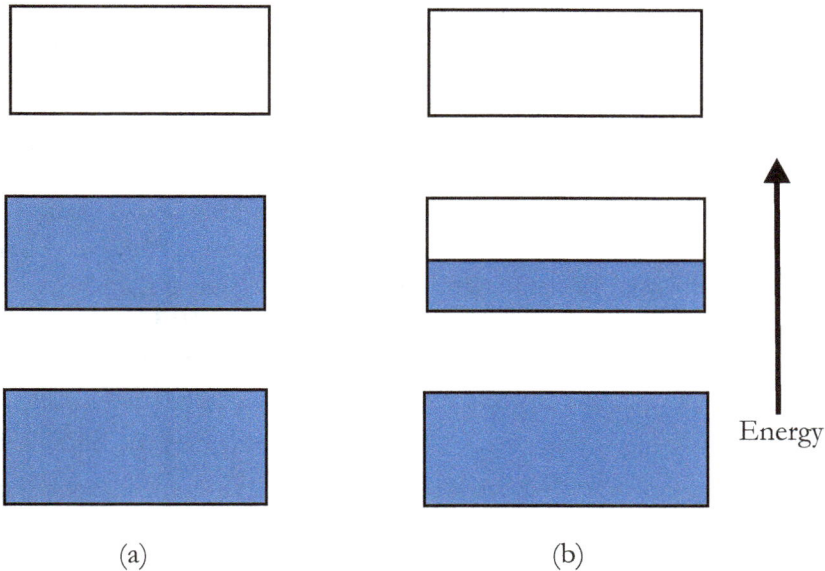

(a)   (b)

What about semiconductors, such as silicon or germanium? In familiar electronics applications, these important elements have their electrical properties radically altered by the addition of small amounts of impurities ("doping"). However, those issues are beyond our present scope. As regards pure semiconductors, at absolute zero they are insulators, i.e., their electron structure is like Figure 24.2a. As the temperature is raised, perhaps to room temperature or higher, they will conduct, albeit far more poorly than proper metals. The reason is that, in semiconductors, the size of the band gap is sufficiently small that thermal energy can result in a small proportion of electrons making the transition from the (full) valence band to the (empty) conduction band above it. In semiconductors the band gap is roughly of the order of an electron-volt. Readers familiar with the Boltzmann factor will readily estimate how tiny the proportion of electrons raised into the conduction band is likely to be at, say, 293K, thus explaining why the conductivity of semiconductors is many orders of magnitude smaller than that of true metals.

Why do crystalline solids exhibit a band gap? The key to understanding the states available to valence electrons is to recall that they are in interaction with a regular lattice of atoms with positive nuclei. Consider, in one spatial dimension, an energy eigenstate wavefunction (i.e., $\langle x|E\rangle$) of the form $exp\{ikx\}$. In the absence of a background potential, this is a solution of the time independent Schrodinger equation for eigenenergy $E = (\hbar k)^2/2m$, with momentum $\hbar k$. This represents a wave of that energy travelling to the right, but there is also a wave of the same energy travelling to the left, $exp\{-ikx\}$, with momentum $-\hbar k$. So, in this 1D case, the energy level is degenerate: there are two states with the same energy.

If we attempt to approximate the state of an electron in a periodic lattice of positive charges as combinations of $exp\{\pm ikx\}$, it can readily be seen why, at certain momenta, the two states cease to be degenerate, corresponding instead to different energies. To see this consider $exp\{ikx\} \pm exp\{-ikx\}$ which, ignoring normalisation factors, are $\cos kx$ and $\sin kx$. Now consider a wavelength of twice the lattice spacing, $\lambda = 2a$, which corresponds to $k = \frac{2\pi}{\lambda} = \frac{\pi}{a}$. (In diffraction theory this wavevector is known as the boundary of the first Brillouin zone, but we shall not need such sophistication here). Assume for sake of argument that $x = 0$ corresponds with the position of a (positively charged) nucleus. The electron density is given by $\cos^2\left(\frac{\pi x}{a}\right)$ for one state and by $\sin^2\left(\frac{\pi x}{a}\right)$ for the other. In the first of these, the electron concentration is greatest near the positive nuclei (at $x = 0$ and $x = a$), and zero midway between them. For the other wavefunction, the electron concentration is zero at the nuclei, and maximum midway between them. It is clear, therefore, that, due to the attraction between the electron and the nuclei, the first state will have reduced energy compared with a wave of uniform density, whilst the second will have raised energy. The pair of degenerate states has split into two states of clearly different energies. This is the simplest picture of how the energy band gap arises. However, let's look with a little more precision (though not much) how one might go about solving the Schrodinger equation in the presence of a periodic potential.

For simplicity I shall confine attention to one dimension. The time-independent Schrodinger equation, expressed in the continuum spatial basis for the wavefunction of an electron is,

$$\left[-\frac{\hbar^2}{2m}\frac{\partial^2}{\partial x^2} + V(x)\right]\psi(x) = E\psi(x) \tag{24.4.1}$$

where $V(x)$ is the potential energy as a function of position. In the periodic potential due to the lattice of positive nuclei we have $V(x + a) = V(x)$, where a is the lattice spacing. Because the potential is periodic it can be written as a Fourier series, thus,

$$V(x) = \sum_{n=-\infty}^{+\infty} V_n \exp\{iG_n x\} \tag{24.4.2}$$

with $G_n = 2\pi n/a$. To keep things simple I shall assume a periodic potential with just two terms, with equal amplitudes, thus,

$$V(x) = V_1(\exp\{iG_1 x\} + \exp\{-iG_1 x\}) \tag{24.4.3}$$

There is a theorem, known as Bloch's Theorem, that the solution of Schrodinger's equation with a periodic potential must be of the form of a plane wave times a periodic function. Thus, in 1D, we can write, in complete generality,

$$\psi(x) = \exp\{ikx\}\sum_{n=-\infty}^{+\infty} A_n \exp\{iG_n x\} \tag{24.4.4}$$

For some constant coefficients, A_n. However, again to keep things simple let's assume that a sufficiently accurate approximation is to retain just three terms in this expansion, giving, $\tag{24.4.5}$

$$\psi(x) = A_{-1}\exp\{i(k - G_1)x\} + A_o\exp\{ikx\} + A_1\exp\{i(k + G_1)x\}$$

Substituting (24.4.3) and (24.4.5) in (24.4.1), the requirement that the coefficients of the three wave terms in (24.4.5) be zero gives three relations between the expansion coefficients A_{-1}, A_0, A_1 which can be written in matrix notation as,

$$\begin{pmatrix} \lambda_k - E & V_1 & V_1 \\ V_1 & \lambda_{k+G_1} - E & 0 \\ V_1 & 0 & \lambda_{k-G_1} - E \end{pmatrix}\begin{pmatrix} A_0 \\ A_1 \\ A_{-1} \end{pmatrix} = 0 \tag{24.4.6}$$

where $\lambda_k \equiv \frac{\hbar^2 k^2}{2m}$. We recognise the condition for a non-trivial solution to (24.4.6) as the secular equation for the energy levels, namely that the determinant of the matrix in (24.4.6) be zero, giving,

$$(\lambda_k - E)(\lambda_{k+G_1} - E)(\lambda_{k-G_1} - E) - V_1^2(\lambda_{k-G_1} + \lambda_{k+G_1} - 2E) = 0$$

$$(24.4.7)$$

Introducing the dimensionless wavevector, energy and potential,

$$\kappa = \frac{k}{G_1} \qquad \varepsilon = \frac{(2\pi\hbar)^2}{2ma^2} = \frac{h^2}{2ma^2} \qquad \tilde{E} = \frac{E}{\varepsilon} \qquad \tilde{V} = \frac{V_1}{\varepsilon} \qquad (24.4.8)$$

(24.4.7) can be written,

$$(\kappa^2 - \tilde{E})((\kappa + 1)^2 - \tilde{E})((\kappa - 1)^2 - \tilde{E}) - \tilde{V}^2[(\kappa + 1)^2 + (\kappa - 1)^2 - 2\tilde{E}] = 0$$

$$(24.4.9)$$

The reader should check he can reproduce the derivation of (24.4.9). This is a cubic equation with three roots for the energy, $\tilde{E}_1 < \tilde{E}_2 < \tilde{E}_3$, but we can confine attention to the smallest two roots. The solutions are readily found numerically. For $\tilde{V} = 0.1$ the lowest two roots are plotted in Figure 24.3, showing the band gap between them at $\kappa = 0.5$ (the boundary of the first Brillouin zone). Figure 24.4 shows how the band gap increases as the amplitude of the periodic potential is increased.

Any real solid-state physicists who have wandered in here will, no doubt, be shuddering at the woeful over-simplification of this derivation. However, I am unapologetic. This is the price one pays for pedagogy. I hope the reader can now follow the trail from the Schrodinger equation to the occurrence of electron energy band gaps in a background with a periodic potential. If so, he will now appreciate how the enormous variation in the electrical conductivity of different materials results from the Schrodinger equation together with the Exclusion Principle – and hence is essentially quantum mechanical in origin. This has been an illustration for one particular material property, conductivity, but it will now be apparent that a material's properties in general arise from the quantum mechanics of their molecular, atomic and electronic constituents. Indeed, the very solidity of solids relies upon the Exclusion Principle, an irreducibly quantum phenomenon.

Figure 24.3: *The energy eigenvalues [roots $\tilde{E}_1$ and $\tilde{E}_2$ of equ.(24.4.9)] plotted against wavevector, κ, illustrated for $\tilde{V} = 0.1$ and showing the band gap at $\kappa = 0.5$*

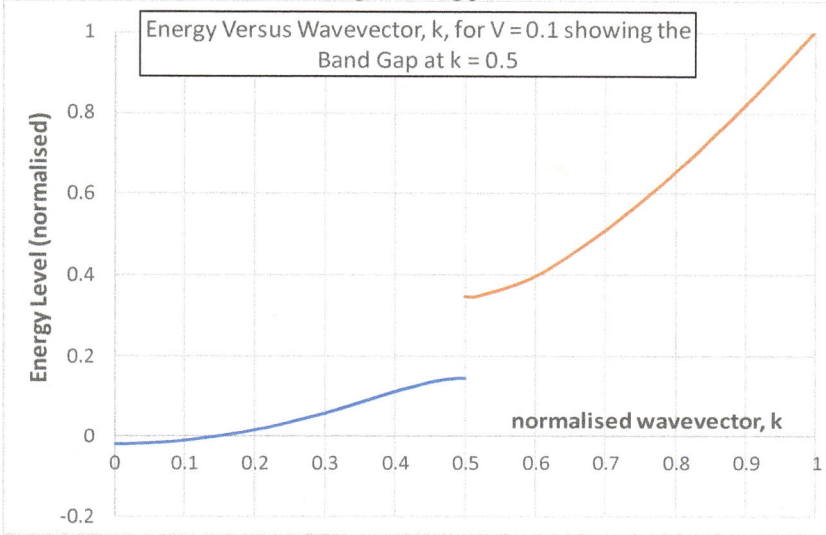

Figure 24.4: *The energy eigenvalues [roots $\tilde{E}_1$ and $\tilde{E}_2$ of equ.(24.4.9)] plotted against potential amplitude, $\tilde{V}$, at the band gap ($\kappa = 0.5$) showing how the band gap increases as the amplitude of the periodic potential increases*

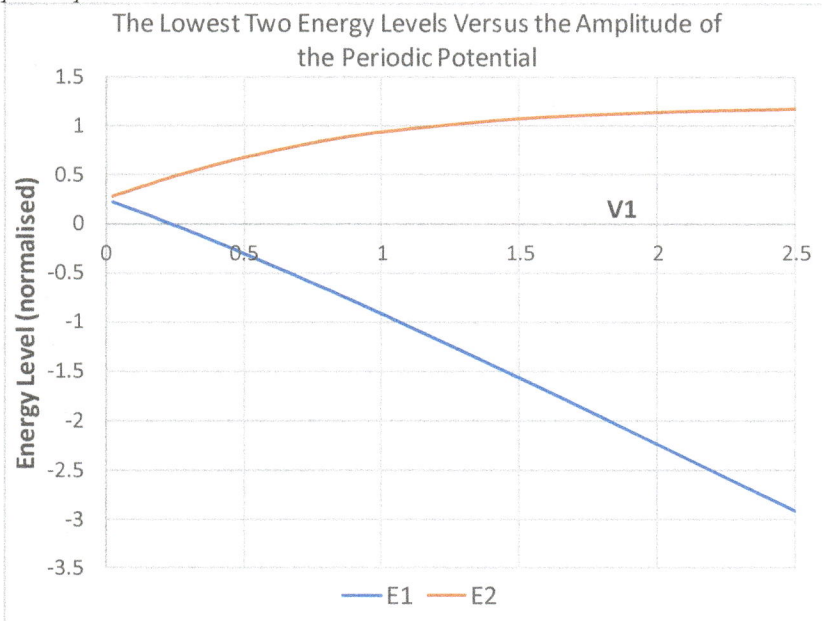

24.5 BCS and the Superconductivity Band Gap

In §24.3 we saw that superfluidity was essentially reliant on the bosonic properties of the atoms in question, resulting (at least heuristically) from a partial Bose-Einstein condensation. Moreover, we saw that superconductivity might be envisaged (again rather heuristically) as superfluidity of the flowing current. But at the end of §24.3 we left unaddressed the conundrum of how electrons, as fermions, can behave like bosons in this respect. The explanation was provided by Bardeen, Cooper and Schrieffer in 1957, Ref.[24.13], for which they won the Nobel Prize in 1972.

Once again the key is to appreciate the significance of electrons moving within the lattice of positive nuclei. An indirect effect of this is to induce a small attractive force between two electrons (which are otherwise repelled by the Coulomb force). The attractive force arises because one electron distorts the lattice of positive nuclei slightly, increasing the local charge density, and thus causing a small attractive force on a second electron. Intuitively one imagines that this effect would be completely swamped by the usual direct inter-electron Coulomb repulsion. But, thanks to quantum collective action, even though the effect is only very slight, it can be sufficient to effectively bind pairs of electrons. These are the famous Cooper Pairs. And a pair of spin half particles constitutes a boson. This is the resolution of the conundrum as to how electrons can exhibit boson-like "superfluid" behaviour. Below I present a highly simplified analysis to illustrate how such a slight effect can result in effective electron pair binding. But, equally importantly, the same analysis shows that the binding phenomenon is intimately related to the production of a band gap – and this band gap is key to the flow of current without resistance.

Consider N electrons, or, perhaps better, N multi-electron quantum states. These quantum states may be defined by the momenta of each contributing electron and also their spin states. All will be affected by the small attractive (negative) potential due to lattice distortions (phonons, actually). Moreover, because we are envisaging states defined by momenta, they will all be delocalised and uniformly spread throughout a large volume. It is reasonable (at least, for illustrative purposes) to assume that the matrix elements of the lattice-induced potential between any two states are all the same, say $-\varepsilon$ where ε may be very small. In matrix form, the interaction Hamiltonian is therefore just the following $N \times N$ lattice induced potential matrix,

$$\begin{pmatrix} -\varepsilon & -\varepsilon & -\varepsilon & -\varepsilon & -\varepsilon & etc \\ -\varepsilon & -\varepsilon & -\varepsilon & -\varepsilon & -\varepsilon & \\ -\varepsilon & -\varepsilon & -\varepsilon & -\varepsilon & -\varepsilon & \\ -\varepsilon & -\varepsilon & -\varepsilon & -\varepsilon & -\varepsilon & \\ -\varepsilon & -\varepsilon & -\varepsilon & -\varepsilon & -\varepsilon & \\ etc & & & & & \end{pmatrix} \qquad (24.5.1)$$

The possible energy levels, with respect to the unperturbed state as datum, are the eigenvalues of the above matrix. But $(N-1)$ of these eigenvalues are zero, and the only non-zero eigenenergy is $-N\varepsilon$. (A reader exercise is to confirm this, see §24.7 for my answer). So we see that the lowest energy level might be arbitrarily large in magnitude, but negative (attractive), even for small ε if the number of electrons involved, N, is sufficiently large. Thus, this quantum collective effect can result in an attractive inter-electron force sufficient for effective electron pair binding.

Actually, this is fortunate because otherwise the above argument would be fatally flawed. Due to the necessity for fermion wavefunctions to be antisymmetric under interchange, if the states implicit in the above derivation were electron states, half the matrix elements in (24.5.1) would be positive and the argument would fall apart. But we can now assume that the states forming the basis in which (24.5.1) is expressed are, in fact, states of electron pairs – and hence bosons. The argument shows this to be a self-consistent assumption.

But note also the band gap. There is just one ground state, with energy $-N\varepsilon$, and this constitutes a macroscopic energy gap below the remaining states with datum energy zero. So, even a tiny attractive effect due to lattice distortion results, not only in Cooper pairs, and hence effective bosonic behaviour of the electrons, but it also results in a substantial band gap.

There remains one thing that might puzzle the reader. In §24.4 the extremely low conductivity of insulators was explained in terms of the existence of a band gap. So how come I am now claiming that superconductivity also relates to the existence of a band gap? How can I argue it both ways? Recall that, in the case of insulators, we were dealing with fermions (single electrons). To get a current to flow – to get the electrons moving – would require giving them more energy. But no such higher energy states are within reach when the Fermi surface is hard up against the top of a band and there is a band gap. Ergo, such a condition produces an insulator.

The difference in the case of superconductors is that, firstly we are dealing with bosons (Cooper pairs of electrons), and secondly the paired electrons in the BCS state experience a substantial negative potential energy, as derived above. Not all the electrons in a sample of superconducting material will be in the BCS state. Not even all the valence electrons. Only a small proportion near the Fermi surface will enter the BCS state. As regards their kinetic energy, the BCS electron pairs lie above the Fermi level – and hence can move so as to create a current. How is this possible if they are also up against a band gap? The answer is that the negative potential energy the Cooper pairs experience more than cancels the increase in kinetic energy. Hence, the BCS state involves a reduction in total energy compared to a non-BCS state electron at the Fermi surface. This is why the superconducting state arises, of course. But it is also why the band gap does not prohibit the BCS electron pairs from carrying large currents: they can increase their kinetic energy whilst not increasing their total energy.

In technological applications, the great boon of superconductors is that wires of small cross-section can carry remarkably large currents – at levels that would otherwise require very heavy cross-sections of normal copper.

Finally, there is the relevance of the band gap to being superconducting. In a normal conductor, resistance occurs because the electrons interact with the rest of the structure (which we can take as manifest as phonons). Such interactions prevent an electron accelerating indefinitely under the action of an applied emf; collisions with phonons take energy away from the electrons and limit their drift speed. But this relies on a lower energy level being available to the electron. In the BCS state, the Cooper pairs are in the ground state, despite their motion. The band gap could be bridged, but only if there was sufficient ambient thermal energy around. Hence, as long as the temperatures are low enough, we get a resistanceless flow of current.

Once again I emphasise the woeful inadequacies of this chapter as a primer on solid state physics – but that was not its purpose. I have failed to distinguish between Type I and Type II superconductors (the latter exhibiting only a partial Meissner effect). I have not explained why some materials superconduct and others do not. I have also failed to mention so-called high temperature superconductors, whose superconducting state arises at the relatively balmy temperature of liquid nitrogen, or higher. At the time of writing the record has been set by lanthanum hydride under high pressure which superconducts at a remarkable 250K, or -23°C, Ref.[24.15]. High temperature superconductivity is not explained by BCS theory. No doubt another Nobel Prize awaits those who ultimately explain it.

24.6 Coda

To reiterate, the purpose of this chapter was to emphasise that properties which are explicable only on the basis of quantum mechanics do indeed occur in "big things", and these quantum behaviours are often related to the occurrence of band gaps in the system's energy spectrum. Some of these examples of behaving quantumly are quite exotic, such as superfluid and superconducting properties, and these tend to occur only at low temperatures when band gaps are not bridged by thermal energies. But many familiar properties of matter at room temperature, or higher temperatures, are also essentially quantum mechanical in origin. This includes many different solid material properties, and we used the conductivity of metals, insulators and semiconductors to illustrate the relevance of quantum mechanics to familiar material properties in §24.4.

Interaction with the environment is unavoidable, and so the mechanisms of decoherence will always be active. The lesson of decoherence theory is that this will tend to drive the system into a pointer state. But whether that means the system ceases to behave "quantumly" begs two questions: what is the pointer state in question, and what does one mean by "quantumly". The fact that you do not fall through your chair – and indeed your whole body and the world around you collapse into an amorphous blob – is actually an example of matter behaving "quantumly", though one regards it as classical merely due to familiarity. As for the pointer states, we have seen in preceding chapters that they depend upon the system's eigen-energy spacing in comparison with the magnitude of the interaction with the environment. It is no accident, then, that the more exotic examples of "behaving quantumly" can often be traced to the existence of band gaps. The broad-brush arguments of chapter 23 identified systems with many degrees of freedom as typically having closely spaced energy levels, and hence that such systems would have pointer states defined by the environmental interaction. But in circumstances where gaps occur in the system's energy spectrum, pointer states can get pinned to the system's internally defined states, resulting in behaviours we more readily describe as "quantumly".

24.7 Reader Exercises

Prove that all but one of the eigenvalues of matrix (24.5.1) are zero, and the one non-zero eigenvalue is $-N\varepsilon$. I shall cheat a little here and appeal to two theorems about determinants. The first is that the sum of the roots of a determinant equals its trace, which in this case is $-N\varepsilon$. The second is that the sum of the squares of the roots equals the sum of the squares of all the components of the matrix, which in this case is $N^2\varepsilon^2$. But it is clear that $-N\varepsilon$ is one of the eigenvalues because each row of the secular matrix then consists of $(N - 1)$ components equal to $-\varepsilon$ and one equal to $(N - 1)\varepsilon$. Hence, a vector whose components are all 1 is the corresponding eigenvector. But that one eigenvalue of $-N\varepsilon$ exhausts the sum-over-the squares-of-roots which must be $N^2\varepsilon^2$, so the remaining eigenvalues must all be zero. **QED**.

24.8 References

[24.1] I.Estermann and O.Stern (1930), *"Beugung von Molekularstrahlen"*, Zeitschrift für Physik, **61**, 95-125.

[24.2] M.Arndt et al, (1999), *"Wave-particle duality of C60 molecules"*, Nature, 401, 680–682. https://www.nature.com/articles/44348/

[24.3] L.Hackermuller, et al, (2003), *"Wave nature of biomolecules and fluorofullerenes"*, Phys. Rev. Lett., 91(90408). https://www.uni-due.de/~hp0198/pubs/prl4.pdf

[24.4] S.Gerlich, *"Quantum interference of large organic molecules"*, Nat Commun. 2011 Apr; 2: 263. https://www.ncbi.nlm.nih.gov/pmc/articles/PMC3104521/

[24.5] W.Pauli (1940), *"The Connection Between Spin and Statistics"*, Physical Review. **58** (8): 716–722.https://doi.org/10.1103/PhysRev.58.716

[24.6] Physics Today Online, *"Cornell, Ketterle, and Wieman Share Nobel Prize for Bose-Einstein Condensates"*, 2001.

[24.7] Einstein, A (1924). *"Quantentheorie des einatomigen idealen Gases"* Königliche Preußische Akademie der Wissenschaften. Sitzungsberichte: 261–267

[24.8] H.Kamerlingh Onnes (1911), Akad. Van Wetenschappen **14**(113), 818.

[24.9] W.Meissner and R.Ochsenfeld (1933), Naturwiss. **21**, 787.

[24.10] S.Balibar, *"The Discovery of Superfluidity"*, https://arxiv.org/abs/physics/0611119

[24.11] I.Carusotto (2010, *"Viewpoint: Sorting superfluidity from Bose-Einstein condensation in atomic gases"*, Physics **3**, 5. https://physics.aps.org/articles/v3/5

[24.12] Press Release, The Nobel Prize in Physics 1996.
https://www.nobelprize.org/prizes/physics/1996/press-release/

[24.13] J.Bardeen, L.N.Cooper, J.R.Schrieffer (1957), Phys.Rev. **106**, 162 and Phys.Rev. 108, 1175.
https://journals.aps.org/pr/abstract/10.1103/PhysRev.108.1175

[24.14] V.F.Weisskopf (1962), "*Superconductivity and the Quantization of Magnetic Flux*", CERN Report 62-30,
https://cds.cern.ch/record/278075/files/p1.pdf

[24.15] A.P. Drozdov et al (2019), "*Superconductivity at 250 K in lanthanum hydride under high pressures*",
https://arxiv.org/ftp/arxiv/papers/1812/1812.01561.pdf

25

Sins of Intention

In chapter 19 on quantum erasure we discussed when a measurement is not a measurement. Here we discuss when a not-measurement is a measurement. Can such a null measurement reduce the state vector? Yes. But it's only the basis vectors that might have been detected, but aren't, that are projected out, whilst the remaining states remain unreduced. Confused? You won't be.

25.1 Null Measurement

The term "null measurement" refers to an experiment which includes an active measuring device but in which the device registers no measurement. Can a null measurement reduce the state vector, just as a completed measurement does? The answer is "yes", though it may be only a partial reduction. Note that state vector reduction by a null measurement is essentially identical to decoherence (see chapter 15) but without an associated change in the state of the measuring device (and hence the apparatus registers no measurement). The phenomenon is closely related to the observability of counterfactuals, as illustrated in chapter 6 by the Elitzur-Vaidman bomb test and the possibility of counterfactual computation. The principle is so important that it bears repeating in this different guise. Indeed, that a null measurement can influence subsequent events is an instance of an observable counterfactual.

25.2 The Gedanken Experiment of Renninger

The gedanken experiment to be considered has been attributed to Renninger, Ref.[25.1]. Suppose the decay of a radioactive nucleus gives rise to a particle (or radiation) in a spherically symmetric state. The particle may be emitted in any direction with equal probability. Now imagine we have two hemispherical shaped perfect detectors, one on the left and one on the right, such that the emitted particle must be detected by one or the other. We can consider the state vector initially to be a superposition of left and right travelling states. If one detector registers a particle, then the state vector is appropriately reduced.

But what if we removed the right detector and we have a means of knowing that the radioactive nucleus has decayed (perhaps through the detection of a different decay product). Suppose the sole remaining (left) detector fails to register a particle. This is equivalent to having detected the particle to be in the right hemisphere – and so the state vector must again be suitably reduced to that state alone. And yet the only detector in use in this scenario failed to detect anything! This is how a null measurement can nevertheless reduce the state vector. Is this correct? It is, and the following detailed analysis elucidates why.

Imagine that we have arranged a large number of particle detectors around the surface of some sphere of radius R, with the radioactive source at its centre. For simplicity we shall assume the detectors cover all 4π steradians of solid angle and are 100% efficient, so that we can guarantee that an emitted particle will be detected in one of them. The spherically symmetric wavefunction representing the emitted particle can be considered as the sum of a large number, N, of isotropically disposed plane waves,

$$|\psi\rangle = \sum_{i=1}^{N}|u_i\rangle \tag{25.2.1}$$

(A reader exercise is to show this – see §25.4 for an answer). Note that the phase relationship between the contributing plane waves in (25.2.1), namely no phase difference, is crucial to the sum being equivalent to a spherical S-wave (i.e., with zero orbital angular momentum). The individual states in (25.2.1) are normalised to $\langle u_i|u_i\rangle = 1/N$ and are orthogonal so that (25.2.1) represents just one particle.

The detectors can be considered as two-state devices, registering 0 when no particle has been detected or 1 when a particle has been detected. Since we shall consider only one particle to be present at a time, at most one detector can register a particle before the whole apparatus is reset (prior to a subsequent event). The state of the set of all detectors is therefore written $|ei\rangle$ where $i = 0$ means that no detectors have yet registered a particle, whereas $i = 1$ means that detector 1 (only) has registered a particle, $i = 2$ means that detector 2 (only) has registered a particle, etc. Moreover we have labelled the detectors so as to correspond to the labelling of the plane waves in (25.2.1), so that plane wave No.1 propagates into detector No.1, etc.

Suppose that the particle is emitted at time 0 and that it takes some time t_1 for the particle to travel from its source to the detectors. Then for times $0 < t < t_1$ the state of the combined particle-plus-apparatus is,

$$|\Psi\rangle = |\psi\rangle|e0\rangle \tag{25.2.2}$$

because the particle and the detectors have not yet had opportunity to interact. The reduced density matrix of the system is trivially just the system's unperturbed density matrix at this time since there is only one apparatus state to trace out, and so it is given by,

$$\hat{\rho}_{sys}(0) = |\psi\rangle\langle\psi| = \left(\Sigma_i|u_i\rangle\right)\left(\Sigma_j\langle u_j|\right) = \Sigma_{i,j}|u_i\rangle\langle u_j| \qquad (25.2.3)$$

In matrix notation the density matrix is thus,

$$(\rho_{sys}) = \frac{1}{N}\begin{pmatrix} 1 & 1 & 1 & 1 & 1 & \dots \\ 1 & 1 & 1 & 1 & 1 & \dots \\ 1 & 1 & 1 & 1 & 1 & \dots \\ 1 & 1 & 1 & 1 & 1 & \dots \\ 1 & 1 & 1 & 1 & 1 & \dots \\ & & & & & etc \end{pmatrix} \qquad (25.2.4)$$

where N is the dimension of the matrix, i.e., the number of detectors and the number of terms in Equ.(25.2.1). The off-diagonal terms in (25.2.4) are witness to the coherence of the initial state. At time t_1, when the interaction has occurred, the state evolves into,

$$|\Psi\rangle \rightarrow |\Psi'\rangle = \Sigma_{i>0}|u_i\rangle\,|ei\rangle \qquad (25.2.5)$$

The density matrix representing the combined particle-plus-apparatus is thus,

$$\hat{\rho} = \left(\Sigma_i|u_i\rangle\,|ei\rangle\right)\left(\Sigma_j\langle u_j|\langle ej|\right) = \Sigma_{i,j}|u_i\rangle\langle u_j| \otimes |ei\rangle\langle ej| \qquad (25.2.6)$$

The reduced density matrix representing the particle alone is obtained by tracing out the apparatus state. Hence,

$$\hat{\rho}^{red} = \Sigma_{ek}\langle ek|\hat{\rho}|ek\rangle = \Sigma_{i,j,k}|u_i\rangle\langle u_j| \otimes \langle ek|ei\rangle\langle ej|ek\rangle$$

$$= \Sigma_{i,j,k}|u_i\rangle\langle u_j|\,\delta_{ki}\delta_{jk} = \Sigma_k|u_k\rangle\langle u_k| \qquad (25.2.7)$$

Notice how different is (25.2.7) from (25.2.3). There are no longer any off-diagonal components. In matrix notation we have,

$$(\rho_{sys}) = \frac{1}{N}\begin{pmatrix} 1 & 0 & 0 & 0 & 0 & \dots \\ 0 & 1 & 0 & 0 & 0 & \dots \\ 0 & 0 & 1 & 0 & 0 & \dots \\ 0 & 0 & 0 & 1 & 0 & \dots \\ 0 & 0 & 0 & 0 & 1 & \dots \\ & & & & & etc \end{pmatrix} \qquad (25.2.8)$$

Entanglement with the apparatus followed by tracing out the apparatus states results in decoherence. The particle is now in a classical mixed state. The transition from (25.2.4) to (25.2.8) is **part of** what is referred to as the reduction of the state vector (or collapse of the wave function). The other part is the selection of one particular state or detector.

So far, so good: this is just the standard decoherence perspective on quantum measurement (as in chapter 15). Things get more interesting, though, if the apparatus is modified by moving one hemisphere of detectors to a different radius (see Figure 25.1). We now consider the left hemisphere to be at a radius R_1, with a time of flight t_1, but the right hemisphere is at a radius $R_2 > R_1$ with time of flight $t_2 > t_1$. The initial state, that is for $0 < t < t_1$, is still as given by (25.2.2-4). Similarly, the final state for $t > t_2$, is still as given by (25.2.5-8). (Actually this is only true in the "geometrical optics" approximation. It is not exactly true due to diffraction effects. This is discussed at the end of this chapter). The question is: what is the state of the particle for $t_1 < t < t_2$, after interaction with the left hemisphere has happened, but interaction with the right hemisphere has not yet happened?

Figure 25.1: Two hemispheres of detectors of different radii

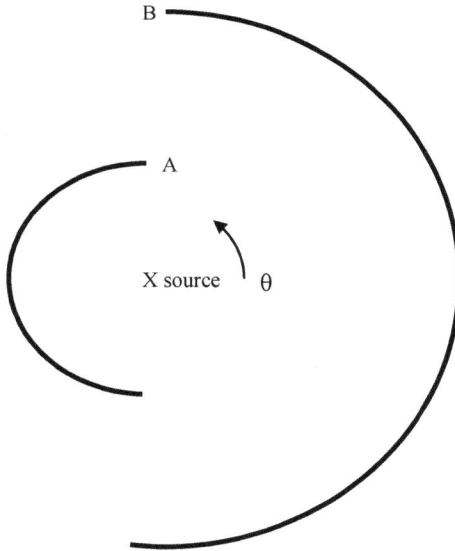

The question is addressed by considering separately the sums over states/detectors in the left and right hemispheres. At the time in question, $t_1 < t < t_2$, the unitary evolution of the initial state has brought only left hemisphere particle states into the left hemisphere detectors, whereas right hemisphere states are still associated with the null reference state of the detectors, i.e.,

$$|\Psi\rangle \rightarrow |\Psi'\rangle = \sum_{i \in L} |u_i\rangle |ei\rangle + \left(\sum_{i \in R} |u_i\rangle\right)|e0\rangle \qquad (25.2.9)$$

where the first sum on the RHS of (25.2.9) is over the left hemisphere whereas the second sum is over the right hemisphere. Note that what this state encodes is that, for particle propagation towards the larger right hemisphere, at times $t_1 < t < t_2$, no detector will have registered a particle. But, crucially, the state is a superposition of propagation towards the left and right.

The density matrix representing the combined particle-plus-apparatus is thus,

$$\hat{\rho} = \sum_{i,j \in L} |u_i\rangle\langle u_j| \otimes |ei\rangle\langle ej| + \left(\sum_{i,j \in R} |u_i\rangle\langle u_j|\right)|e0\rangle\langle e0| +$$
$$\sum_{i \in L}\sum_{j \in R} |u_i\rangle\langle u_j| \otimes |ei\rangle\langle e0| + \sum_{i \in R}\sum_{j \in L} |u_i\rangle\langle u_j| \otimes |e0\rangle\langle ej| \qquad (25.2.10)$$

Tracing out the environment states now means summing not just over left-hemisphere detector states $\{|ei\rangle, i = 1,2,3...\}$ but also including the state $|e0\rangle$ in the sum, since that remains a possible state of the detectors at this time. The second pair of terms (the cross terms) in (25.2.10) contribute nothing when traced out, whereas both the first two terms contribute, i.e.,

$$\hat{\rho}^{red} = \sum_{k=0}^{N}\langle ek|\hat{\rho}|ek\rangle = \sum_{k \in L} |u_k\rangle\langle u_k| + \left(\sum_{i,j \in R} |u_i\rangle\langle u_j|\right) \qquad (25.2.11)$$

Hence decoherence has taken place over the states corresponding to the left hemisphere, whereas coherence remains for the right hemisphere states. The signature of the former is that this part of the reduced density matrix has become diagonal, whereas the signature of the latter is that the remaining part of the density matrix still has off-diagonal components. In matrix notation,

$$\left(\rho^{red}\right) = \frac{1}{N}\begin{pmatrix} 1 & 0 & 0 & 0 & 0 & \cdots & & & & & \\ 0 & 1 & 0 & 0 & 0 & \cdots & & & & & \\ 0 & 0 & 1 & 0 & 0 & \cdots & & & 0 & & \\ 0 & 0 & 0 & 1 & 0 & \cdots & & & & & \\ 0 & 0 & 0 & 0 & 1 & \cdots & & & & & \\ & & & & & etc & & & & & \\ & & & & & & 1 & 1 & 1 & 1 & 1 & \cdots \\ & & & & & & 1 & 1 & 1 & 1 & 1 & \cdots \\ & & 0 & & & & 1 & 1 & 1 & 1 & 1 & \cdots \\ & & & & & & 1 & 1 & 1 & 1 & 1 & \cdots \\ & & & & & & 1 & 1 & 1 & 1 & 1 & \cdots \\ & & & & & & & & & & & etc \end{pmatrix}$$

$$(25.2.12)$$

where the left hemisphere corresponds to the first $N/2$ components and the right hemisphere to the second $N/2$ components. The density matrix (25.2.11-12) is the full description of the state of the particle at time $t_1 < t < t_2$ assuming that we have not yet looked at the left hemisphere detectors to see whether they have registered a particle. If we did, and they had, then the actual $|u_i, i \in L\rangle$ state would have been found. If we looked and found the left hemisphere detectors had not registered a particle, then the state of the particle would then be given by the density matrix for the remaining (right hemisphere) degrees of freedom alone, i.e.,

$$\hat{\rho}^{red} = 2\left(\sum_{i,j \in R} |u_i\rangle\langle u_j|\right) \tag{25.2.13}$$

The factor of 2 is to re-establish normalisation given that the sum in (25.2.13) now extends only over $N/2$ detectors. In matrix notation, if there is no particle registered on the left hemisphere the particle state becomes,

$$(\rho^{red}) = \frac{2}{N}
\begin{pmatrix}
0 & 0 & 0 & 0 & 0 & \cdots & & & & & & \\
0 & 0 & 0 & 0 & 0 & \cdots & & & & 0 & & \\
0 & 0 & 0 & 0 & 0 & \cdots & & & & & & \\
0 & 0 & 0 & 0 & 0 & \cdots & & & & & & \\
0 & 0 & 0 & 0 & 0 & \cdots & & & & & & \\
 & & & & & etc & & & & & & \\
 & & & & & & 1 & 1 & 1 & 1 & 1 & \cdots \\
 & & & & & & 1 & 1 & 1 & 1 & 1 & \cdots \\
 & & & 0 & & & 1 & 1 & 1 & 1 & 1 & \cdots \\
 & & & & & & 1 & 1 & 1 & 1 & 1 & \cdots \\
 & & & & & & 1 & 1 & 1 & 1 & 1 & \cdots \\
 & & & & & & & & & & & etc
\end{pmatrix}$$

$$\tag{25.2.14}$$

What does this mean? It means that with the left hemisphere present but failing to detect a particle, the particle state is nevertheless changed from (25.2.4) to (25.2.14). The state vector is partially reduced in that it is now concentrated on the right hemisphere states. However, full coherence is retained amongst the right hemisphere states, in contrast to the fully decohered density matrix (25.2.8). For example, if the right hemisphere of detectors was replaced (after suitable focussing) by a screen with a pair of slits, a diffraction pattern could be developed.

25.3 Diffraction Effects

To avoid misleading the reader we must now mention in passing that the above analysis is incomplete. The reason is that the presence of two hemispherical boundaries means that neither plane waves nor the Bessel function "hemispherical" wave,

$$0 < \theta < \pi/2: \qquad \psi(\vec{r}, t) \propto e^{-i\omega t} j_0(kr) \qquad\qquad (25.3.15a)$$

$$\pi/2 < \theta < \pi: \qquad \psi(\vec{r}, t) = 0 \qquad\qquad (25.3.15b)$$

are correct solutions to the Schrodinger equation. Note that in the absence of boundaries, the spherical wave (25.3.15a) would be a valid solution to the Schrodinger equation over the whole 4π steradians (see §25.4). However, (25.3.15a) and (25.3.15b) are discontinuous over their mutual boundary at $\theta = \pi/2$. So the true solution must smear (25.3.15a,b) so as to produce a continuous function (with continuous derivative). In the optical analogy, the physically effect which causes this modification to (25.3.15a,b) is diffraction from the edge of the inner hemisphere. Hence the detectors near the edge of the outer hemisphere would record a diffraction pattern. Moreover, diffraction would deflect some of the particles to angles $\theta > \pi/2$ so that they would miss the outer hemisphere entirely. The total signal received by the two hemispheres would not sum to unity (there's a hole in our bucket!). The lost signal could be captured by putting another arc of detectors beyond the right hemisphere, thus…

Figure 25.2: Additional detector to capture diffracted particles

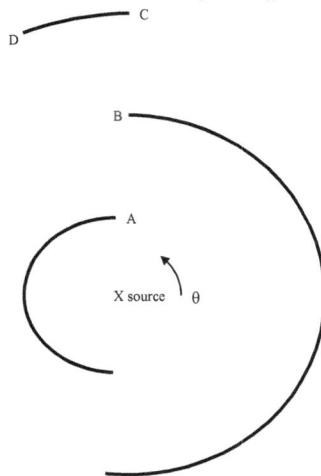

Hence the diffraction around edge A causes some particles to pass to the left of point B, missing the right-hand hemisphere, but these particles are captured by the additional detectors placed on the arc CD. Of course, there will also be diffraction from edge B, so the new detector CD would need to extend clockwise beyond the 12 o'clock position. However these diffraction effects are mentioned only for completeness. They do not detract from the moral of the example: null measurements can partially decohere the density matrix (partially reduce the state vector).

25.4 Reader Exercise

What is a "spherical wave" and why can it be written in terms of plane waves by (25.2.1)? The spherically symmetric S-wave solution to the free Schrodinger equation, for a state of definite energy, is $\psi(\bar{r}, t) \propto e^{-i\omega t} j_0(kr)$ where the energy is $\hbar\omega = (\hbar k)^2/2m$ and j_0 is the first spherical Bessel function, which is just the sinc function, i.e., $j_0(x) \equiv \sin x / x$. The plane wave solutions of the same energy are $\psi(\bar{r}, t) \propto e^{-i\omega t} e^{i\bar{k}\cdot\bar{r}}$ so that an isotropic, in-phase sum of plane waves is,

$$\frac{1}{4\pi} \int e^{i\bar{k}\cdot\bar{r}} d\Omega = -\frac{1}{2} \int_0^\pi e^{ikr\cos\theta} d(\cos\theta) = \frac{1}{2} \cdot \frac{e^{ikr} - e^{-ikr}}{ikr} = j_0(kr)$$

The integral on the LHS, above, is the continuum version of the sum over states in (25.2.1). So the states $|u_i\rangle$ in (25.2.1) can be identified, modulo normalisation, with plane waves of the same energy as the spherical wave (i.e., the same scalar k), noting that the plane waves in all directions contribute with the same amplitude and phase. There are therefore no expansion coefficients in (25.2.1). Obviously (25.2.1) becomes exact only in the limit of the subscript i becoming a continuous variable, namely the solid angle.

25.5 Reference

[25.1] Mauritius Renninger, *Messungen ohne Storung des Messobjekts (Measurement without disturbance of the measured objects)*, Zeitschrift für Physik, 1960; **158**(4): 417-421. An English translation by W. De Baere is available as arXiv:physics0504043.

26

Superdense Coding and Quantum Security

A single qubit can communicate only one classical bit of information. It takes two qubits to communicate two classical bits. However, if A wants to transmit to B two classical bits of information, there is a way of doing so which involves the transmission of only one quantum qubit – although another qubit, which is not actually transmitted from A, must also be involved. The way this works requires the two qubits to be entangled. This is superdense coding. And there is another benefit, which introduces the reader to the world of quantum cryptography, namely that an eavesdropper can learn nothing of the intended message by intercepting the one qubit transmitted from A to B.

26.1 How Much Information Can One Qubit Communicate?

A single qubit is defined by a pair of complex numbers, $\alpha|0\rangle + \beta|1\rangle$. Due to normalisation, $|\alpha|^2 + |\beta|^2 = 1$, this reduces to three real numbers, $re^{i\theta}|0\rangle + \sqrt{1-r^2} \cdot e^{i\varphi}|1\rangle$ where $0 \leq r \leq 1$, $0 \leq \theta < 2\pi$, $0 \leq \varphi < 2\pi$. As long as interference with another state is not envisaged, the overall phase is unimportant and we may reduce the general state to the specification of just two real numbers. Nevertheless, as the decimal expansion of a real number may be arbitrarily long, in principle it requires an infinite amount of information to specify the state of one qubit!

However, this is not the question being asked. The question is how much information can be **conveyed** by one qubit? Providing Alice and Bob, the sender and receiver, share an understanding of the polarization direction they are to use, then Alice can prepare an up or a down spin state to code one classical bit of information, and Bob can "read" that information by measuring the spin wrt the agreed axis. So, it is clear that one qubit can convey one classical bit. So far, so good.

If all Alice has is one qubit, the best she can do as an alternative to the above scheme is to prepare it as "spin up" wrt some other direction in 3D space – and this is just the same as saying she can prepare an arbitrary qubit state, $\alpha|0\rangle + \beta|1\rangle$. Can Bob contrive to "read" more than one bit of information from this state, for example by obtaining some knowledge, however incomplete, of the complex numbers α and β? Clearly not, because all Bob can do with a single qubit is to obtain an "up" or a "down" measurement with respect to an axis of his choosing. He remains ignorant of α and β. A quantum state is not, itself, an observable.

If Bob was sent a large number, N, of identical states, he could build up a statistical picture of the magnitudes of α and β. But he would only be able to deduce their magnitudes roughly, with an uncertainty of order $1/\sqrt{N}$. So the amount of information obtained by this would be of order $\sqrt{N}$, but it would have taken N qubits to convey it. Clearly, not an efficient coding procedure as N qubits could easily be used to convey N bits in a straightforward fashion.

You will not be surprised, then, that the maximum amount of classical information that can be communicated by N qubits is N bits. This is true no matter how devious you may be. This result – or rather a more complicated relationship between entropy expressions which implies it – is known as Holevo's Theorem. You may be forgiven, therefore, if you are confused regarding the claim made in chapter 1 that there is a way of conveying *two* classical bits via one qubit: superdense coding. But to do so, we require an additional resource, which is why Holevo's bound is not violated.

26.2 Superdense Coding

In superdense coding we see how it is possible for the ubiquitous Alice to transmit just one qubit to the equally inescapable Bob and yet to convey two bits of classical information. To do this, however, Alice and Bob must initially have received a pair of maximally entangled qubits, one of the bits accessible to Alice and the other to Bob, just as in an EPR type scenario. This pair of entangled qubits could be supplied by some central facility which plays no part in what information is to be conveyed. The pair of entangled qubits which start the process initially have no message content.

Alice wishes to convey two bits of information, which we will write as B_1 and B_2, both these independently taking the value 0 or 1. But recall that Alice has only one of the entangled pair of qubits available to her. She has to encode her two bits of information into this one accessible qubit. She does this by modifying her qubit using a physical interaction which corresponds to the operator $\sigma_x^{B_1}\sigma_z^{B_2}$ where $\sigma_x, \sigma_y, \sigma_z$ are the Pauli matrices which implement spin measurements on a spin-half particle (see §2.9 and §2.10). If $B_1 = 0$ then $\sigma_x^{B_1}$ becomes the identity operator, and similarly $\sigma_z^0 = 1$. Taking the states $|0\rangle$ and $|1\rangle$ to represent respectively spin up and spin down wrt the z-axis, the actions of the operators (see §2.10) are,

$$\sigma_z|0\rangle = |0\rangle, \quad \sigma_z|1\rangle = -|1\rangle, \quad \sigma_x|0\rangle = |1\rangle, \quad \sigma_x|1\rangle = |0\rangle \tag{26.2.1}$$

But the initial state of the entangled pair of particles is (say),

$$|\omega\rangle = (|00\rangle + |11\rangle)/\sqrt{2} \tag{26.2.2}$$

where the first particle is that accessible to Alice and the second particle is that accessible to Bob. Consequently, the four possible operations which Alice may perform on her qubit, corresponding to the two bits of information she wishes to communicate, create the following two-particle states,

$$\sigma_x^0 \sigma_z^0 |\omega\rangle = (|00\rangle + |11\rangle)/\sqrt{2} \qquad (26.2.3a)$$

$$\sigma_x^1 \sigma_z^0 |\omega\rangle = (|10\rangle + |01\rangle)/\sqrt{2} \qquad (26.2.3b)$$

$$\sigma_x^0 \sigma_z^1 |\omega\rangle = (|00\rangle - |11\rangle)/\sqrt{2} \qquad (26.2.3c)$$

$$\sigma_x^1 \sigma_z^1 |\omega\rangle = (|10\rangle - |01\rangle)/\sqrt{2} \qquad (26.2.3d)$$

Having performed the physical transformation corresponding to whichever of (26.2.3a-d) Alice wants, she then sends her qubit to Bob. The joint state of the pair of particles is one of the above four states, and which state they are in encodes the two bits of information. But how can Bob decode it?

It is actually immediately obvious that Bob can, in principle, decode it because the four states in (26.2.3a-d) are mutually orthogonal and hence perfectly distinguishable – and recall that Bob now has access to both qubits (though he has received only one of them from Alice). To do so he needs to construct an Hermitian matrix with the four states (26.2.3a-d) as eigenvectors, and with distinct eigenvalues, to complete the job. Physically measuring the observable corresponding to that Hermitian operator will then yield one of its eigenvalues, $\lambda_1, \lambda_2, \lambda_3, \lambda_4$. As we know that the state received by Bob is one of the corresponding eigenstates, by construction, the measured eigenvalue then uniquely and definitely determines which state Bob received, provided that we ensure the eigenvalues, $\lambda_1, \lambda_2, \lambda_3, \lambda_4$ are all different. For example, if we work in the basis defined by $|00\rangle, |10\rangle, |01\rangle, |11\rangle$ in that order, then the matrix of required eigenvectors, i.e., the states (26.2.3a-d), is,

$$(E) = \frac{1}{\sqrt{2}} \begin{pmatrix} 1 & 0 & 1 & 0 \\ 0 & 1 & 0 & 1 \\ 0 & 1 & 0 & -1 \\ 1 & 0 & -1 & 0 \end{pmatrix} \qquad (26.2.4)$$

A suitable matrix defining our required observable is then given by,

$$(Q) = (E)(\Lambda)(E)^+ \qquad (26.2.5)$$

where (Λ) is the diagonal matrix with elements $\lambda_1, \lambda_2, \lambda_3, \lambda_4$. For example, if we arbitrarily assign values -3/2, -1/2, 1/2, 3/2 to the eigenvalues then we find,

$$(Q) = \frac{1}{2} \begin{pmatrix} -1 & 0 & 0 & -2 \\ 0 & 1 & -2 & 0 \\ 0 & -2 & 1 & 0 \\ -2 & 0 & 0 & -1 \end{pmatrix} \qquad (26.2.6)$$

It can readily be confirmed that (26.2.3a-d) are the eigenvectors of (26.2.6) with eigenvalues -3/2, -1/2, 1/2, 3/2 respectively. Thus, realising a physical measurement of the observable defined by (26.2.6) yields the two bits of information. A measurement result of -3/2 means the message is $B_1 = 0$, $B_2 = 0$; a measurement result of -1/2 means the message is $B_1 = 1, B_2 = 0$, and so on.

Do not misinterpret this purely algebraic process to imply that two spin-half particles are physically the same as a single spin 3/2 particle. In fact, two spin-half particles actually combine to form a singlet (spin zero) state and a triplet (spin one) state, see §2.10. This way of looking at the state which Bob receives provides a physically more appealing schema for its decoding. Consider the total (vectorial) spin operator,

$$\hat{\vec{S}} = \frac{1}{2}(\hat{\vec{\sigma}}_1 + \hat{\vec{\sigma}}_2) \qquad (26.2.7)$$

Its scalar-square $\hat{S}^2$ takes the values $s(s + 1)$ for a system of spin s. Measuring this observable, i.e., the total spin of the pair of particles, reveals state (26.2.3d) to be the singlet with $s = 0$, whereas the other three states all have $s = 1$. (An exercise for the reader is to show this). Hence, measuring zero total spin reveals the message to be $B_1 = 1, B_2 = 1$. Otherwise, as states (26.2.3a-c) are all eigenstates of the operator $\hat{S}^2$ they are left unchanged by its measurement and so are still available to be decoded without having been corrupted by this first measurement step. Note that we cannot yet distinguish between states (26.2.3a-c) as they all have eigenvalue 2 wrt the measurement of $\hat{S}^2$ (i.e., $s = 1$).

Next, if the result of the first measurement was $s = 1$, we can measure the product of the two x-spin operators, $\hat{\sigma}_{1x}\hat{\sigma}_{2x}$ where the subscripts indicate that one operator acts on the first particle and the other on the second particle. Another exercise for the reader is to show that all three of the states (26.2.3a-c) are eigenstates of this operator, but state (26.2.3c) has eigenvalue -1 whereas states (26.2.3a,b) both have eigenvalues +1. Hence measuring the value -1 for this observable reveals the message to be $B_1 = 0, B_2 = 1$.

The remaining two states which have not yet been distinguished, (26.2.3a,b), have remained unchanged by both measurements so far as they were both eigenstates of both observables measured. They can be distinguished by measuring the product observable $\hat{\sigma}_{1y}\hat{\sigma}_{2y}$. Both states are again eigenstates of this operator, but (26.2.3a) has eigenvalue -1 whereas (26.2.3b) has eigenvalue +1. The former result reveals the message to be $B_1 = 0, B_2 = 0$ whilst the latter result reveals the message to be $B_1 = 1, B_2 = 0$. This completes the decoding.

Whilst the above protocol requires three measurements, it is possible to decode the two bits in just two measurements. Texts on quantum computing prefer to couch the process in terms which are significant in the context of computers, i.e., in the language of "gates". Thus Ref.[26.1], for example, provides a decoding scheme using a CNOT gate and a so-called Hadamard gate (see that reference, or an equivalent quantum computing text, for the meaning of these terms, or look ahead to §27.2).

Superdense coding has been (partially) demonstrated using photons in Ref.[26.2]. Rather than two bits being transmitted per qubit, the authors achieved transmission of a "trit", or three-state message per qubit. Full implementation of superdense coding was claimed using an NMR technique in Ref.[26.3].

26.3 Eavesdropping

The beauty of the superdense coding scheme outlined in §26.2 lies not only in the fact that only one qubit need be physically transmitted by Alice, but also that this transmitted qubit can provide an eavesdropper with no information about the message at all. The whole of the message – both bits of information – resides in the bipartite two-qubit state. Without possession of the other qubit, which Bob has, intercepting the transmitted qubit is of no value to the would-be spy. And recall that the other qubit was sent to Bob by some third party facility, not Alice, and this third party has no knowledge of the message.

To see why possession of Alice's qubit alone is useless, note that all the eavesdropper can do with it is measure it. But if the result is "spin up", this is equally likely to have resulted from any of the four states (26.2.3a-d), and the same is true if the result is "spin down".

26.4 Exercises for the Reader

[1] Show that states (26.2.3a-c) are all eigenstates of $\hat{S}^2$ with eigenvalue 2.

[2] Show that state (26.2.3d) is an eigenstate of $\hat{S}^2$ with eigenvalue 0.

[3] Show that states (26.2.3a-b) are both eigenstates of $\hat{\sigma}_{1x}\hat{\sigma}_{2x}$ with eigenvalue 1.

[4]   Show that state (26.2.3c) is an eigenstate of $\hat{\sigma}_{1x}\hat{\sigma}_{2x}$ with eigenvalue -1.
[5]   Show that state (26.2.3a) is an eigenstate of $\hat{\sigma}_{1y}\hat{\sigma}_{2y}$ with eigenvalue -1.
[6]   Show that state (26.2.3b) is an eigenstate of $\hat{\sigma}_{1y}\hat{\sigma}_{2y}$ with eigenvalue 1.

I make use of (26.2.1) together with $\sigma_y|0\rangle = i|1\rangle$ and $\sigma_y|1\rangle = -i|0\rangle$. Hence we get the following,

$$\hat{\sigma}_{1z}\hat{\sigma}_{2z}|00\rangle = |00\rangle \qquad\qquad \hat{\sigma}_{1z}\hat{\sigma}_{2z}|11\rangle = |11\rangle$$

$$\hat{\sigma}_{1z}\hat{\sigma}_{2z}|01\rangle = -|01\rangle \qquad\qquad \hat{\sigma}_{1z}\hat{\sigma}_{2z}|10\rangle = -|10\rangle$$

$$\hat{\sigma}_{1x}\hat{\sigma}_{2x}|00\rangle = |11\rangle \qquad\qquad \hat{\sigma}_{1x}\hat{\sigma}_{2x}|11\rangle = |00\rangle$$

$$\hat{\sigma}_{1x}\hat{\sigma}_{2x}|01\rangle = |10\rangle \qquad\qquad \hat{\sigma}_{1x}\hat{\sigma}_{2x}|10\rangle = |01\rangle$$

$$\hat{\sigma}_{1y}\hat{\sigma}_{2y}|00\rangle = -|11\rangle \qquad\qquad \hat{\sigma}_{1y}\hat{\sigma}_{2y}|11\rangle = -|00\rangle$$

$$\hat{\sigma}_{1y}\hat{\sigma}_{2y}|01\rangle = |10\rangle \qquad\qquad \hat{\sigma}_{1y}\hat{\sigma}_{2y}|10\rangle = |01\rangle$$

The required results then follow. Note in particular,

$$\left(\hat{\sigma}_{1x}\hat{\sigma}_{2x} + \hat{\sigma}_{1y}\hat{\sigma}_{2y} + \hat{\sigma}_{1z}\hat{\sigma}_{2z}\right)|00\rangle = |00\rangle$$

$$\left(\hat{\sigma}_{1x}\hat{\sigma}_{2x} + \hat{\sigma}_{1y}\hat{\sigma}_{2y} + \hat{\sigma}_{1z}\hat{\sigma}_{2z}\right)|11\rangle = |11\rangle$$

$$\left(\hat{\sigma}_{1x}\hat{\sigma}_{2x} + \hat{\sigma}_{1y}\hat{\sigma}_{2y} + \hat{\sigma}_{1z}\hat{\sigma}_{2z}\right)|01\rangle = 2|10\rangle - |01\rangle$$

$$\left(\hat{\sigma}_{1x}\hat{\sigma}_{2x} + \hat{\sigma}_{1y}\hat{\sigma}_{2y} + \hat{\sigma}_{1z}\hat{\sigma}_{2z}\right)|10\rangle = 2|01\rangle - |10\rangle$$

And that $\hat{S}^2 = \frac{3}{2} + \frac{1}{2}\left(\hat{\sigma}_{1x}\hat{\sigma}_{2x} + \hat{\sigma}_{1y}\hat{\sigma}_{2y} + \hat{\sigma}_{1z}\hat{\sigma}_{2z}\right)$ which yields [1] and [2].

26.5 References

[26.1] Joachim Stolze, Dieter Suter, *"Quantum Computing: A short course from theory to experiment"*, Wiley-VCH, 2004.

[26.2] Klaus Mattle, Harald Weinfurter, Paul G. Kwiat, and Anton Zeilinger, "*Dense Coding in Experimental Quantum Communication*", Phys.Rev.Lett. **76**(25) 4656-4659, June 1996

[26.3] Fang, X., X. Zhu, M. Feng, X. Mao, and F. Du, (2000), *"Experimental Implementaton of Dense Coding Using Nuclear Magnetic Resonance"*, Phys.Rev. A 61, 022307. https://arxiv.org/abs/quant-ph/9906041v1

27

Quantum Teleportation

Quantum Teleportation: What it is and what it is not. What has to be transported conventionally through space, and why it does not provide a means of faster-than-light travel or communication. As with superdense coding, the resource which makes quantum teleportation work is a bipartite entangled state shared by sender and recipient.

27.1 It's Teleportation, Jim, but Not as We Know It

Of course, it's a cheat, really. The classical equivalent of what passes for teleportation in the quantum communication fraternity would be "here's an instruction book for how to construct Captain Kirk atom by atom – the rest is up to you". Any matter or energy which is transported in the process has to be done by conventional means. Do not expect the whole of Captain Kirk to be teleported anywhere. All that is teleported is….well, I'm tempted to say "information", but that is not right either. You get more than just information. What you get is the exact quantum state, which is something rather ineffable beyond just its possible information content.

To be impressed by this, you first need to appreciate the no-cloning theorem (see chapter 4). This says that there is no process that will produce an exact copy of an arbitrary quantum state whilst preserving the original. But quantum teleportation **does** provide the ubiquitous Bob with an exact replica of Alice's quantum state. How does this get around the no-cloning theorem? Simple – Alice's original of the quantum state is destroyed in the process. This is why it is called "teleportation". In a sense, one might claim that Bob does not receive a copy of the quantum state, he gets the original because the original has vanished and been replaced with Bob's copy (potentially at some great distance). Hence the designation as "teleportation". But neither matter nor energy is teleported, and really the process would be better described as remote-copying-with-destruction-of-the-original. Not as snappy as "teleportation", admittedly.

The important thing to appreciate is that Alice has no knowledge herself of the quantum state which is to be teleported. Say it's a qubit state, $\alpha|0\rangle + \beta|1\rangle$. Alice does not know α and β. Otherwise, of course, it would be trivial. She would merely transmit the values of these complex numbers to Bob who could then construct the corresponding state in his own lab. The point here is that α and β are not observable (not measurable), so Alice cannot discover what they are even given the state itself.

Does the quantum state get transmitted to Bob transluminally? No. It is true that entangled quantum states do genuinely exhibit non-local characteristics, the spooky action-at-a-distance which perturbed Einstein so. And, as we shall see below, there is a key moment when Alice performs a measurement which provides Bob instantly with a certain quantum state, one out of several which were previously in superposition. But this quantum state is not *the* quantum state. To acquire *the* quantum state, Bob must perform an operation on the precursor state himself – and the operation he must perform requires information from Alice to specify it. This key information is just ordinary, classical, information which must be transmitted in the normal way. So, not only has no matter or energy been transmitted faster than light, but nor does any information or the quantum state itself, get transmitted transluminally by the teleportation process. Causality is respected as regards matter, energy *and* information and quantum states.

The possibility of quantum teleportation, and the entanglement protocol to achieve it, was first described by Bennett et al, Ref.[27.1]. It is convenient to express the protocol in the language of quantum computing, using quantum "gates", so these are described first.

27.2 Preliminaries – Quantum Gates

Before describing the teleportation process, it is necessary to appreciate what physical interactions Bob and Alice can perform on qubits in their possession. In the context of computation or communication the elementary interactions with qubits are referred to as 'gates', in analogy with classical computing. Also as with classical computing, there are usually only one or two qubits which enter the gate, and only one or two qubits which are output from the gate.

The simplest example is the controlled-not gate, or CNOT. This takes two qubits as input and two qubits as output. Its operation is as follows: if the first qubit is in state $|0\rangle$ then the CNOT gate passes both qubits through to the output unchanged. But if the first qubit is $|1\rangle$, then the first qubit is again unchanged but the second qubit is changed to the opposite state, i.e., $|0\rangle$ becomes $|1\rangle$ and vice-versa. Using matrix notation the states are written,

$$1 = |0\rangle|0\rangle;\ 2 = |0\rangle|1\rangle;\ 3 = |1\rangle|0\rangle;\ 4 = |1\rangle|1\rangle \tag{27.2.1}$$

CNOT can therefore be represented by,

$$\text{CNOT} = \begin{pmatrix} 1 & 0 & 0 & 0 \\ 0 & 1 & 0 & 0 \\ 0 & 0 & 0 & 1 \\ 0 & 0 & 1 & 0 \end{pmatrix} \tag{27.2.2}$$

CNOT is the only two-state gate we shall need. But we also need some single qubit gates. Firstly there are the gates defined by the Pauli matrices,

$$\bar{\sigma} = \left[\begin{pmatrix} 0 & 1 \\ 1 & 0 \end{pmatrix}, \begin{pmatrix} 0 & -i \\ i & 0 \end{pmatrix}, \begin{pmatrix} 1 & 0 \\ 0 & -1 \end{pmatrix} \right] \tag{27.2.3}$$

Hence, in the z-representation in which $|0\rangle \rightarrow \begin{pmatrix} 1 \\ 0 \end{pmatrix}$ and $|1\rangle \rightarrow \begin{pmatrix} 0 \\ 1 \end{pmatrix}$ we have,

$$\sigma_z|0\rangle = |0\rangle; \ \sigma_z|1\rangle = -|1\rangle; \ \sigma_x|0\rangle = |1\rangle; \ \sigma_x|1\rangle = |0\rangle;$$
$$\sigma_y|0\rangle = i|1\rangle; \ \sigma_y|1\rangle = -i|0\rangle \tag{27.2.4}$$

Finally, we shall need the Hadamard gate which is defined by,

$$\hat{H}ad = \frac{1}{\sqrt{2}}(\sigma_x + \sigma_z) \tag{27.2.5}$$

and hence, $\hat{H}ad|0\rangle = \frac{1}{\sqrt{2}}(|0\rangle + |1\rangle)$ and $\hat{H}ad|1\rangle = \frac{1}{\sqrt{2}}(|0\rangle - |1\rangle)$.

All the above transformations are unitary and represent operations which can be realised physically on suitable systems.

27.3 The Quantum Teleportation Process

The simplest example is the teleportation of a single qubit state. Call the qubit to be teleported, initially in Alice's possession,

$$|\psi\rangle = \alpha|0\rangle + \beta|1\rangle \tag{27.3.1}$$

Recall that neither Bob nor Alice know what state this is, i.e., they do not know α and β. The resource needed for teleportation to be possible is that Bob and Alice must share a maximally entangled state of two other qubits. Call this two-qubit entangled state $|\varphi\rangle$. For definiteness we specify it as,

$$|\varphi\rangle = \frac{1}{\sqrt{2}}(|0\rangle|0\rangle + |1\rangle|1\rangle) \tag{27.3.2}$$

The first of the qubits in $|\varphi\rangle$ is accessible to Alice, whereas the second is accessible to Bob. For example, they may be a pair of spin ½ particles which were entangled and then propagated in opposite directions, one towards Alice and one towards Bob. The initial state of all three qubits is thus,

$$|\psi\rangle|\varphi\rangle = (\alpha|0\rangle + \beta|1\rangle)\frac{1}{\sqrt{2}}(|0\rangle|0\rangle + |1\rangle|1\rangle)$$

$$= \frac{1}{\sqrt{2}}[\alpha|0\rangle|0\rangle|0\rangle + \alpha|0\rangle|1\rangle|1\rangle + \beta|1\rangle|0\rangle|0\rangle + \beta|1\rangle|1\rangle|1\rangle] \qquad (27.3.3)$$

where the 1st (leftmost) state refers to the qubit to be teleported. Hence the 1st and 2nd qubits are initially accessible to Alice, but only the 3rd qubit is accessible to Bob. Alice now carries out some operations on the qubits she can influence. Firstly she carries out a CNOT, which has the following effect,

$$CNOT|\psi\rangle|\varphi\rangle = \frac{1}{\sqrt{2}}[\alpha|0\rangle|0\rangle|0\rangle + \alpha|0\rangle|1\rangle|1\rangle + \beta|1\rangle|1\rangle|0\rangle + \beta|1\rangle|0\rangle|1\rangle]$$
$$(27.3.4)$$

Next Alice carries out a Hadamard operation on the first qubit, the one that initially contains the state to be teleported. This gives,

$$\hat{H}ad.\,CNOT|\psi\rangle|\varphi\rangle = \frac{1}{2}\left[\begin{array}{l} \alpha(|0\rangle + |1\rangle)\{|0\rangle|0\rangle + |1\rangle|1\rangle\} \\ +\beta(|0\rangle - |1\rangle)\{|1\rangle|0\rangle + |0\rangle|1\rangle\} \end{array}\right] \qquad (27.3.5)$$

This superposition of eight states can be re-arranged as follows,

$$\hat{H}ad.\,CNOT|\psi\rangle|\varphi\rangle = \frac{1}{2}\left[\begin{array}{l} |0\rangle|0\rangle(\alpha|0\rangle + \beta|1\rangle) + |0\rangle|1\rangle(\alpha|1\rangle + \beta|0\rangle) \\ +|1\rangle|0\rangle(\alpha|0\rangle - \beta|1\rangle) + |1\rangle|1\rangle(\alpha|1\rangle - \beta|0\rangle) \end{array}\right]$$
$$(27.3.6)$$

But the third qubit, the one accessible to Bob, can be written as $|\psi\rangle$, $\sigma_x|\psi\rangle$, $\sigma_z|\psi\rangle$ and $\sigma_x\sigma_z|\psi\rangle$ in the four terms respectively, i.e.,

$$\hat{H}ad.\,CNOT|\psi\rangle|\varphi\rangle = \frac{1}{\sqrt{2}}[|0\rangle|0\rangle + |0\rangle|1\rangle\sigma_x + |1\rangle|0\rangle\sigma_z + |1\rangle|1\rangle\sigma_x\sigma_z]|\psi\rangle$$
$$(27.3.7)$$

Let us review the situation now from the perspective of Alice. Her two qubits, the 1st and 2nd, occur with equal weight in the above superposition. Rather miraculously, the α and β parameters have switched from being associated with her qubits to being associated with Bob's, the 3rd qubit. From Bob's perspective, his qubit could be in any of the states $|\psi\rangle$, $\sigma_x|\psi\rangle$, $\sigma_z|\psi\rangle$ or $\sigma_x\sigma_z|\psi\rangle$. So, at this point, teleportation has not happened.

Alice now performs a measurement in the z-basis of the 2 x 2 dimensional bipartite Hilbert space accessible to her, i.e., she measures the 'spins' (if that is what they are) of her two qubits in the z-direction. The state vector collapses and just one of the four superposed states in (27.3.6) is projected out. Alice obtains the result 00, or 01, or 10, or 11. Correspondingly the quantum state of the qubit in Bob's possession projects out to be either $|\psi\rangle$, or $\sigma_x|\psi\rangle$, or $\sigma_z|\psi\rangle$ or $\sigma_x\sigma_z|\psi\rangle$ respectively. This happens non-locally in the manner characteristic of entangled bipartite states when the parts are spacelike separated. (Note that these four states in Bob's possession cannot be mutually orthogonal, but nevertheless these are the four states projected out because this comes about due to Alice's measurements, and **her** four states in (27.3.7) **are** mutually orthogonal).

The important thing to realise is that, after Alice's measurement, Bob's quantum state is *deterministically* in one of the states $|\psi\rangle$, or $\sigma_x|\psi\rangle$, or $\sigma_z|\psi\rangle$ or $\sigma_x\sigma_z|\psi\rangle$, though he does not yet know which. Prior to Alice's measurement, Bob's quantum state was a superposition of these states (with 'coefficients' entangled with the quantum states of Alice's qubits). This is crucial. The collapse of the state vector, (27.3.7), which led to this deterministic state of affairs is the key step in the teleportation.

However, Bob does not yet possess the state $|\psi\rangle$. But in order for him to do so he only needs to know which of the four states, $|\psi\rangle$, or $\sigma_x|\psi\rangle$, or $\sigma_z|\psi\rangle$ or $\sigma_x\sigma_z|\psi\rangle$, he now has in his possession. This is because each of the operators σ_x, σ_z and $\sigma_x\sigma_z$ is unitary and hence reversible. So, if Bob knew, say, that he currently had state $\sigma_x|\psi\rangle$, he would only need to carry out the physical operation corresponding to the inverse of σ_x (which happens to be just σ_x itself) in order to recover the state $|\psi\rangle$, and hence to succeed in completing the teleportation. Similarly, the states $\sigma_z|\psi\rangle$ and $\sigma_x\sigma_z|\psi\rangle$ can also be used to recover $|\psi\rangle$ if Bob knows which state he now possesses.

But it is a simple matter for Bob to find out which state he has. He just needs to ask Alice! Her measurement has provided the result 00, or 01, or 10, or 11, and these correspond one-to-one with Bob's four quantum states $|\psi\rangle$, or $\sigma_x|\psi\rangle$, or $\sigma_z|\psi\rangle$ or $\sigma_x\sigma_z|\psi\rangle$. So, a simple classical communication of two bits of information from Alice provides Bob with the information he needs to carry out the final step of deconvolution to obtain the successfully teleported state, $|\psi\rangle$. Applause!

Note that Alice no longer has the state $|\psi\rangle$. She is left with one of the product states, $|0\rangle|0\rangle$ or $|0\rangle|1\rangle$ or $|1\rangle|0\rangle$ or $|1\rangle|1\rangle$ with equal probability. Hence, she is left with not the least vestige of information about her original quantum state $|\psi\rangle$, thus respecting the no-cloning theorem.

Also note that it has required the transmission of *two* bits of classical information to successfully teleport a one qubit quantum state. In chapter 26 we saw that superdense coding is essentially the reverse of this teleportation process in which the transmission of one qubit suffices to communicate two bits of classical information. It is tempting to conclude that one qubit equals two bits, but no, not really. Because in both cases there were other qubits crucially involved in the process, and entanglement is the required resource. So neither superdense coding nor quantum teleportation violate the Holevo bound that says one qubit can, at most, convey only one classical bit of information.

It is surprising that the arbitrary quantum state $|\psi\rangle$ can be teleported using only two classical bits of information. Recall that $|\psi\rangle$ is specified by two complex numbers, or three real numbers taking normalisation into account. This reduces to two real numbers if we ignore an arbitrary overall phase (though we could not ignore the overall phase if we wished to teleport more than one state and retain their relative phase, which would be necessary to correctly reproduce subsequent superpositions and interference). But even two real numbers "contain" an infinite amount of information (i.e., an infinite number of decimal digits). So it is surprising that the state can be teleported using just two bits of information. The resolution of this paradox is that the coefficients α and β which specify $|\psi\rangle$ are not observable, i.e., the state itself is not an observable. One can regard the possibility of teleporting a qubit using just two bits of information as a proof that the quantum state of the qubit is not observable – because if it were we would have a means of transmitting a potentially infinite amount of information in just two classical bits, which makes no sense. It also raises ontological questions about the nature of quantum states.

Quantum teleportation in the sense discussed here has been demonstrated experimentally many times, over increasingly large distances and at increasingly fast bit-rates. The first demonstrations were in 1997/98. Some of the earliest papers are Refs.[27.2-8].

27.4 References

[27.1] C. H. Bennett, G. Brassard, C. Crépeau, R. Jozsa, A. Peres, W. K. Wootters: (1993), "*Teleporting an Unknown Quantum State via Dual Classical and Einstein-Podolsky-Rosen Channels*", Phys. Rev. Lett. 70, 1895-1899 (1993).

[27.2] Bouwmeester, D., J.-W. Pan, K. Mattle, M. Eibl, H. Weinfurter, and A. Zeilinger, "*Experimental quantum teleportation*" 1997, Nature (London) 390, 575.

[27.3] Bouwmeester, D., J.-W. Pan, H. Weinfurter, and A. Zeilinger, "*High-Fidelity Teleportation of Independent Qubits*" 2000, J. Mod. Opt. 47, 279.

[27.4] Boschi, D., S. Branca, F. De Martini, L. Hardy, and S. Popescu, "*Experimental realization of teleporting an unknown pure quantum state via dual classical and Einstein-Podolsky-Rosen channels*" 1998, Phys. Rev. Lett. 80, 1121.

[27.5] Braunstein, S.L., C.A. Fuchs, and H.J. Kimble, "*Criteria for Continuous-Variable Quantum Teleportation*" 2000, J. Mod.Opt. 47, 267.

[27.6] Vaidman, L., 1998, "*Teleportation: Dream or Reality?*", *Talk at the Conference: Mysteries, Puzzles and Paradoxes in Quantum Mechanics*, Gargano, Italy; arXiv:quant-ph/9810089.

[27.7] Pan, J-W., D. Bouwmeester, H. Weinfurter, and A. Zeilinger, "*Experimental entanglement swapping: entangling photons that never interacted*" 1998, Phys. Rev. Lett. 80, 3891

[27.8] Furuzawa, A., J.L. Sørensen, S.L. Braunstein, C.A. Fuchs, H.J. Kimble, and E.S. Polzik, "*Unconditional Quantum Teleportation*" 1998, Science 282, 706.

28

Waves Making Tracks

The straight tracks of alpha or beta particles radiating out from a radioactive source in a cloud chamber give every impression that the agency causing them is particle-like. How is this consistent with the quantum description of particles in terms of Hilbert space vectors or wavefunctions? Why should a delocalised wave make a straight track?

28.1 Wave Functions

I introduced the idea of wavefunctions, and how they relate to Hilbert space vectors, in §2.6, but have made relatively little use of them in this book. However we must consider wavefunctions in this chapter because the subject matter is explicitly about continuous trajectories in space, so it is unavoidable to introduce the continuum of spatial basis states. Let's first recap what is meant by a wavefunction.

The state of a stationary particle at a well-defined position in 3D space may be written $|\bar{r}\rangle$ where $\bar{r}$ is its position vector in (real physical) space. For some other state of the particle, expressed as normal as a Hilbert space vector, $|\psi\rangle$, the wavefunction of the particle in that state is defined as the inner product of $|\psi\rangle$ with $|\bar{r}\rangle$. The wavefunction is just that – a function of the vector $\bar{r}$ – and it is written,

$$\psi(\bar{r}) = \langle \bar{r} | \psi \rangle \tag{28.1.1}$$

There are certain niceties regarding normalisation and orthogonality. We shall not need to go into those except to note that we can choose some arbitrary volume, V, within which the states are normalised. I choose unit volume.

I shall now derive the Schrodinger equation in configuration space form for a free particle of well defined energy, E. We have already seen in §2.4 that the Schrodinger equation is,

$$\hat{H}|\psi\rangle = i\hbar\partial_t|\psi\rangle \tag{2.4.1}$$

where the Hamiltonian, $\hat{H}$, is the operator for energy. The formal solution of (2.4.1) is,

$$|\psi(t)\rangle = exp\left\{-i\frac{\hat{H}t}{\hbar}\right\}|\psi(0)\rangle \tag{2.4.2}$$

So, if the initial state is an energy eigenstate, then the state at subsequent times is given by,

$$|\psi(t)\rangle = exp\left\{-i\frac{Et}{\hbar}\right\}|\psi(0)\rangle \tag{28.1.2}$$

The Schrodinger equation, (2.4.1), then becomes independent of time (the stationary Schrodinger equation),

$$\hat{H}|\psi\rangle = E|\psi\rangle \tag{28.1.3}$$

which, of course, is just the energy eigen-equation. Adopting the configuration basis, $|\bar{r}\rangle$, the operator $\hat{H}$ becomes a differential operator acting on the variable $\bar{r}$ and the Schrodinger equation becomes,

$$\hat{H}\psi(\bar{r}) = E\psi(\bar{r}) \tag{28.1.4}$$

The energy of a free particle is just its kinetic energy (i.e., there is no potential energy if the particle is free in the sense of not being acted upon by an external force). In terms of the particle's moment, P, and mass, m, the non-relativistic kinetic energy is just $P^2/2m$. But we deduced in chapter 5 that the operator which represents momentum for a particle moving in 3D continuum space is,

$$\hat{\bar{P}} \rightarrow -i\hbar\bar{\nabla} \tag{5.6.4}$$

where $\bar{\nabla}$ is the gradient operator with components $(\partial_x, \partial_y, \partial_z)$. In one spatial dimension (5.6.4) is just,

$$\hat{P}_x \rightarrow -i\hbar\partial_x \tag{5.6.1}$$

Putting these together we get the stationary Schrodinger equation for a free particle expressed in the continuum spatial basis,

$$-\frac{\hbar^2}{2m}\nabla^2\psi(\bar{r}) = E\psi(\bar{r}) \tag{28.1.5}$$

The operator on the LHS of (28.1.5) is the explicit form of the Hamiltonian for a free particle in differential form. In one spatial dimension it reduces to,

$$-\frac{\hbar^2}{2m}\partial_x^2\psi(x) = E\psi(x) \tag{28.1.6}$$

The purpose of dragging you through these derivations is to motivate introducing the plane wave and spherical wave solutions to these equations. The reader should be able to check that the following are the plane wave solutions to equs.(28.1.5,6),

3D Plane Wave: $\qquad \psi(\bar{r}) = exp\left(i\bar{k}\cdot\bar{r}\right)$ $\qquad\qquad$ (28.1.7a)

1D Plane Wave: $\qquad \psi(x) = exp(ikx)$ $\qquad\qquad\qquad$ (28.1.7b)

In the 3D case the (vectorial) momentum is $\bar{p} = \hbar\bar{k}$, and in the 1D case it is $p_x = \hbar k$. In order to satisfy the Schrodinger equation, (28.1.6), the energy must be given by $E = (\hbar k)^2/2m$. This may be inverted to give the magnitude of the wavevector in (28.1.7a,b) in terms of the energy, noting that its direction is arbitrary.

Since (28.1.7a,b) are states of well-defined momentum, the uncertainty relation implies that the particle's spatial position should be completely uncertain (see chapter 5). Indeed it is, as the probability of finding the particle at position $\bar{r}$ is $|\langle \bar{r} | \psi \rangle|^2 = |\psi(\bar{r})|^2$ and for the plane wave wavefunctions (28.1.7a,b), $|\psi(\bar{r})|^2$ is a constant, meaning the probability of finding the particle in any region of space is the same as in any other region: it is uniformly spread over space.

Another exact solution of the 3D Schrodinger equation for a free particle is the spherical wave. In this case it is assumed that the wavefunction depends upon the radial distance from the origin, r, only, i.e., there is no dependence upon the angular parts of the spherical polar coordinates, θ, ϕ. In this case the del-squared operator can be written,

$$\nabla^2 \to \frac{1}{r^2} \partial_r (r^2 \partial_r) \tag{28.1.8}$$

An exercise for the reader is to confirm that the following is a solution to (28.1.5) when (28.1.8) holds,

3D Spherical Wave: $\quad \psi(r) = \frac{exp(ikr)}{r}$ (28.1.9)

for which the relationship with energy is again given by $E = (\hbar k)^2/2m$. Just as a plane wave is the correct wavefunction for a particle with a well-defined momentum vector (magnitude and direction), so a spherical wave is the correct wavefunction for a particle whose magnitude of momentum is well defined, but whose direction is completely unknown and isotropically distributed. For a spherical wave, the probability of finding the particle at a given spatial location falls off inversely as the square of the radial distance from the origin, as follows from taking the absolute square of (28.1.9). However, the angular location of the particle is completely indeterminate.

Thus (28.1.7a,b) and (28.1.9) are solutions which describe free particles with definite energy. Bear this in mind as we turn to the behaviour observed in cloud chambers.

28.2 The Conundrum

One gets used to measurement causing the 'reduction of the state vector' (or 'collapse of the wavefunction') despite its nature still being mysterious. One becomes accustomed to a particle with well-defined momentum, and hence approximating to a plane wave, nevertheless being detected as if it were particle-like. Thus, a plane wave quantum state normalised to unity (within some finite volume) is detected by a photographic screen as a single point (to within the resolving power of the emulsion). Or, impinging upon an array of detectors, a normalised plane wave state causes only one detector to click. Having accepted these instances of the collapse of the wavefunction, there is still another type of detection which is more unsettling. This is particle detection using a cloud chamber or bubble chamber. In these detectors the whole track of the particle is made visible. How can a plane wave give rise to a very particle-like straight track? It's tempting to think that all this talk of waves is so much moonshine when confronted with the ostensibly obvious evidence of particle tracks.

The resolution of this problem was first provided by Mott, Ref.[28.1] and it has been discussed often since, see for example Refs.[28.2-3]. A cloud chamber consists of a volume of supercooled vapour in a container which is either transparent or has a transparent window. Alcohol is a convenient working vapour in the lab. The passage of a radioactive decay product, an alpha or beta particle, causes ionisation of molecules within the air/vapour mixture, and these act as nucleation points for tiny droplets of liquid phase to form. A string of such tiny droplets is formed and traces the trajectory of the alpha or beta particle.

If the interaction between the particle and the air/vapour molecules, which results in their ionisation and hence droplet formation, is not to interfere too much with the trajectory of the particle, the energy of the particle must be very large compared with the energy required to cause ionisation. That is indeed the case for radioactive decays. Beta particles are typically emitted with kinetic energies between 0.1 and 0.5 MeV, whilst alpha particle kinetic energies are typically ~0.5 MeV. In comparison, ionisation energies are in the order of a few eV. Hence, the interaction leading to ionisation disturbs the alpha/beta particle only very little, and from the perspective of classical physics it is to be expected that the track of the particle would be straight (just as a tennis ball is not noticeably affected by Brownian motion). But this does not address the conundrum raised by the quantum mechanical perspective. Suppose the alpha/beta particle were emitted from the radioactive decay as a so-called S-wave, that is without any angular momentum and in a spherically symmetric state. Its wavefunction is (28.1.9). It is far from obvious that this should give rise to a straight, particle-like track. The next two sections provide the resolution to this seeming paradox; the first is simple, the second not so easy but more complete.

28.3 Argument from the Uncertainty Principle

From the classical perspective, the fundamental reason for the straight track is the high energy of the particle compared with the ionisation energy of the air/vapour. The fundamental reason in quantum mechanics is similar, but better argued from momenta rather than energy. The formation of a droplet is effectively a position measurement. What is the uncertainty in that position measurement? That would be the size of the droplet – which, for sake of argument we assume is about 0.1mm (10^{-4} m). But, one could argue, in principle one could detect a droplet forming when it was far smaller, by using a suitable microscope. In the limit, the best we could possibly do would be to locate the particle by this means to a region approaching the size of an atom, say $\sim 10^{-10}$ m. Clearly, that's really pushing it.

So what does that tell us about the uncertainty in the particle's momentum? The uncertainty principle (chapter 5) tells us that the product of the uncertainties in position and momentum is at least $\hbar/2 \sim 0.5 \times 10^{-34}$ Js. Hence for position measurements whose uncertainty is 10^{-4} m the uncertainty in momentum is at least 0.5×10^{-30} kgms^{-1}, whilst for position measurements with uncertainty 10^{-10} m the uncertainty in momentum is at least 0.5×10^{-24} kgms^{-1}.

How do those momentum uncertainties compare with the momentum of the particles? A beta particle (i.e., an electron) with kinetic energy 0.1 MeV is travelling at a bit above half the speed of light and has momentum of $\sim 1.8 \times 10^{-22}$ kgms^{-1} (reader exercise to confirm). Hence, even assuming the upper limit on the momentum uncertainty, it is still only $\sim 0.3\%$ of the particle's momentum. For lower energy beta particles, say 0.01 MeV, the upper limit is a momentum uncertainty of $\sim 1\%$ of the total momentum.

For alpha particles the situation is even clearer. A 5 MeV alpha particle has a momentum of about 10^{-19} kgms^{-1}, so the upper limit of momentum uncertainty is five orders of magnitude smaller.

The significance of the ratio of momentum uncertainty to total momentum is that it equals the angular deviation of the particle track from a straight line resulting from each ionisation event. To see this, note that the momentum uncertainty applies also to the direction transverse to the particles main motion. For an alpha particle this deviation is so slight that we expect tracks which look completely straight – and that is indeed what is found. For electrons (beta particles) the situation is not so clear. Even a rather low energy beta particle (0.01 MeV) only deflects by, at worst, 0.01 radians ~ 0.6 degrees on each interaction. But many such interactions occur to form the whole track. Thus, for electrons – especially those of rather low energy – we expect to get some random kinks or curviness in their cloud chamber tracks – and that is indeed what is found.

This argument goes a long way to rationalise what is seen in cloud chambers, but one is still left with some unresolved questions regarding how a wavefunction like (28.1.9) ends up creating a straight track. This is addressed head-on by scattering theory, to which I turn next.

28.4 Argument from Scattering Theory

This section is more mathematically demanding than the rest of this book, as we plough headlong into some algebraic properties of wavefunctions which arise from the scattering of a particle off another particle (in this case, off molecules of air/vapour). It is worth the effort to understand it, though, as it puts to rest the conceptual conundrum raised by particle tracks in quantum theory. As a by-product you learn a little about scattering theory.

The fact that each ionisation event absorbs a negligible fraction of the particle's energy allows us to approximate the scattering between the particle and the air/vapour molecules as elastic scattering (i.e., that it conserves kinetic energy). We can assume that the particle is emitted in an isotropic (angle-independent) spherical wave, (28.1.9). Consider the first ionisation event which can potentially scatter the incident particle. It can occur with equal probability in any direction in 4π steradians. This is consistent with cloud chamber observations in which the tracks fan out in all directions from the source as centre. The energy of the particle together with the direction defined by the position of this first scattering event wrt the source defines an incident wavevector, $\bar{k}_{in}$. At this point the initial spherical wave, (28.1.9), has become a plane wave, (28.1.7a), by virtue of its direction of propagation having been defined. (This is a particular example of collapse of the state vector as the initial spherical wave can be considered as composed of a superposition of plane waves, see §25.4).

The wave outgoing from the scattering event consists of a plane wave with wavevector $\bar{k}_{in}$, which is the part unaffected by scattering, plus the scattered wave. It is only the latter we are concerned about, as this may potentially cause the particle track to deviate from a straight course. The scattered wave is an anisotropic out-going spherical wave,

$$\psi_{scattered} \propto f(\theta) \frac{exp\{ikr\}}{r} \qquad (28.4.1)$$

where we have redefined the radial coordinate r to be centred on the scattering atom in the air/vapour. This spherical wave is not isotropic (the same in all directions) but has an angular dependence, θ. Without loss of generality we can take this angle to be defined wrt the direction of the incident wavevector, $\bar{k}_{in}$, as polar axis. The scattered wave, (28.4.1), must satisfy the 3D Schrodinger equation, but that leaves broad scope for the angular dependence of the amplitude function, $f(\theta)$. This scattering amplitude will depend upon the interaction between the particle and the atom. This interaction can be represented, at least approximately, by a potential energy which is a function of position, $V(\bar{r})$ (the gradient of which, in classical physics, would be the magnitude of the force acting between the particle and the molecule). Recalling that the Hamiltonian represents total energy, and that the LHS of (28.1.5) is only the kinetic energy, we can write the Schrodinger equation including a potential energy function as,

$$\left\{-\frac{\hbar^2}{2m}\nabla^2 + V(\bar{r})\right\}\psi(\bar{r}) = E\psi(\bar{r}) \qquad (28.4.2)$$

Finally, in our 28th chapter, we meet the Schrodinger equation in the form that is most familiar from standard QM texts. I will now state without proof that an approximate expression for the scattering amplitude in (28.4.1) is,

$$f(\theta) \propto \int V(\bar{r}) \, exp\{i\bar{q} \cdot \bar{r}\} \cdot d^3r \qquad (28.4.3)$$

where $\bar{q} = \bar{k}_{in} - \bar{k}_{out}$ and $\bar{k}_{out}$ is the vector defined by the direction given by θ and by $k_{out} = k_{in} = k$. Hence $q = 2k \sin(\theta/2)$. (28.4.3) is known as the Born approximation (see, for example, Ref.[28.3]).

The potential function is significant in magnitude over distances at least comparable with some size scale, a, roughly the "size of the atom" (or larger). This, together with (28.4.3), is the key to understanding why the scattered wave remains tightly collimated around the "forward" direction, near $\theta = 0$, and hence that, despite being a wave, it leaves a virtually straight track.

For particle energies of the order of 0.1MeV or higher, $ak \gg 1$ (for either alpha or beta particles). Consequently, so long as θ is not very small we also have $aq = 2ka \sin(\theta/2) \gg 1$. Consequently the exponential in (28.4.2) will oscillate rapidly over most of the volume integral, resulting in strong cancellation. The exceptional case is for very small scattering angles, θ, for which the exponential factor can be near unity over a substantial spatial region. Consequently, it is clear that the scattering amplitude will be large for near-forward scattering ($\theta \approx 0$) but diminish rapidly at larger angles. A specific illustration of this is provided by a Coulomb potential with an exponential shielding decay (due to the overall neutrality of the atom), $V \propto e^{-r/a}/r$. Explicit integration yields $f \propto 1/(1 + a^2 q^2)$, so the amplitude does indeed reduce steeply for $aq > 1$, i.e., away from the near-forward direction.

The occurrence of a straight track can be argued in several different ways. The simplest is by appeal to the strongly forward-focussed nature of $f(\theta)$. This is sufficient to produce a tightly collimated beam in typical cases. Taking $a \sim 1$ Angstrom and a particle energy of the order of 0.1 MeV, the angular spread of the scattered beam is $\Delta\theta \sim 1/ak \sim 10^{-2}$ (electrons) or, for alpha particles, $\Delta\theta \sim 1/ak \sim 10^{-4}$. Note that this relationship of scattering angle to target size and particle energy (or de Broglie wavelength) is essentially the same as would apply if the particle-wave were diffracted from an aperture of size $\sim a$. This is not coincidence. The Born approximation, (28.4.3), shows that the scattering amplitude is proportional to the spatial Fourier transform of the potential function. But diffraction patterns are the Fourier transform of the aperture function, so the potential is analogous to an aperture function.

This example may leave you feeling slightly uneasy because it appears to rely upon strongly biased forward scattering, which in turn appears to rest upon assumptions regarding the scattering potential and hence upon the details of the particle-molecule interaction. This unease is admittedly ameliorated by the fact that a strongly collimated $f(\theta)$ arises even from the spherically symmetric Coulomb potential, the collimation therefore resulting purely from the initial plane wave, $\bar{k}_{in}$. Actually the only requirement is that the potential be confined to some spatially small region $\sim a$. Nevertheless a more convincing argument follows.

In reality, the wave resulting from the first scattering event does not then propagate freely. To form the particle track there will be a large number of nucleation events at a sufficiently close spacing to give the illusion of a continuous track. Consider the wave resulting from scattering the first scattered wave a second time. Assuming this occurs at a distance of b from the first event we have,

$$\psi_{scattered} \propto f(\theta') \frac{exp\{ikb\}}{b} f(\theta'') \frac{exp\{ikr\}}{r} \qquad (28.4.4)$$

where the total deflected angle is now $\theta = \theta' + \theta''$. The probability amplitude for scattering through this total angle, due to just two scattering events separated by distance b, is the sum of all expressions like (28.4.4) with differing individual angles θ' and θ'' but the same sum $\theta = \theta' + \theta''$. Hence, ignoring the effect of the relative phases, the overall amplitude cannot exceed,

$$\tilde{f}(\theta) \propto \int f(\theta - \theta') f(\theta') d\theta' \qquad (28.4.5)$$

The initial momentum $\bar{k}_{in}$ becomes $\bar{k}'$ after the first scattering event and $\bar{k}''$ after the second. The Born approximation gives us, $\qquad (28.4.6)$

$$f(\theta') f(\theta'') \propto \int V(\bar{r}') exp\{i\bar{q}' \cdot \bar{r}'\} \cdot d^3 r' \int V(\bar{r}'') exp\{i\bar{q}'' \cdot \bar{r}''\} \cdot d^3 r''$$

The two spatial integrals in (28.4.6) are over widely separated regions, because the atoms are small compared with the spacing between the two scattering events, $a << b$. The momentum transfers are given by,

$$\bar{q}' = \bar{k}_{in} - \bar{k}' \; ; q' \approx 2k \sin \frac{\theta'}{2} \qquad (28.4.7)$$

$$\bar{q}'' = \bar{k}' - \bar{k}'' \; ; q'' \approx 2k \sin \frac{\theta''}{2} \qquad (28.4.8)$$

The argument is now identical to that following (28.4.3). The potential function extends over distances at least of order $\sim a$. Assuming this is of atomic dimensions or greater, then for particle energies of the order of 0.1 MeV or higher we will have $ak >> 1$. Consequently, we also have $aq' >> 1$ and $aq'' >> 1$ providing that θ' and θ'' are not too small. The exponentials in (28.4.6) will then oscillate rapidly over most of the volume integrals, resulting in strong cancellation. The exceptional case is for very small scattering angles for which the exponentials can be nearly unity. But **both** θ' and θ'' must be small for the overall amplitude to be large. Consequently, the integrand in (28.4.6) will be significant only for those cases in which $\bar{k}_{in}, \bar{k}'$ and $\bar{k}''$ are all nearly colinear.

If we suppose the particle undergoes just these two nucleation events, then $\bar{k}'' \equiv \bar{k}_{out}$. So the argument implies that the outgoing direction of motion is parallel to the incoming direction, to within some small tolerance,

$\bar{k}_{in} \approx \bar{k}_{out}$. Moreover the two nucleation sites are displaced along the direction of $\bar{k}'$ which is also parallel to the incoming and outgoing directions of motion. So the direction of the particle track is faithful to the particle's classical velocity.

However, if we extend this argument to N successive scattering events we run into a problem for large N. The momentum transfer between successive events $\bar{q}_i = \bar{k}_{i-1} - \bar{k}_i$ is again required to be small so that the scattering angle θ_i between $\bar{k}_{i-1}$ and $\bar{k}_i$ is generally within $\Delta\theta \sim 1/ak$. But over a large number, N, of uncorrelated scattering events the standard deviation of the overall angle of scattering would be $\sim\sqrt{N} \cdot \Delta\theta$, which could be a large angle if N is sufficiently large. This would suggest that interactions with very large numbers of atoms would cause the track to drift seriously away from the direction of the initial momentum. But provided that the particle's energy remains large, with relatively little energy being lost in the collisions, we suspect that this cannot be right. There is something wrong with the argument based solely on $f(\theta_i)$. What has been forgotten is the role of the phase of the wavefunction, and the importance of this when expressions like (28.4.4) are summed over all possible particle paths. The generalisation of (28.4.4) to many scattering events is,

$$\psi_{scattered} \propto \prod_{j=in}^{out} f(\theta_j) \frac{exp\{ikb_j\}}{b_j} = \left(\prod_{j=in}^{out} \frac{f(\theta_j)}{b_j} \right) exp\{ik \sum_{j=in}^{out} b_j\}$$

(28.4.9)

where b_j is the distance between the $j-1^{th}$ and j^{th} scatterings, and the product and the sum in (28.4.9) extend over the whole track. The total amplitude will be the integral of (28.4.9) over both the intermediate scattering angles and the distances b_j. Both the direction of scattering and the distance to the next event can be represented by a vector $\bar{b}_i$, so the overall amplitude can be written,

$$\psi_{scattered} \propto \int exp\{ik \sum_{j=in}^{out} b_j\} \left(\prod_{j=in}^{out} \frac{f(\theta_j)}{b_j} d^3 b_j \right)$$

(28.4.10)

Now the distance between scattering events is large compared with the atomic size, a, so we have $bk >> ak >> 1$. So we can deploy the same rapid-phase-variation argument again in the context of the term $exp\{ik \sum_{j=in}^{out} b_j\}$ in (28.4.10). In this context the argument is just the method of stationary phase: the integral will be small unless $\sum_{j=in}^{out} b_j$ is stationary with respect to variations in the individual b_j - and subject to fixed starting and finishing points on the track. But this simply means that the track must be straight, because $\sum_{j=in}^{out} b_j$ is the total track length and this is minimum when the track is straight. The phase is stationary when the total path length is minimum, and hence when the track is straight. This is the best argument for the track being straight.

The overall wavefunction in (28.4.10) is actually a path integral, and we see that the reason for the straight track is essentially the same as the reason why the classical limit of a free particle's trajectory is straight. This follows from applying the method of stationary phase to the Feynman path integral, Ref.[28.4]. The global term $exp\{ik \sum_{j=in}^{out} b_j\}$ prevents the particle drifting away from the straight track which would occur for a large number of uncorrelated classical scattering events. Nevertheless, the argument still requires $bk \gg ak \gg 1$ and will become weaker as the number of ionisation events becomes large. Hence, whilst alpha particle tracks are indeed very straight, beta particle tracks – especially those of lower energy – can be curved or kinked in a random manner.

28.5 Reader Exercises

Find the momentum of an electron with kinetic energy 0.1 MeV
The mass of an electron in MeV is 0.511. Energy in the relativistic regime is $E = \gamma mc^2$, so the ratio of the total energy to the rest mass energy gives $\gamma = \frac{0.611}{0.511} = 1.20$. But $\gamma = \left(1 - \frac{v^2}{c^2}\right)^{-\frac{1}{2}}$ which gives $v = 0.55c = 1.65 \times 10^8$ m/s. The electron mass in kg is 9.1 x 10^{-31} and the momentum in the relativistic regime is γmv which is 1.8 x 10^{-22} kgm/s. Hence an uncertainty in momentum of 0.5 x 10^{-24} kgms^{-1} is ~0.3% of this momentum. If the electron energy were 10 times smaller, i.e., 0.01 MeV, then the corresponding calculation gives the momentum to be 5.4 x 10^{-23} kgm/s, so the uncertainty is ~1% of this. QED.

Find the momentum of an alpha particle with kinetic energy 5 MeV
5 MeV = 8 x 10^{-13} J. In the non-relativistic regime this can be equated to $\frac{1}{2}mv^2$ and the mass of the alpha is 6.6 x 10^{-27} kg, hence $v = 1.5 \times 10^7$ m/s. Hence the momentum is $mv \approx 10^{-19}$ kgm/s which is about 5 orders of magnitude larger than the uncertainty in momentum stated in the text of 0.5 x 10^{-24} kgms^{-1}. QED.

28.6 References

[28.1] N.F. Mott, Proc. Roy. Soc. Al26, 79 (1929).

[28.2] J.S. Bell (1971) *"On the Hypothesis that the Schrodinger Equation is Exact"*, Ref.TH.1424-CERN, *Contribution to the international colloquium on issues in contemporary physics and philosophy of science, and their relevance for our society, Penn State University, September 1971.*

[28.3] L.I.Schiff (1968), *"Quantum Mechanics"*, 3rd ed, McGraw-Hill (see §38).

[28.4] R.P.Feynman (1948), *"Space-Time Approach to Non-Relativistic Quantum Mechanics"*, Rev.Mod.Phys. **20**, 267.

29

Algebraic Methods

In general the state structure of a given system will not be easy to calculate, e.g., the eigenenergy states for an arbitrary Hamiltonian. However, there are some examples where the state structure can be derived in an elementary fashion which is purely algebraic rather than requiring the states to be expressed in the spatial basis, i.e., as wavefunctions requiring the solution of differential equations. All these cases can also be solved via wavefunctions by solving the associated differential equations, but the algebraic method is so elegant – and so important – that all students of quantum mechanics should be familiar with them. The examples given here are Fock space, angular momentum states, and the solution of the non-relativistic hydrogen atom.

29.1 Fock Space

Students are generally introduced to Fock space in the context of the quantum mechanics of the non-relativistic harmonic oscillator. It is later used as the platform on which is built relativistic quantum field theory. However, I'd like to stress here that the structure of the Fock states does not need the harmonic oscillator to be introduced at all. It so happens that the Fock states are indeed the eigenstates of the harmonic oscillator Hamiltonian, but that is by way of being a fluke. The Fock state structure actually arises from the canonical commutation relation between momentum and position observables, $\hat{p}$ and $\hat{q}$, as we derived in chapter 5, i.e.,

$$[\hat{p}, \hat{q}] = \frac{\hbar}{i} \tag{29.1.1}$$

Being observables, the operators $\hat{p}$ and $\hat{q}$ are Hermitian. Define a dimensionless operator, $\hat{a}$, as follows,

$$\hat{a} = \frac{1}{\sqrt{2\dot{m}\hbar}}(\dot{m}\hat{q} + i\hat{p}) \tag{29.1.2}$$

where $\dot{m}$ is some constant with the dimensions of mass/time (kgs⁻¹) whose purpose is to make $\hat{a}$ dimensionless. Note that it is natural to introduce a quantity with such dimensions if we wish to consider a linear combination of $\hat{p}$ and $\hat{q}$ which is dimensionally consistent. Operator $\hat{a}$ is not Hermitian, it's conjugate is,

$$\hat{a}^{+} = \frac{1}{\sqrt{2\dot{m}\hbar}}(\dot{m}\hat{q} - i\hat{p}) \tag{29.1.3}$$

An exercise for the reader is to show, from (29.1.1), that the commutator between $\hat{a}$ and its conjugate is,

$$[\hat{a}, \hat{a}^+] = 1 \qquad (29.1.4)$$

However, $\hat{a}^+\hat{a}$ is Hermitian. Consider its eigenstates, which we may write as,

$$\hat{a}^+\hat{a}|\lambda\rangle = \lambda|\lambda\rangle \qquad (29.1.5)$$

The eigenvalues, λ, are necessarily real, but for this operator they must also be non-negative. This can be seen by noting that if we write $|\psi\rangle = \hat{a}|\lambda\rangle$ then it follows that $\langle\psi|\psi\rangle = \langle\lambda|\hat{a}^+\hat{a}|\lambda\rangle = \lambda\langle\lambda|\lambda\rangle$ and the fact that vector norms are positive, or zero for the null vector, implies $\lambda \geq 0$.

Theorem: $\hat{a}^+|\lambda\rangle$ is also an eigenvector of $\hat{a}^+\hat{a}$ with eigenvalue $(\lambda + 1)$

Proof: $\hat{a}^+\hat{a}|\lambda\rangle = \lambda|\lambda\rangle$ hence $\hat{a}^+(\hat{a}^+\hat{a}|\lambda\rangle) = \lambda\hat{a}^+|\lambda\rangle$

Hence, using (29.1.4), $\hat{a}^+(\hat{a}\hat{a}^+ - 1)|\lambda\rangle = \lambda\hat{a}^+|\lambda\rangle$

Hence, $\hat{a}^+\hat{a}(\hat{a}^+|\lambda\rangle) = (\lambda + 1)\hat{a}^+|\lambda\rangle$

Hence, $\hat{a}^+|\lambda\rangle$ is an eigenvector of $\hat{a}^+\hat{a}$ with eigenvalue $(\lambda + 1)$. Since eigenvectors are determined only up to a multiplicative constant, we can write,

$$\hat{a}^+|\lambda\rangle = k_\lambda|\lambda + 1\rangle \qquad (29.1.6)$$

for some function of λ written as k_λ.

Operators with the property expressed in (29.1.6) are known as "ladder operators" because they convert one eigenvector into another eigenvector with an eigenvalue one unit higher (i.e., figuratively speaking, they climb one rung up the ladder of eigenvalues).

Theorem: $\hat{a}|\lambda\rangle$ is an eigenvector of $\hat{a}^+\hat{a}$ with eigenvalue $(\lambda - 1)$ if $\lambda \geq 1$

Proof: $\hat{a}^+\hat{a}|\lambda\rangle = \lambda|\lambda\rangle$ hence $\hat{a}(\hat{a}^+\hat{a}|\lambda\rangle) = \lambda\hat{a}|\lambda\rangle$

Hence, using (29.1.4), $(\hat{a}^+\hat{a} + 1)\hat{a}|\lambda\rangle = \lambda\hat{a}|\lambda\rangle$

Hence, $\hat{a}^+\hat{a}(\hat{a}|\lambda\rangle) = (\lambda - 1)\hat{a}|\lambda\rangle$

Hence, if $\lambda \geq 1$ so that $(\lambda - 1)$ is non-negative, then $\hat{a}|\lambda\rangle$ is an eigenvector of $\hat{a}^+\hat{a}$ with eigenvalue $(\lambda - 1)$. Since eigenvectors are determined only up to a multiplicative constant, we can write,

For $\lambda \geq 1$ $\qquad \hat{a}|\lambda\rangle = \tilde{k}_\lambda|\lambda - 1\rangle \qquad (29.1.7)$

for some function of λ written as $\tilde{k}_\lambda$.

Theorem: The smallest eigenvalue of $\hat{a}^+\hat{a}$ is zero.

Proof: We already know there is a smallest eigenvalue, which cannot be less than zero, because $\lambda \geq 0$. Suppose there is an integer eigenvalue, $\lambda = n$. By repeated application of (29.1.6) we eventually arrive at the state whose eigenvalue is zero, i.e., $\hat{a}^n|\lambda\rangle \propto |\lambda - n\rangle = |0\rangle$. On the other hand suppose there is a state whose eigenvalue is not integral. By repeated application of (29.1.6) we eventually arrive at a state with eigenvalue between 0 and 1, i.e., we can take this state to have $0 < \lambda < 1$. However, we have shown that $\hat{a}^+\hat{a}(\hat{a}|\lambda\rangle) = (\lambda - 1)\hat{a}|\lambda\rangle$, but as $(\lambda - 1)$ would now be negative, by construction, this cannot be a valid eigen-equation, despite being true, and the contradiction is avoided only if $\hat{a}|\lambda\rangle = 0$, and so is not actually an eigenvector. But in that case $\hat{a}^+\hat{a}|\lambda\rangle = \lambda|\lambda\rangle$ is also zero, despite λ having been assumed non-zero. But this implies that our postulated eigenvector $|\lambda\rangle$ with $0 < \lambda < 1$ must be the zero vector, which is a contradiction thus proving that there is no non-integral eigenvalue.

Hence, we conclude that the possible eigenvalues are 0, 1, 2, 3.... and that this sequence does not terminate. Representations of the position and momentum operators obeying the canonical commutation relation, (29.1.1), are therefore necessarily infinite dimensional (i.e., Fock space is an infinite dimensional Hilbert space). This is why we have not dealt with the canonical commutation relation and the spatial representation very much in this book, because we have focused instead on finite dimensional Hilbert spaces (because they are pedagogically simpler).

We can now write the basis states of Fock space as $|n\rangle$ for $n = 1, 2, 3...$ and we shall assume these states are normalised to unity. We can now find the multiplicative factors which occur in (29.1.6,7). $\hat{a}^+|n\rangle = k_n|n + 1\rangle$ gives $\langle n|\hat{a}\hat{a}^+|n\rangle = |k_n|^2 = \langle n|(1 + \hat{a}^+\hat{a})|n\rangle = 1 + n$. Without loss of generality we can assume the constants to be real, and hence $k_n = \sqrt{n + 1}$. Similarly, $\hat{a}|n\rangle = \tilde{k}_n|n - 1\rangle$ gives $\langle n|\hat{a}^+\hat{a}|n\rangle = |\tilde{k}_n|^2 = n$ and hence $\tilde{k}_n = \sqrt{n}$. So equs.(29.1.6,7) become,

$$\hat{a}^+|n\rangle = \sqrt{n + 1}|n + 1\rangle \qquad (29.1.8a)$$

For $n \geq 1$ $\qquad\qquad \hat{a}|n\rangle = \sqrt{n}|n - 1\rangle \qquad\qquad (29.1.8b)$

In matrix notation the Fock states can be written,

$$|0\rangle = \begin{pmatrix} 1 \\ 0 \\ 0 \\ 0 \\ etc \end{pmatrix}, \quad |1\rangle = \begin{pmatrix} 0 \\ 1 \\ 0 \\ 0 \\ etc \end{pmatrix}, \quad |2\rangle = \begin{pmatrix} 0 \\ 0 \\ 1 \\ 0 \\ etc \end{pmatrix}, \text{ etc} \qquad (29.1.9)$$

So in this matrix representation, (29.1.8a) gives the operator $\hat{a}^+$ as,

$$\hat{a}^+ = \begin{pmatrix} 0 & 0 & 0 & 0 & 0 & 0 \\ 1 & 0 & 0 & 0 & 0 & 0 \\ 0 & \sqrt{2} & 0 & 0 & 0 & 0 \\ 0 & 0 & \sqrt{3} & 0 & 0 & 0 \\ 0 & 0 & 0 & \sqrt{4} & 0 & 0 \\ 0 & 0 & 0 & 0 & \sqrt{5} & etc \end{pmatrix} \qquad (29.1.10)$$

where it is understood that the representation is actually infinite, extending to $|n \to \infty\rangle$, so that (29.1.10) shows only the first 6 x 6 terms. The operator $\hat{a}$ is simply the transpose of (29.1.10). Note that we can invert (29.1.1,2) to give,

$$\hat{q} = \sqrt{\frac{\hbar}{2\dot{m}}}(\hat{a}^+ + \hat{a}) \quad \text{and} \quad \hat{p} = i\sqrt{\frac{\dot{m}\hbar}{2}}(\hat{a}^+ - \hat{a}) \qquad (29.1.11)$$

These operators can also be written in matrix notation as follows,

$$\hat{q} = \sqrt{\frac{\hbar}{2\dot{m}}} \begin{pmatrix} 0 & 1 & 0 & 0 & 0 & 0 \\ 1 & 0 & \sqrt{2} & 0 & 0 & 0 \\ 0 & \sqrt{2} & 0 & \sqrt{3} & 0 & 0 \\ 0 & 0 & \sqrt{3} & 0 & \sqrt{4} & 0 \\ 0 & 0 & 0 & \sqrt{4} & 0 & \sqrt{5} \\ 0 & 0 & 0 & 0 & \sqrt{5} & etc \end{pmatrix} \qquad (29.1.11)$$

$$\hat{p} = i\sqrt{\frac{\dot{m}\hbar}{2}} \begin{pmatrix} 0 & -1 & 0 & 0 & 0 & 0 \\ 1 & 0 & -\sqrt{2} & 0 & 0 & 0 \\ 0 & \sqrt{2} & 0 & -\sqrt{3} & 0 & 0 \\ 0 & 0 & \sqrt{3} & 0 & -\sqrt{4} & 0 \\ 0 & 0 & 0 & \sqrt{4} & 0 & -\sqrt{5} \\ 0 & 0 & 0 & 0 & \sqrt{5} & etc \end{pmatrix} \qquad (29.1.12)$$

The reader may check that the matrices (29.1.11,12) obey the commutation relation (29.1.1), but only provided that they extend to infinity. If we truncate the matrix representation at some maximum n, say at $n = 5$ this giving matrices which are 6 x 6, as above, we get instead of the exact commutator the following,

$$[\hat{p}, \hat{q}]_{6\times6} = \frac{\hbar}{i} \begin{pmatrix} 1 & 0 & 0 & 0 & 0 & 0 \\ 0 & 1 & 0 & 0 & 0 & 0 \\ 0 & 0 & 1 & 0 & 0 & 0 \\ 0 & 0 & 0 & 1 & 0 & 0 \\ 0 & 0 & 0 & 0 & 1 & 0 \\ 0 & 0 & 0 & 0 & 0 & -5 \end{pmatrix} \tag{29.1.13}$$

So truncation at a finite number of basis states fails to exactly respect the desired commutation relation, (29.1.1), but deviates from it only in the last diagonal term, which in general is $-n$, where n is the largest Fock state, i.e., in a representation which is $(n+1) \times (n+1)$.

The representation of the position and momentum operators in the form (29.1.11,12) receives little attention in standard texts which invariably favour the continuum representation in which $\hat{q}$ is represented by the real continuous variable x, and $\hat{p}$ is represented by the differential operator $-i\hbar\partial_x$. These have the advantage of being an exact representation of the desired commutator, (29.1.1), but so is the discrete matrix representation of equs.(29.1.10-12) as long as it is understood that the matrices are infinite.

Finally, the one-dimensional harmonic oscillator Hamiltonian can be written,

$$\hat{H} = \frac{1}{2m}\hat{p}^2 + \frac{1}{2}m\omega^2\hat{q}^2 \tag{29.1.14}$$

where m is the mass of the oscillator and ω is its classical natural angular frequency. Setting $\dot{m} = m\omega$ the reader can check that this can be re-written as,

$$\hat{H} = \hbar\omega\left(\hat{a}^+\hat{a} + \frac{1}{2}\right) \tag{29.1.15}$$

Consequently, the Fock states, $|n\rangle$, which have been constructed to be eigenstates of $\hat{a}^+\hat{a}$ are also the eigenstates of the oscillator Hamiltonian which is thus revealed to have energy levels,

$$E_n = \hbar\omega\left(n + \frac{1}{2}\right) \tag{29.1.16}$$

for $n = 0,1,2,3$ .... Hence, the ground state has non-zero energy, namely $\hbar\omega/2$, the so-called zero-point energy. This becomes a bit of an embarrassment if space is considered to consist of an infinite array of oscillators (e.g., of the electromagnetic field) as the zero-point energy is infinite. In quantum field theory this awkwardness is swept under the carpet by the simple expedient of ignoring it (thinly disguised by the use of terms like "normal ordering"). It is worth noting in this context that if we use the discrete representation truncated as some maximum finite n then the canonical commutator, (29.1.1), is replaced by,

$$[\hat{p}, \hat{q}]_{n+1 \times n+1} = \frac{\hbar}{i} \hat{\xi} \tag{29.1.17}$$

where $\hat{\xi}$ is the matrix appearing in (29.1.13), up to the maximum n considered. (19.1.4) then becomes,

$$[\hat{a}, \hat{a}^+] = \hat{\xi} \tag{29.1.18}$$

and the Hamiltonian becomes,

$$\hat{H} = \hbar \omega \left(\hat{a}^+ \hat{a} + \frac{1}{2} \hat{\xi} \right) \tag{29.1.15}$$

The expectation value of the energy wrt every state $|i\rangle$ is therefore the same zero-point energy as before, i.e., $\hbar\omega/2$, except for the state of maximum $i = n$ for which the expectation value of the energy is $-n\hbar\omega/2$. Hence, the zero point energy considered as the sum of the expectation values over all states is zero (put another way, $Tr(\hat{\xi}) = 0$). Hence, the truncated discrete representation has no overall zero-point energy.

29.2 Angular Momentum

Don't worry if you find this opening paragraph rather opaque. You don't need to understand it on first reading.

We have already met the commutation relations between the three components of spin angular momentum in §2.9, namely,

$$\left[\hat{S}_x, \hat{S}_y\right] = i\hbar\hat{S}_z \quad \left[\hat{S}_y, \hat{S}_z\right] = i\hbar\hat{S}_x \quad \left[\hat{S}_z, \hat{S}_x\right] = i\hbar\hat{S}_y \tag{2.9.7}$$

However, in §2.9 the connection between operators with these commutation properties and angular momentum was only hinted at. We saw that the Pauli matrices ($\times \hbar/2$) obey (2.9.7), and that the Pauli matrices are a representation of Hamilton's quaternion algebra and that the resulting algebra provides a means of implementing rotations in three-dimensional (Euclidean) space. Moving from 19th century to 20th century mathematics, (2.9.7) define the properties of the generators of the Lie algebra, which in turn defines, by exponentiation, the Lie group SU(2), the universal covering of the group SO(3) of rotations in three-dimensional (Euclidean) space. There is no need for the reader to appreciate these more sophisticated mathematical constructions. I am merely reiterating the connection between (2.9.7) and rotations. The connection with angular momentum is by analogy with position and linear momentum. These are conjugate variables, and, as we saw in chapter 5, the generator of translational displacement (∂_x) is a representation of the momentum operator (after multiplying by $-i\hbar$). So, in the same way, the generators of rotational displacements $(\times \hbar)$ are representations of angular momentum.

The reader can be forgiven for finding that last paragraph rather abstruse. There is a far simpler way of seeing how the commutator structure of (2.9.7) arises by considering, not spin, but orbital angular momentum. Orbital angular momentum is the product of linear momentum and its perpendicular distance from the point about which the angular momentum is being defined. The vectorial angular momentum can thus be written $\bar{L} = \bar{r} \times \bar{p}$. This is not specific to quantum mechanics. However, in quantum mechanics, if we adopt the continuum spatial representation, then the momentum operator can be written $\bar{p} = -i\hbar\bar{\nabla}$. Putting these together we get a differential representation of the orbital angular momentum operators as follows,

$$\hat{L}_x = -i\hbar(y\partial_z - z\partial_y), \quad \hat{L}_y = -i\hbar(z\partial_x - x\partial_z), \quad \hat{L}_z = -i\hbar(x\partial_y - y\partial_x)$$

(29.2.1)

These operators have the same commutation relations as (2.9.7). For example,

$$\begin{aligned}
\left[\hat{L}_x, \hat{L}_y\right] &= (-i\hbar)^2\left[y\partial_z - z\partial_y, z\partial_x - x\partial_z\right] \\
&= (-i\hbar)^2\{[y\partial_z, z\partial_x] + [z\partial_y, x\partial_z]\} \\
&= (-i\hbar)^2\{y\partial_x - x\partial_y\} = i\hbar\hat{L}_z
\end{aligned}$$

(29.2.2)

and similarly,

$$\left[\hat{L}_x, \hat{L}_y\right] = i\hbar\hat{L}_z \quad \left[\hat{L}_y, \hat{L}_z\right] = i\hbar\hat{L}_x \quad \left[\hat{L}_z, \hat{L}_x\right] = i\hbar\hat{L}_y \qquad (29.2.3)$$

which are identical to (2.9.7). Similarly, we can define the total angular momentum as the sum of the spin and the orbital angular momentum,

$$\hat{J}_i = \hat{L}_i + \hat{S}_i \qquad (29.2.4)$$

Spin and orbital angular momentum commute, $\left[\hat{L}_i, \hat{S}_k\right] = 0$, so it immediately follows that the components of $\bar{J}$ have the same commutation properties. To drive this home, here they are again,

$$\left[\hat{J}_x, \hat{J}_y\right] = i\hbar\hat{J}_z \quad \left[\hat{J}_y, \hat{J}_z\right] = i\hbar\hat{J}_x \quad \left[\hat{J}_z, \hat{J}_x\right] = i\hbar\hat{J}_y \qquad (29.2.5)$$

(From here on I shall drop the caret, it being understood that the above symbols relate to operators). We shall work with $\bar{J}$ and that will suffice to address both $\bar{L}$ and $\bar{S}$. The question to be addressed is: what are the eigenvalues and eigenvector structure of $\bar{J}$?

We know, from §2.5.5, that the three components of $\bar{J}$ cannot have the same eigenvectors because they do not commute. By convention, J_z is the component chosen to develop the eigenvectors. However, the square "magnitude" operator defined by $J^2 = J_x^2 + J_y^2 + J_z^2$ does commute with J_z, as can be seen as follows,

$$[J_z, J^2] = [J_z, J_x^2 + J_y^2] = J_z J_x^2 - J_x^2 J_z + J_z J_y^2 - J_y^2 J_z$$
$$= (J_x J_z + i\hbar J_y)J_x - J_x(J_z J_x - i\hbar J_y) + (J_y J_z - i\hbar J_x)J_y - J_y(J_z J_y + i\hbar J_x)$$
$$= 0 \tag{29.2.6}$$

Consequently, there are states which are eigenstates simultaneously of both J_z and J^2. Suppose $|a, b\rangle$ is such a state, where,

$$J^2|a,b\rangle = a|a,b\rangle \qquad J_z|a,b\rangle = b|a,b\rangle \tag{29.2.7}$$

Define $\qquad J_\pm = J_x \pm i J_y \tag{29.2.8}$

So, $\qquad [J_z, J_\pm] = [J_z, J_x] \pm i[J_z, J_y] = i\hbar J_y \pm i(-i\hbar J_x)$
$$= \pm\hbar J_x + i\hbar J_y = \pm\hbar(J_x \pm i J_y) = \pm\hbar J_\pm \tag{29.2.9}$$

The operators $J_\pm$ are ladder operators, similar to those we met in §29.1, which change one eigenstate into another, as follows,

$$J_z J_+|a,b\rangle = (J_+ J_z + \hbar J_+)|a,b\rangle$$
$$= (bJ_+ + \hbar J_+)|a,b\rangle = (b + \hbar)J_+|a,b\rangle \tag{29.2.10}$$

Hence, given an eigenstate $|a, b\rangle$ of J_z with eigenvalue b, we can find another eigenstate of J_z namely $J_+|a, b\rangle$ with eigenvalue $(b + \hbar)$. In the same way the operator J_- forms a new eigenstate with eigenvalue $b - \hbar$. A sequence of eigenstates with eigenvalues spaced by $\hbar$ is therefore constructed by repeating these operations, $(J_+)^n|a, b\rangle$ has eigenvalue $(b + n\hbar)$ and $(J_-)^n|a, b\rangle$ has eigenvalue $(b - n\hbar)$. Note that we have not claimed that $(J_\pm)^n|a, b\rangle$ are normalised, and in fact they will not be.

We have, $\tag{29.2.11}$
$$\langle a, b|J^2|a, b\rangle = a = \langle a, b|(J_x^2 + J_y^2 + J_z^2)|a, b\rangle \geq \langle a, b|J_z^2|a, b\rangle = b^2$$

From which it follows that $a \geq 0$ and that, for a given finite value for a there is an upper bound to b^2 as it cannot exceed a. It follows that the process of finding ever higher eigenvalues of J_z by repeatedly applying J_+ to states with a fixed a cannot go on for ever. So where does the reasoning break down? The only possibility is that, for some largest possible $b = b_{max}$ we get $J_+|a, b_{max}\rangle = 0$ so that (29.2.10) fails to determine a new eigenstate. Now consider, (29.2.12)

$$J_-J_+ = (J_x - iJ_y)(J_x + iJ_y) = J_x^2 + J_y^2 + i[J_x, J_y] = J^2 - J_z^2 - \hbar J_z$$

Hence,

$$J_-J_+|a, b\rangle = (J^2 - J_z^2 - \hbar J_z)|a, b\rangle = (a - b^2 - \hbar b)|a, b\rangle \qquad (29.2.13)$$

So, if we choose $b = b_{max}$ in (29.2.13), (29.2.10) tells us $J_+|a, b_{max}\rangle = 0$ so we must have $a - b^2 - \hbar b = 0$, i.e.,

$$a = b_{max}(b_{max} + \hbar) \qquad (29.2.14)$$

Similarly, the process of finding ever lower eigenvalues of J_z by repeatedly applying J_- cannot go on for ever because, for a sufficiently large and negative eigenvalue, this eigenvalue squared will exceed a. So, there is a minimum $b = b_{min}$ such that $J_z J_-|a, b\rangle = (b - \hbar)J_-|a, b\rangle$ must instead imply $J_-|a, b_{min}\rangle = 0$. So consider, (29.2.15)

$$J_+J_- = (J_x + iJ_y)(J_x - iJ_y) = J_x^2 + J_y^2 - i[J_x, J_y] = J^2 - J_z^2 + \hbar J_z$$

Hence,

$$J_+J_-|a, b\rangle = (J^2 - J_z^2 + \hbar J_z)|a, b\rangle = (a - b^2 + \hbar b)|a, b\rangle \qquad (29.2.16)$$

So that choosing $b = b_{min}$ means we require,

$$a = b_{min}(b_{min} - \hbar) \qquad (29.2.17)$$

But we also know that the eigenvalues are spaced apart by $\hbar$, so that the maximum and minimum values for b must differ by an integral multiple of $\hbar$, i.e., there must be a positive integer n such that $b_{max} - b_{min} = n\hbar$.

Equating (29.2.14) and (29.2.17) and using this relation between b_{max} and b_{min} gives,

$$b_{max}(b_{max} + \hbar) = (b_{min} + n\hbar)(b_{min} + (n + 1)\hbar) = b_{min}(b_{min} - \hbar) \qquad (29.2.18)$$

Re-arranging and simplifying this gives simply,

$$b_{min} = -\frac{n}{2}\hbar \qquad\qquad b_{max} = \frac{n}{2}\hbar \qquad (29.2.19)$$

Substitution in (29.2.14) or (29.2.17) gives,

$$a = \frac{n}{2}\left(\frac{n}{2} + 1\right)\hbar^2 \qquad (29.2.20)$$

where $n = 0,1,2,3, ...$ It is usual to write $\frac{n}{2} = j$, so that the possible values of the quantum number j are $0, \frac{1}{2}, 1, \frac{3}{2}, 2, \frac{5}{2}, ...$ And what we have denoted by b it is usually to write as $m\hbar$, where m is the azimuthal quantum number, which, we have seen, takes values from $-j$ to $+j$, inclusive. The conventional notation for the angular momentum states is thus $|j, m\rangle$.

From here on we shall set $\hbar = 1$. Our eigenvalue equations can now be written in standard notation as,

$$J^2|j, m\rangle = j(j + 1)|j, m\rangle \qquad J_z|j, m\rangle = m|j, m\rangle \qquad (29.2.21)$$

We now return to the issue of the normalisation of states like $J_+|j, m\rangle$. Due to (29.2.10) we can write,

For $m < j$; $\qquad\qquad J_+|j, m\rangle = k_m|j, m + 1\rangle \qquad\qquad (29.2.22a)$

For $m > -j$; $\qquad\qquad J_-|j, m\rangle = \tilde{k}_m|j, m - 1\rangle \qquad\qquad (29.2.22b)$

The first of these gives $k_m = \langle j, m + 1|J_+|j, m\rangle$. Replacing m with $m + 1$ in the second gives,

$$\tilde{k}_{m+1} = \langle j, m|J_-|j, m + 1\rangle = \langle j, m + 1|J_+|j, m\rangle^* = k_m^* \qquad (29.2.23)$$

Multiplying (29.2.22b) by J_+ and (29.2.22a) by J_- gives, ignoring the validity limits,

$$J_-J_+|j, m\rangle = k_m\tilde{k}_{m+1}|j, m\rangle = |k_m|^2|j, m\rangle \qquad (29.2.24a)$$

$$J_+J_-|j, m\rangle = \tilde{k}_m k_{m-1}|j, m\rangle = |k_{m-1}|^2|j, m\rangle \qquad (29.2.24b)$$

But subtracting these two equations and using (29.2.12,15) which give the commutator $J_+J_- - J_-J_+ = 2J_z$ yields,

$$|k_{m-1}|^2 - |k_m|^2 = 2m \qquad (29.2.25)$$

Noting that $k_j = 0$ so that (29.2.22a) yields zero when $m = j$, we can thus find all the coefficients from the difference equation, (29.2.25), by assuming, without loss of generality, that the coefficients are real. Thus, setting $m = j$, then $m = j - 1$, then $m = j - 2$, etc., in (29.2.25),

$$k_{j-1}^2 = 2j$$

$$k_{j-2}^2 = 2j + 2(j - 1)$$

$$k_{j-3}^2 = 2j + 2(j - 1) + 2(j - 2), \text{etc.} \qquad (29.2.26)$$

Modulo the factor of 2, the general expression is seen to be the sum of the integers starting at $m + 1$ and ending at j. But this is the sum of the integers from 1 to j, which is $j(j + 1)/2$, minus the sum of the integers from 1 to m, which is $m(m + 1)/2$. Hence we find,

$$k_m = \sqrt{j(j + 1) - m(m + 1)} \tag{29.2.27a}$$

And
$$\tilde{k}_m = \sqrt{j(j + 1) - m(m - 1)} \tag{29.2.27b}$$

Note that these respect $k_j = 0$ and $\tilde{k}_{-j} = 0$ and $\tilde{k}_{m+1} = k_m$, as required.

With (29.2.27a,b) we have the final piece in the jigsaw that provides the complete structure of the angular eigenvectors. For any given half-integral value for j this has provided a representation of the commutators, (29.2.5), and their eigen-structure of dimension $(2j + 1)$, this subspace being spanned by the $(2j + 1)$ different m-states, $\{|j, m\rangle, m \epsilon [-j, +j, step\ 1]\}$.

It is important to appreciate that (29.2.27a,b) provide the matrix representations of these finite dimensional representations of the angular momentum operators. As a concrete example, suppose $j = \frac{3}{2}$. Working in the conventional representation in which,

$$\left|\frac{3}{2}, -\frac{3}{2}\right\rangle = \begin{pmatrix} 1 \\ 0 \\ 0 \\ 0 \end{pmatrix}, \left|\frac{3}{2}, -\frac{1}{2}\right\rangle = \begin{pmatrix} 0 \\ 1 \\ 0 \\ 0 \end{pmatrix}, \left|\frac{3}{2}, \frac{1}{2}\right\rangle = \begin{pmatrix} 0 \\ 0 \\ 1 \\ 0 \end{pmatrix}, \left|\frac{3}{2}, \frac{3}{2}\right\rangle = \begin{pmatrix} 0 \\ 0 \\ 0 \\ 1 \end{pmatrix} \tag{29.2.28}$$

This makes J_z diagonal with elements which are its eigenvalues, namely $-\frac{3}{2}, -\frac{1}{2}, \frac{1}{2}, \frac{3}{2}$ in that order. J^2 is the unit matrix multiplied by $j(j + 1) = \frac{15}{4}$. However, less obviously, (29.2.27a) gives,

$$J_+ = \begin{pmatrix} 0 & 0 & 0 & 0 \\ \sqrt{3} & 0 & 0 & 0 \\ 0 & 2 & 0 & 0 \\ 0 & 0 & \sqrt{3} & 0 \end{pmatrix} \tag{29.2.29}$$

And J_- is the transpose of J_+. From (29.2.8) it is readily found that the x and y components of angular momentum are represented by,

$$J_x = \frac{1}{2} \begin{pmatrix} 0 & \sqrt{3} & 0 & 0 \\ \sqrt{3} & 0 & 2 & 0 \\ 0 & 2 & 0 & \sqrt{3} \\ 0 & 0 & \sqrt{3} & 0 \end{pmatrix} \tag{29.2.30}$$

$$J_y = \frac{i}{2} \begin{pmatrix} 0 & \sqrt{3} & 0 & 0 \\ -\sqrt{3} & 0 & 2 & 0 \\ 0 & -2 & 0 & \sqrt{3} \\ 0 & 0 & -\sqrt{3} & 0 \end{pmatrix} \qquad (29.2.31)$$

$$J_z = \frac{1}{2} \begin{pmatrix} -3 & 0 & 0 & 0 \\ 0 & -1 & 0 & 0 \\ 0 & 0 & 1 & 0 \\ 0 & 0 & 0 & 3 \end{pmatrix} \qquad (29.2.32)$$

The reader can readily confirm that (29.2.30-32) obey the commutations relations (29.2.5). The above development for total angular momentum can be interpreted to apply for spin alone, or for orbital angular momentum alone, or for their sum, the total angular momentum. This follows from the fact they all have the same commutation relations, (2.9.10), (29.2.5), (29.2.5), and the above derivation has only used these commutation relations and nothing else. However, one final thing should be emphasised. Whilst spin and total angular momentum including spin can take any of the half-integral values $j = 0, \frac{1}{2}, 1, \frac{3}{2}, 2, \frac{5}{2}, \ldots$ note that orbital angular momentum alone can take only integral values. This is associated with the fact that systems of half-integral spin change sign under a rotation by 2π, whilst orbital angular momentum, familiar in classical mechanics, has not such property. Another way of seeing that orbital angular momentum takes only integral values is to note that the eigenfunctions in the spatial-differential representation, (29.2.1), are the spherical harmonic functions, Y_L^m, and take only integral L values (see any standard QM text for this conventional development).

29.3 The Algebraic Hydrogen Atom

In chapter 28 we derived Schrodinger's equation in its spatial-differential form,

$$\left\{ -\frac{\hbar^2}{2m} \nabla^2 + V(\bar{r}) \right\} \psi(\bar{r}) = E\psi(\bar{r}) \qquad (29.3.1)$$

In undergraduate QM courses, solutions to Schrodinger's equation in this form are usually illustrated by a standard set of cases, including potential wells, the harmonic oscillator and – most importantly – the hydrogen atom. It was the success of the "new quantum theory", as it was in 1926, in deriving the principal structure of the energy levels of the hydrogen atom that started to cement its place as the underpinning of the physics of the very small.

We want to find the energy levels of the non-relativistic hydrogen atom Hamiltonian, which is the same thing as the time-independent Schrodinger equation,

$$\left\{-\frac{\hbar^2}{2m}\nabla^2 - \frac{q^2}{r}\right\}\psi(\bar{r}) = E\psi(\bar{r}) \tag{29.3.2}$$

where q is the proton charge (and the magnitude of the negative electron charge). The $-q^2/r$ term is therefore the electrostatic Coulomb potential energy of the electron in the electric field of the proton, negative because attractive. (You may prefer to replace q^2 with $q^2/4\pi\varepsilon_0$, depending which units you prefer).

In 1926, Schrodinger solved his equation for the hydrogen atom, Ref.[29.1], following the same approach that is standard in most QM texts, namely by seeking the configuration-space wavefunction, $\psi(\bar{r})$, which satisfies (29.3.2). One proceeds by assuming the wavefunction is separable in spherical polar coordinates and so find that the answer is obtained as a product of a spherical harmonic for the angular dependency and a radial function. The latter is a product of an exponential term and a Laguerre polynomial. That the last factor is a polynomial rather than an infinite series is required to ensure that the wavefunction converges to zero at spatial infinity, rather than diverging. The discreteness of the Laguerre polynomials introduces the principal quantum number, n, which parameterises these polynomials and provides an explicit expression for the energy levels, E_n.

However, Schrodinger was not the only person who derived the principal energy levels of hydrogen. Pauli also did so in the same year, 1926, and Pauli has precedence. Rather remarkably, Pauli did not need the configuration-space Schrodinger equation, (29.3.2). His method started from the commutation relations between the various relevant observables and derived the energy levels as a purely algebraic consequence of them, without needing the spatial-basis representation at all. This is the method that I now present.

Pauli started with the commutation relations between the observables of momentum and spatial position, which we discussed in chapter 5. The full set is,

$$\left[x_i, x_j\right] = 0 \quad \text{and} \quad \left[p_i, p_j\right] = 0 \quad \text{and} \quad \left[x_i, p_j\right] = i\hbar\delta_{ij} \tag{29.3.3}$$

In terms of these observables, and in the continuum, the orbital angular momentum observables are given by (29.2.1), which can be written more compactly as $\bar{L} = \bar{r} \times \bar{p}$, and these then provide the commutation relations between the $\bar{L}$ given by (29.2.3). The key to Pauli's treatment was then to introduce the Runge-Lenz vector, in operator form, as follows,

$$\bar{M} = \frac{1}{2m}(\bar{p} \times \bar{L} - \bar{L} \times \bar{p}) - e^2\hat{r} \qquad (29.3.4)$$

where the operator $\hat{r}$ is shorthand for the operator $\bar{r}/r$. This does not come from nowhere. To explain where it comes from, let us make a little excursion into classical mechanics, in which application (29.3.4) appears as an ordinary vector, and simplifies to $\bar{M} = \bar{p} \times \bar{L}/m - e^2\hat{r}$.

Whilst the $\bar{L}, \bar{r}, \bar{p}$ in (29.3.4) are all operators, if we interpret them instead as classical (vector) quantities, then (29.3.4) was already well known in the classical mechanics of the gravitational two-body problem and known as the Runge-Lenz vector. Note that the gravitational potential is also attractive and proportional to $1/r$ so the gravitational problem and the electrostatic problem are mathematically equivalent. The importance of the Runge-Lenz vector in the classical two-body gravitational problem lies in the fact that it is a constant of the motion ($\frac{d\bar{M}}{dt} = 0$). It is simple to prove in classical mechanics that the angular momentum is conserved for any central force, i.e., any force which is in the direction of the radial vector and whose magnitude is a function of the radial coordinate only necessarily gives $\frac{d\bar{L}}{dt} = 0$ in classical mechanics. But note that the instantaneous plane of motion is the plane which contains both the vectors $\bar{r}$ and $\bar{p}$. That plane is therefore equivalently defined as being perpendicular to $\bar{L} = \bar{r} \times \bar{p}$. It follows that if $\bar{L}$ is a constant of the motion, the motion must be confined to a plane. In particular, this is why planetary orbits are in a fixed plane (at least, if the effects of the other planets is neglected).

It turns out that, for the particular case of an inverse square law force, i.e., a potential energy proportional to $-1/r$, the Runge-Lenz vector is also a constant of the motion (as we will show in the QM context below). But consider what this implies if we assume that the orbiting body, starting at some arbitrary position at orientation $\hat{r}$, sweeps out a full revolution, i.e., that it is gravitationally bound. The angle defined by unit vector $\hat{r}$ changes monotonically until it arrives back at the starting angle, i.e., the same $\hat{r}$. But as $\bar{M} = \bar{p} \times \bar{L}/m - e^2\hat{r}$ is constant, and $\bar{L}$ is constant, and we are considering the same $\hat{r}$ at the beginning and the end of this revolution, it follows that the vectorial momentum $\bar{p}$ must also have returned to its starting value. What this means physically is that the orbit is closed.

To summarise, the fact that the Runge-Lenz vector is a constant of the motion is what leads to the classical two-body Kepler-Laplace bound orbits being closed loops (actually ellipses, of course). If the gravitational force were to deviate from being an inverse-square law – and indeed it does when general relativistic effects are taken into account – then the Runge-Lenz vector is no longer a constant of the motion and the bound orbits not only deviate from the elliptical, but are not closed loops either. The result is the precession of Mercury's perihelion, for example.

That lengthy digression was merely to motivate the introduction of (29.3.4), which is the QM generalisation of the Runge-Lenz vector, so that it did not seem to have appeared from nowhere. We can now derive the commutation properties of the vector-operator (29.3.4) from those of the other observables. In fact, the only purpose of introducing the momentum and position observables is so that their commutation properties, (29.3.3), can be used for this purpose, after which we shall need only the following,

$$[L_j, L_k] = i\hbar \, \epsilon_{jkn} \, L_n \tag{29.3.5}$$

$$[M_j, L_k] = i\hbar \, \epsilon_{jkn} \, M_n \tag{29.3.6}$$

$$[M_j, M_k] = -\frac{2i\hbar}{m} \hat{H} \, \epsilon_{jkn} \, L_n \tag{29.3.7}$$

$$[\bar{M}, \hat{H}] = [\bar{L}, \hat{H}] = 0 \tag{29.3.8}$$

where ϵ_{jkn} is the alternating tensor which is +1 when jkn is a positive permutation of xyz, and -1 if it is a negative permutation, and zero otherwise. It is straightforward, if tedious, to derive these from the definitions and the commutation relations between $\bar{L}, \bar{r}, \bar{p}$. In (29.3.7,8), $\hat{H}$ is the Hamiltonian operator from (29.3.2), i.e.,

$$\hat{H} = \frac{\hat{p}^2}{2m} - \frac{q^2}{r} \tag{29.3.9}$$

Consequently, (29.3.8) shows that both vector operators $\bar{L}$ and $\bar{M}$ are constants of the motion, as they were classically, as this is what commuting with the Hamiltonian means (see chapter 5). In the context of the Kepler problem, $\bar{M}$ lies in the plane of the orbit whereas $\hat{L}$ is perpendicular to the plane of the orbit. Corresponding to this we find that the operators obey,

$$\bar{L} \cdot \bar{M} = \bar{M} \cdot \bar{L} = 0 \tag{29.3.10}$$

Finally, the counterpart of the classical expression for the square of the Runge-Lenz operator, $M^2 = \frac{2E}{m} L^2 + K^2$, where the gravitational potential energy is $-K/r$, can be derived to be,

$$M^2 = \frac{2\hat{H}}{m} \left(L^2 + \hbar^2 \right) + q^4 \tag{29.3.11}$$

We now confine attention to,

- a degenerate sub-space of Hilbert space for which all states have the same energy, E;

- bound states so that $E < 0$.

So long as we confine attention to this sub-space we can replace the Hamiltonian operator on the RHS of (29.3.7) simply with E and we can normalised the Runge-Lenz vector operator such that,

$$\bar{M}' = \sqrt{-\frac{m}{2E}} \cdot \bar{M} \tag{29.3.12}$$

(recalling that $E < 0$ by assumption). So (29.3.5-7) simplify to,

$$\left[L_j, L_k \right] = i\hbar \, \epsilon_{jkn} \, L_n \tag{29.3.12}$$

$$\left[M'_j, L_k \right] = i\hbar \, \epsilon_{jkn} \, M'_n \tag{29.3.13}$$

$$\left[M'_j, M'_k \right] = i\hbar \, \epsilon_{jkn} \, L_n \tag{29.3.14}$$

The substitution,

$$N_j^{\pm} = \left(L_j \pm M'_j \right)/2 \tag{29.3.15}$$

leads to a still simpler set of commutators,

$$\left[N_j^+, N_k^+ \right] = i\hbar \, \epsilon_{jkn} \, N_n^+ \tag{29.3.16}$$

$$\left[N_j^-, N_k^- \right] = i\hbar \, \epsilon_{jkn} \, N_n^- \tag{29.3.17}$$

$$\left[N_j^+, N_k^- \right] = 0 \tag{29.3.18}$$

So the algebra decomposes into two closed but independent algebras, one for $\bar{N}^+$ and one for $\bar{N}^-$, these two sub-algebras being identical and equal to that for angular momentum. [Specifically the commutator structure is the Lie algebra of the Lie group SU(2), the universal covering group of the group of rotations in 3-dimensions, SO(3)]. The utility of recasting the algebra in the form of (29.3.16-18) is that we already understand the eigenvalue structure of the algebra of each of (29.3.16,17) from §29.2 in the context of angular momentum. Writing the squared-scalars $N_+^2 \equiv \bar{N}^+ \cdot \bar{N}^+$ and $N_-^2 \equiv \bar{N}^- \cdot \bar{N}^-$ we know from §29.2 that their eigenvalues are,

$$N_+^2 = n_+(n_+ + 1)\hbar^2 \quad \text{and} \quad N_-^2 = n_-(n_- + 1)\hbar^2 \qquad (29.3.19)$$

where n_+ and n_- take half-integral values 0, ½ , 1, ³/₂ , 2, ⁵/₂,... Now it follows from (29.3.15) that,

$$N_+^2 + N_-^2 = (L^2 + M'^2)/2 \qquad (29.3.20)$$

And
$$N_+^2 - N_-^2 = (\bar{L} \cdot \bar{M}' + \bar{M}' \cdot \bar{L})/2 \qquad (29.3.21)$$

But from (29.3.10) tells us that (29.3.21) is zero and hence $N_+^2 = N_-^2$, so in this particular application (the hydrogen atom) we require $n_+ = n_-$. Hence, taking the expectation value in an energy eigenstate, (29.3.20) gives,

$$\langle (L^2 + M'^2)/2 \rangle = \frac{1}{2}\langle L^2 - \frac{m}{2E} M^2 \rangle = 2n_+(n_+ + 1)\hbar^2 \qquad (29.3.22)$$

But from (29.3.11) we have,

$$L^2 - \frac{m}{2E} M^2 = -\left(\frac{q^4 m}{2E} + \hbar^2\right) \qquad (29.3.23)$$

So that (22) becomes,

$$\frac{q^4 m}{2E} + \hbar^2 = -4n_+(n_+ + 1)\hbar^2 \qquad (29.3.24)$$

Hence,
$$\frac{q^4 m}{2E} = -(2n_+ + 1)^2 \hbar^2 \qquad (29.3.25)$$

Hence,
$$E = -\frac{q^4 m}{2\hbar^2 (2n_+ + 1)^2} \equiv -\frac{1}{2}\frac{\alpha^2 mc^2}{n^2} \qquad (29.3.26)$$

where
$$n = 2n_+ + 1 \qquad (29.3.27)$$

Equ.(29.3.26) is the familiar non-relativistic solution for the energy levels of a hydrogen atom, where $\alpha = q^2/\hbar c$ is the fine structure constant ($\approx 1/137$) and n is the principal quantum number, which, since n_+ takes half-integral values 0, ½ , 1, ³/₂ , 2, ⁵/₂,... we see that n takes integral values 1, 2, 3, 4, 5...

29.4 References

[29.1]   E.Schrodinger (1926), Ann.Physik, 79, 361. In German, https://onlinelibrary.wiley.com/doi/abs/10.1002/andp.19263840404

[29.2]   W.Pauli (1926) Z.Physik, 36, 336-363. Published May 1926,. https://link.springer.com/article/10.1007/BF01450175. English: https://www.worldscientific.com/doi/abs/10.1142/9789814542319_0008

30

The Pitfalls of Popularisation

Popularisers of physics have a tough job. We must allow them considerable latitude as regards strict accuracy. Rigour is often the enemy of comprehensibility. Nevertheless, it is instructive to examine some sins that have been committed because there is also an obligation not to mislead the public.

Presenters of popular science programmes on TV can sometimes get a little carried away. Quantum mechanics is particularly easy to misrepresent in the heat of the moment. Here we examine and moderate some of the claims which have been made.

30.1 Can a Macroscopic Object Spontaneously Escape from a Box?

The presenter shows us a 60g object which he places inside a box which is roughly a 50mm cube, complete with lid. This purports to be an illustration of Heisenberg's uncertainty principle. How long, he asks, will we have to wait before the object spontaneously appears outside the box? He tells us that the answer is as follows,

$$t \sim \frac{LSm}{\pi h} \tag{30.1.1}$$

where $L = 0.05$m is the size of the box, $S \sim 0.04$m is the size of the object, and $m = 0.06$ kg is the mass of the object. As an illustration of the uncertainty principle this is fine. As an estimate of the time to spontaneously escape the box, it is not.

Neglecting factors of order unity, equ.(30.1.1) may be derived as follows. Confining the object in the box implies a minimum momentum of $p \sim \pi \hbar / L$. If the object is to appear outside the box, the uncertainty in its position must be of the order of its size, so that the uncertainty in its momentum is $\Delta p \sim \hbar / 2S$. The length of time we must wait is related to the uncertainty in the object's energy by $t \sim \hbar / 2\Delta E$. Finally, the uncertainty in momentum and energy are related by $\Delta E = p\Delta p / m$. Putting these together does indeed produce (30.1.1).

Inserting the numerical values in (30.1.1) gives a comfortingly long time before the object (ostensibly) materialises outside the box, of order 10^{29} s or about 10^{12} times the age of the universe. So everyone is happy that the probability of the object spontaneously escaping is sufficiently small that, for all intents and purposes, it will just not happen, thus conforming to our experience.

Well, no. Actually I'm not very happy that a probability of $\sim 10^{-22}$ per year is small enough. After all there are $\sim 10^{24}$ stars in the observable universe, so there are an awful lot of places where such behaviour could occur. Recall that the first nuclear reaction in solar mass stars, $p + p \rightarrow D + e^+ + v$, typically has a reaction time of $\sim 10^{10}$ years. The protons have a probability of only $\sim 10^{-17}$ of undergoing this reaction each time they collide. Yet this reaction is responsible for the sequence of events which leads to the formation of all the chemical elements in the universe (beyond lithium). My suspicion is that the probability for the object to escape the box is actually far smaller than 10^{-22} per year. It is, as we shall now see.

Something crucial has been missed in the estimate given by (30.1.1) – namely, the box! What has been calculated is the probability for spontaneous relocation over a distance of 40mm *through empty space*. But really the object must get through the wall of the box. This makes the probability far, far smaller. Of course it does. If we make the box strong enough, the object is not going to get out – just as you'd expect. In fact, even a remarkably feeble box will prevent the object escaping for much, much longer times than implied by Equ.(29.1.1), even though that is $\sim 10^{12}$ times the age of the universe.

Heisenberg's uncertainty principle is no longer sufficient to calculate the confinement time when the object has a box wall to penetrate. We are now in the realms of quantum tunnelling. An order of magnitude estimate of the tunnelling probability is easily made. The two key parameters are the thickness of the box wall (a) and the energy which would be required to push the object into the wall (V). The latter is a measure of the strength of the wall. In reality this potential energy barrier is going to be very large. Two solids do not usually superimpose to occupy the same space. The attempt to force them to do so would usually result in one of them breaking up. This illustrates the very substantial energy required. Even if we assume there was enough interstitial space within the atomic lattice of the two solids, the lattice disruption of both the object and the box wall would lead to a very large energy requirement. Fortunately, though, it is not necessary to attempt any realistic estimate of this barrier energy. Denote by ΔV the amount by which the potential energy barrier exceeds the kinetic energy of the object. (If the kinetic energy exceeded the size of the potential barrier the object could just bust straight through the wall). We shall assume that ΔV is a paltry 1 Joule, which is certainly many orders of magnitude smaller than the realistic magnitude for even a flimsy box wall.

The relationship between momentum and non-relativistic kinetic energy is $E = p^2/2m$. In terms of the de Broglie wavenumber, $p = \hbar k$. When within the wall, the object has a kinetic energy which is negative, namely $-\Delta V$, and hence impossible classically. In quantum mechanics this means that the wave-vector is imaginary, $k = i\sqrt{2mV}/\hbar$. The normally oscillatory wave $exp\{ikx\}$ therefore becomes an exponential decay $exp\{-\sqrt{2mV} \cdot x/\hbar\}$. Consequently, the probability amplitude for finding the object on the outer surface of the box is smaller than that on the inner surface of the box by a factor $exp\{-\sqrt{2mV} \cdot a/\hbar\}$. But probabilities are the absolute squares of probability amplitudes, so the probability of the object tunnelling through the wall is roughly,

$$\text{Tunnelling probability} \sim exp\left\{-\frac{2\sqrt{2mV}}{\hbar} \cdot a\right\} \qquad (30.1.2)$$

Substituting $V \sim 1$ Joule and, say, $a \sim 0.001$ m, gives, for our 0.06 kg object,

$$\text{Tunnelling probability} \sim exp\{-6.6 \times 10^{30}\} \qquad (30.1.3)$$

Now **that** is a small number! To drive home just how small, consider this. Really (30.1.3) gives the probability of tunnelling through the wall at every attempt to do so. But how frequently does the object make the attempt? A crude approximation is to assume that an attempt to penetrate the wall is being made on every 'cycle', where the cycles in question relate to the frequency derived from the object's energy. Let's give the object lots of energy by raising it to 3000K. Its typical thermal energy is $k_B T = 4 \times 10^{-20}$ J (where k_B is Boltzmann's constant). The associated frequency is $\sim 6 \times 10^{13}$ Hz. So we must multiply (30.1.3) by 6×10^{13} to get the probability of escape per second, or by 2×10^{21} to get the probability of escape per year. So we get,

$$\text{Tunnelling probability} \sim 2 \times 10^{21} \times exp\{-6 \times 10^{30}\} \text{ per year} \qquad (30.1.4)$$

How much difference has it made to have multiplied (30.1.3) by 2×10^{21}? Effectively no difference at all because (30.1.4) is the same as,

$$\text{Tunnelling probability} \sim exp\{49 - 6 \times 10^{30}\}$$

$$\approx exp\{-6 \times 10^{30}\} \text{ per year} \qquad (30.1.5)$$

which is identical to (30.1.3)!

Note that the "$\approx$" sign in (30.1.5), whilst strictly an approximation, is extremely accurate in the sense that the logarithm of the tunnelling probability is accurate to 29 decimal places.

Unlike the comparatively huge probability of $\sim 10^{-22}$ per year given by (30.1.1), the probability given by (30.1.5) really *is* effectively zero. Even if there were 10^{500} universes in the multiverse (enough perhaps to realise every possible string theory) and even if they all survived for 10^{500} years (they wouldn't), this would make negligible impact on the probability for the object to tunnel through the box wall within the life of any of these universes. It would remain as given by (30.1.5) to an accuracy (in its logarithm) of 26 decimal places.

So when people say, "in principle, the object could spontaneously appear outside the box if you waited long enough" I think it is best to simply regard this as untrue for macroscopic objects, for all practical purposes. The trouble is that people do not have sufficient respect for probabilities which are truly, prodigiously, stupendously, small.

30.2 Is Everything Connected to Everything Else?

In the later parts of this section I address – with some trepidation – issues which are not properly understood (at least, not by me). I think it is fitting to close a book on quantum mechanics on issues which are still puzzling, because that is an honest reflection of its status on some matters. Let's start, though, with our hapless hypothetical TV populariser.

The presenter explains the exclusion principle to the audience. "Two electrons", he says, "can never share the same quantum state". He takes up an object and says that the electrons in the object cannot share a quantum state with any other electrons in the universe. He warms the object in his hand, remarking that this will cause some electrons to move to a higher energy quantum state. He then confounds the audience by claiming that, as a result, every electron in the universe has had to react to this change of state within the object. Is this true?

Actually there are several things worrying about the claim, or at least the way it was expressed. Firstly, and the least concerning, is a technicality. Warming an object at room temperature by just a few degrees is manifest at the atomic level by increasing the mean kinetic energy of the atoms (or molecules) of which the object is composed. It does not primarily raise electrons to higher atomic orbitals. However, I suppose one has to concede that a few electrons may be raised in energy, so I'll give this one a pass.

Of greater concern is that the phrasing of the presentation suggests that changing the state of an electron here will change the state of electrons everywhere as a consequence of the exclusion principle (that no two electrons can be in the same quantum state). Why? There appears to be a suggestion that the electron we have directly influenced must have moved to a state already occupied by some other electron, which therefore needs to move to another state so as not to violate the exclusion principle. And also that the displaced electron then displaces another in turn. And so on. Across the entire universe. But this is a wildly false assumption. There are far more quantum states available to electrons in the universe than there are electrons, and so most are unoccupied.

A semantic confusion can arise because, in many contexts, the electrons in a number of unexcited hydrogen atoms are often said to be in "the same state", namely the ground state with principal quantum number 1. But there is semantic confusion over what is meant by "the same state". Leave spin out of it: consider all the electrons to have the same spin state, there is still no difficulty. The electrons in two different unexcited hydrogen atoms are not in the same quantum state. They are merely in corresponding, or equivalent, states, not literally the same quantum state. Energy alone does not uniquely specify a quantum state. Forget energy and focus on states defined by position (consider a distributed location defined by a wavepacket if it gives you more comfort). There is plenty of space available in the universe which is unoccupied by electrons, and hence plenty of states available. Two electrons in two different, well-separated, hydrogen atoms are not in the same quantum state, despite both being in (say) the $n = 1$, $L = 0$, spin-up state.

One could try to make the TV presenter's claim sensible by considering the electromagnetic interaction between charges. Thus, if the change of state of an electron is considered to alter its mean position, then the electrostatic Coulomb energy wrt other electrons will all be changed. But this does not seem to be what the presenter had in mind as it has nothing to do with the exclusion principle. In any case, a physical interaction of any kind cannot justify a claim that there is an immediate effect on every other electron in the universe. The limitation of the maximum propagation speed of any physical influence to the speed of light prohibits that. It will be north of 40 billion years, at least, before any physical influence can propagate to the furthest reaches of the observable universe (let alone whatever lies beyond that).

So far my counterarguments have been, I believe, sound. But there is one slant on the issue which gives me pause. We have seen that entanglement can give rise to genuinely non-local effects. The reduction of the state vector applicable to a particle currently in Andromeda can be accomplished, apparently instantly, by a measurement on an entangled partner here on Earth. Moreover, the exclusion principle guarantees that electrons are always entangled! How come? Recall that the theorem behind the exclusion principle is the spin-statistics theorem, see §24.2. In the context of quantum field theory, this theorem is rigorously provable as a consequence of Lorentz invariance (i.e., as a consequence of special relativity). It states that indistinguishable particles with integral spin (integral multiples of $\hbar$) must have joint quantum states which are fully symmetric under interchange of any pair of particles, whereas indistinguishable particles with half-integral; spin $(\frac{1\hbar}{2}, \frac{3\hbar}{2}, ...)$ must have joint quantum states which are fully antisymmetric under interchange of any pair of particles. Thus, if there are two electrons and one of them is in state 1 and the other is in state 2 it is meaningless to ask which is which. Hence writing the state as $|1\rangle|2\rangle$ where the first ket refers to electron No.1 and the second to electron No.2 is wrong. This would imply a distinction between this state and the state $|2\rangle|1\rangle$, but there cannot be such a distinction because the electrons are indistinguishable. Since electrons have spin half, the correct state is actually,

$$|\psi\rangle = \tfrac{1}{\sqrt{2}}[|1\rangle|2\rangle - |2\rangle|1\rangle] \qquad (30.2.1)$$

This is where the exclusion principle, which applies only for half-integral spin particles (fermions) comes from. If we attempt to put both electrons in state 1, or both in state 2, then (30.2.1) becomes zero – there is no valid state of two electrons in which both are in the same state, and this arises from the required antisymmetry of joint fermion states.

But (30.2.1) is an entangled state. Is this, finally, a way of salvaging the presenter's reputation? We know that state vector reduction can occur over spacelike separations between entangled particles. So does the automatic entanglement of fermions via the antisymmetry rule provide a means by which one electron can affect all other electrons in the universe? Is everything connected after all?

Let me emphasise that the antisymmetry of the wavefunctions of electrons can have observable effects. For example, the first excited state of the helium atom has an energy level whose energy is shifted from where it would otherwise be by an effect of the antisymmetry, the shift being referred to as the "exchange energy". So, antisymmetry is real and has observable effects. However, and for this very reason, one cannot tell if it is electron No.1 or electron No.2 which is in the excited state. They both are, and both are not, in a state like (30.2.1).

Consider, then, two hydrogen atoms in their ground state (and assume both electrons are in the same spin state). Suppose one atom is in Alice's possession and the other in Bob's. The atoms are denoted accordingly by A and B, and we denote the ground state by the letter G. As usual we use a notation in which the leftmost ket refers to electron No.1 and the ket on the right to electron No.2. The combined electron state, with antisymmetry imposed, is thus,

$$|\psi\rangle = \frac{1}{\sqrt{2}} [|A, G\rangle|B, G\rangle - |B, G\rangle|A, G\rangle] \qquad (30.2.2)$$

We are happy to imagine that Alice and Bob and their corresponding hydrogen atoms are some prodigious distance apart. Now let us interact with Alice's atom and raise its electron to an excited state. Take careful note: we are raising the state of Alice's atom, not specifically the state of No.1 or No.2 electron. The latter is impossible as we don't know which is which. So, in obvious notation, (30.2.2) becomes,

$$|\psi\rangle = \frac{1}{\sqrt{2}} [|A, E\rangle|B, G\rangle - |B, G\rangle|A, E\rangle] \qquad (30.2.3)$$

Have we succeeded in changing the state of Bob's atom? No. Whether we regard Alice's atom as containing electron No.1 (left term) or electron No.2 (right term), the state of Bob's atom is unchanged. So, it looks like the TV presenter cannot save face by claiming this "exclusion principle entanglement" loophole either.

But wait a minute. What if the two atoms were "really" entangled. That is, entangled in the usual way, not simply because of the requirement for antisymmetry? For example, ignoring the requirement for antisymmetry under interchange for a moment, suppose one of the pair of hydrogen atoms is excited initially, but in such a way that we do not know which and the state is entangled thus,

$$|\psi\rangle = \frac{1}{\sqrt{2}} [|A, E\rangle|B, G\rangle + |A, G\rangle|B, E\rangle] \qquad (30.2.4)$$

This assumes electron No.1 is on atom A, and electron No.2 is on atom B. We are not allowed to assume this and must make the state antisymmetric under interchange, so we get, (30.2.5)

$$|\psi\rangle = \frac{1}{2}[|A,E\rangle|B,G\rangle - |B,G\rangle|A,E\rangle + |A,G\rangle|B,E\rangle - |B,E\rangle|A,G\rangle]$$

Now we can exploit the usual non-local entanglement connection between the two atoms. For example, if Alice measures the energy of her atom A and finds it to be in the ground state then we have forced the state of Bob's atom B from a superposition of energy states into a definite excited state. (It does not matter if we consider (30.2.4) or the more correct (30.2.5) to conclude this). So our struggling TV presenter might be able to save face if he can claim that all electrons are entangled with all other electrons in the universe (in the usual, non-exclusion principle, sense). But why should they be? Especially if they are in distant regions which have never been causally connected to permit an entanglement to arise?

At this point cosmology raises its head. (This is not getting any simpler, is it?). You see, the universe is remarkably isotropic, which is rather a conundrum given that diametrically opposite points appear (naively[5]) never to have been in causal contact. But the standard cosmological model gets around this problem by assuming that, in fact, there was a causal connection between all parts of the universe in the first stupendously tiny fraction of a second ($\sim 10^{-34}$ s) after the big band. So, if entanglement of all electrons arose then….but hold on, did all the electrons now present in the universe even exist then? I think I have taken this far enough. Our unfortunate TV presenter would be grasping at the most water-logged of straws at this point.

Is everything connected to everything else? If you think so, it's up to you to present an argument; I have failed to do so. And on that note of admitted ignorance, I close the book. I hope you enjoyed it.

---

[5] Amusingly this implicitly accuses general relativistic cosmology of being naïve – if inflation is ignored. Not *that* naïve, then.

Index

A searchable pdf of this book may downloaded here,
http://rickbradford.co.uk/theunweirding.pdf

Internet links in the pdf may facilitate access to the references.

Any errors or other comments welcome, email rawbradford@gmail.com

www.ingramcontent.com/pod-product-compliance
Lightning Source LLC
Chambersburg PA
CBHW071325210326
41597CB00015B/1357